Allelopathy
New Concepts and Methodology

Allelopathy

New Concepts and Methodology

Editors

Yoshiharu Fujii

Syuntaro Hiradate

National Institute for Agro-environmental Sciences
Tsukuba
Japan

CRC Press
Taylor & Francis Group
Boca Raton London New York

CRC Press is an imprint of the
Taylor & Francis Group, an **informa** business
A SCIENCE PUBLISHERS BOOK

CRC Press
Taylor & Francis Group
6000 Broken Sound Parkway NW, Suite 300
Boca Raton, FL 33487-2742

ISBN-13: 9781578084463 (hbk)

Visit the Taylor & Francis Web site at
http://www.taylorandfrancis.com

and the CRC Press Web site at
http://www.crcpress.com

Library of Congress Cataloging-in-Publication Data

World Congress on Allelopathy (3rd : 2002 : Tsukuba Kenkyu Gakuen Toshi, Japan)
Allelopathy : new concepts and methodology/editors, Yoshiharu Fujii, Syuntaro Hiradate
p. cm.
Papers presented at the Third World Congress on Allelopathy, held in Tsukuba, Japan, 2002
Includes bibliographical references and index.
ISBN 1-57808-446-6
1. Allelopathy--Congresses. 2. Allelopathic agents--Congresses. I. Fujii, Yoshiharu. II. Hiradate, Syuntaro. III. Title.

SB732.75.W67 2002
571.9'2--dc22

2006045069

Foreword

The technological advances made in allelopathy research in recent years have been created, analyzed, and developed by scientific establishments throughout the world. They present exciting and intellectually challenging problems which are solvable using modern techniques. These modern and advanced techniques as described in the chapters presented in this volume are representative of the exciting research and development approaches today.

The principal goal of allelopathy is to foster sustainable agriculture, forestry, and environment. The objective is to minimize the industrial chemicals and to maximize the use of natural resources locally available while improving crop productivity, forestry and the environment. According to Elroy Rice, the public concern for environmental pollution due to the use of pesticides has made a dramatic impact on challenging allelopathy scientists; they, now, have some of the technological tools and equipment that are so necessary to solve these problems.

This book represents primarily a collection of papers that was presented at the Third World Congress on Allelopathy held in Tsukuba, Japan. The theme was "Challenge of the New Millennium". Distinguished scientists from more than 30 countries presented their research results at the Congress. Their presentations were thoughtful, deliberate and challenging to the allelopathy community. It is expected that some of these results will be put into practice by the agricultural, forestry and environmental systems.

The International Allelopathy Society was founded in Old Delhi, India in 1994 by 31 distinguished scientists representing about 20 countries of the world. The IAS has held three World Congresses on Allelopathy the first being in Cadiz, Spain in 1996, the second in Thunderbay, Canada in 1999, and the third was held at Tsukuba Japan in 2002.

The Distinguished Awards presented at the Third World Congress were: the Molisch Award for outstanding academic achievement and/or service relating to the field of allelopathy-Steven O. Duke; the Grodzinski Award for the best single publication or book relating to allelopathy-Azim Mallik; the Rice Award for the best single paper or article presented at a symposium sponsored by the IAS-R.G. Belz, and the IAS Sustained Support Award-George R. Waller. These awards are presented every third year coinciding with the meeting of the International Allelopathy Society.

Interactions between participants at this Congress help develop and sustain cooperative relationships among scientists. Allelopathy scientists are using techniques to solve problems throughout the world. The industrial demand for allelopathic chemicals and practices, while still in its infancy, is increasing in its demand for improvement of agriculture, forestry and environment.

The International Allelopathy Society acknowledges support from the following organizations: Food and Agriculture Organization [FAO], Commemorative Association for the Japanese World Exposition of 1970, Tsukuba City, The Weed Science Society of Japan [WSSJ], Japanese Allelopathy Society [JAS], Noda Plants Ltd., Field Science Co. Ltd., Japan Turfgrass II, Caro Trading Co. Ltd., and National Institute for Agro-Environmental Sciences [NIAES]. Sincere appreciation is due to the authors of these papers.

George Waller
Professor Emeritus
Department of Biochemistry and
Molecular Biology
Oklahoma State University
Room 246, Noble Research Center
Stillwater, OK 74078 USA

Preface

The term 'Allelopathy' is already a time-honored concept, starting from the definition of Dr. Molisch in 1937. Allelopathy is an action of natural bioactive chemicals produced by plants to other life. The related concept of 'Antibiotics', which is an action of natural chemicals produced by microorganism to other biota, began being investigated almost at the same time as 'Allelopathy', but developed rapidly and was soon able to make a major impact on medical science. Compared to antibiotics, the development of allelopathy was very slow, owing to the ambiguity of the phenomena in the field.

Allelopathy is now becoming increasingly important. One reason is that this concept helps in organic or natural farming with or without limited use of synthetic agrochemicals (herbicide, insecticide, fungicides, etc.). Another reason is the understanding of allelopathy in natural ecosystems. Allelochemicals belong to 'secondary metabolites'. Secondary metabolites indicate those without indispensable constituents in plants and exist only in the plant kingdom. In the past, the meaning of these chemicals in plants seemed to be a pool of energy or reducing agents, or simple wastes. But recently, the 'Allelopathy' hypothesis describes the real meaning of these secondary metabolites as a tool of immobile plants to protect themselves from surrounding plants or other life that might attack them, or a tool to communicate with each other or to communicate with other life for their survival. It has been commonly assumed that there are more than 500,000 plant species and more than 30,000 secondary natural chemicals in this world. However, we are sure that there are still many natural chemicals unknown to us. The third importance of allelochemicals is their use as a source of new agrochemicals.

The title of this book 'Allelopathy: New Concepts and Methodology' is a challenge for new research in allelopathy. As for new methodology and concepts, Dr. Duke's group introduces a 'Dose/Response

Relationship' concept. They point out that all chemicals are neither always toxic nor stimulatory but are dependent on the dose. Dr. Blum explains the discrepancy between laboratory data and field data. Our group on the other hand, is exploiting new bioassay for allelopathy and explains the 'Plant-Box Method' as a bioassay tool to evaluate allelopathy by root exudates.

As for new challenges in allelochemicals, the possibility finding of biologically active natural chemicals from allelochemicals as agrochemicals is discussed by Dr. Cutler's group. Wheat is one of the most important grains in the world. Allelochemicals of wheat and their ecological degradation, are reported by Dr. Haig and Dr. Macias' group. Reflecting on the importance of allelopathy in the field, the interaction of allelochemicals to soil and soil components is significant. Dr. Hiradate explains the mode of action of 'Mugineic acid', which is a unique amino acid isolated by Dr. Takagi in Japan (Mugi means wheat in Japanese) as a specific mechanism, of a ferrous iron acquiescing tool in some wheat and other graminaceous plants. Allelopathy is a multidisciplinary concept and plant growth promotion is a new challenging field. Dr. Hasegawa's group explains their new findings regarding lepidomoidic acids.

'Invasive alien plants' is a serious weed problem in new countries or areas, and is now becoming an increasingly important topic. Dr. Qasem explains white rocket as a potentially invasive weed and Dr. Kazinczi reports on the problem weeds of Hungary.

As for practical weed suppression, cover crop or cover plants with allelopathic activity are the most ideal tools. Buckwheat, sunflower, hairy vetch, *Astragalus adsurgens* (new ground cover plants related to Chinese Milk Vetch), and potential cover plants from ornamental plants are reviewed in the following section.

Rice allelopathy is a newly developing area because of the importance of rice and the growing need for direct seeding or large-scale farming. Reviews concerning the allelochemicals and genetic approach are compiled in the next two chapters.

Trees are perennial plants and stand in the same place for long periods. Thus, the accumulation of allelochemicals is a problem for regeneration. But owing to the difficulty of research and its time-consuming nature, there is not much data available in this field. Dr. Ito postulates the importance of 'Stem Flow' in tree ecosystems in his chapter.

Other more challenging and newly developing areas in allelopathy such as aquatic plants, mushrooms and plant insect interactions, are reviewed in the last section. The editors specially asked Dr. Yamaoka, a medical doctor, to review their research on the influence of plant volatile allelochemicals to animals including humans.

This book is a result of the Third World Congress on Allelopathy held in Tsukuba, Japan, 2002. The chapters by Dr. Belz and Dr. Blum were the winners of the Best Poster Award. Other chapters were either based on invited talks or are selected papers from the posters. There are additional interesting topics related to new concepts and methodologies. 'Function of Flavonoids as Allelopathic Substances', by Mohammad Masud Parvez *et al.*, was selected for the Gold Poster Award, and 'Biological Phenomenon of Nitrification Inhibition: A Case Study with *Brachiaria humidicola*' by Dr. Takayuki Ishikawa, and 'Effects of 2-Benzoxazolinone on Seedlings and Plants of *Lactuca sativa* L'. by Dr. Manuel J. Reigosa *et al.*, were selected for the Silver Poster Award. 'Weed Suppression and Rice Yield in the Hairy Vetch Introduced Paddy Field', by Sakae Horimoto *et al.*, and 'Allelopathy of Sweet Vernalgrass (*Anthoxanthum odoratum* L.) for Invading Zoysia-grassland in Japan' by Dr. Yoshito Yamamoto are interesting practical reports awarded with the Bronze Poster Award. Finally, the IAS Sustained Support Award was given to Dr. George R. Waller for his outstanding accomplishments and contributions in the field of allelopathy.

Finally, the editors thank Dr. Mohammad Masud Parvez, as a technical editor, for his support. We hope this book will help develop new paradigms in the field of allelopathy, contribute to the peaceful existence of human beings, and the harmonious coexistence of all living organisms in the world.

Tsukuba, Japan
December 2005

Yoshiharu Fujii
Syuntaro Hiradate

Acknowledgment

All the chapters included in this book have been peer-reviewed. The editors of this book wish to express their sincere thanks to the following individuals for their generous support to make this book possible.

Stephen Duke
Udo Blum
Horace Cutler
Stephen Cutler
Terry Haig
Francisco Macias
Mohammad Masud Parvez
Jamal Qasem
Zahida Iqbal
Helena Gawronska
John Teasdale
Yong-qing Ma
Chuihua Kong
Hiroshi Araya

Tsukuba, Japan
December 2005

Yoshiharu Fujii
Syuntaro Hiradate

Contents

SECTION 1

New Methodology and Approach (Dose Response, Bioassay)

1

Dose/Response Relationships in Allelopathy Research

Regina G. Belz,[1] Edivaldo D. Velini[2] and Stephen O. Duke[3]

[1]University of Hohenheim, Institute of Phytomedicine 360, Department of Weed Science, 70593 Stuttgart, Germany

[2]Univerity of São Paulo State, Botucatu, Brazil

[3]Natural Products Utilization Research Unit, USDA, ARS, P.O. Box 8048, University, MS 38677, USA

Understanding the relationships between the dose of a phytotoxin and its effects is crucial in many types of allelopathy studies. However, realistic considerations of dose/response relationships and time of exposure are too infrequently considered in allelopathy studies. Examples are given of the proper methods for conducting dose/response studies. Synergism has been invoked in many allelopathy papers, without sufficient proof. The use of dose/response studies in determination of interactions (both antagonistic and synergistic) between different compounds is discussed. Dose/Response studies are also crucial to understand the modes of action of phytotoxins. For example, the dose/response relationships and their timing with different effects can strongly suggest that certain responses are closer to the molecular target site than others. Differences in slopes of dose/response curves suggest differences in modes of action, whereas similar slopes suggest similar modes of action. Very accurate dose/response curves are usually needed to prove hormesis (stimulatory effects of phytotoxins at subtoxic doses). Little is known about hormesis of allelochemicals, a phenomenon that might be very important in allelopathy. Finally, the effects of different target plant densities on dose/response relationships, as well as dose/response relationships generated by different allelochemical-producing donor plant densities are discussed. These methods can be crucial to proving allelopathy and to understand the mechanisms of allelopathy.

Keywords: *allelochemical, dose/response, herbicide, hormesis, phytotoxin*

INTRODUCTION

Theophrastus Bombastus von Hohenheim (1493-1541), otherwise known as Paracelsus, is often paraphrased as saying "the poison is in the dose". For this concept, he is considered the father of toxicology. In other words, a sufficient dose of almost anything can be toxic, but smaller doses are either benign or, in some cases, beneficial (hormesis). This dictum has been clear to animal and plant toxicologists dealing with pesticides and other xenobiotics, but realistic considerations of dose, in terms of the interactions of dose and duration of exposure, have too infrequently been the case in allelopathy studies.

In traditional plant toxicology, dose/response relationships are used for many purposes, of which we will list only a few. First, they are necessary to compare the toxicity of different phytotoxins and to determine safe levels of potential phytotoxins. Such information is absolutely necessary in conducting mode of action studies, in order to use appropriate doses of the toxicant and to compare different modes of action. Dose/response curves are required to compare the level of toxicity of compounds to different plants of the same species (e.g., susceptible and resistant) or to different plant species. Without this information, few conclusions can be made with certainty.

Since allelopathy is predicated on the assumption that a natural compound from a plant acts as a phytotoxin in nature, understanding dose/response relationships between the allelochemical and the target plant is essential to prove the role of a natural phytotoxin as an allelochemical. For example, if significant phytotoxicity is found only at doses above those found in soil, the role of a natural phytotoxin as an allelochemical is questionable. This chapter is meant to review the topic of dose/response relationships of phytotoxins, with emphasis on natural product-based phytotoxins, some of which are allelochemicals.

CLASSICAL DOSE/RESPONSE RELATIONSHIPS WITH PURE COMPOUNDS AND MIXTURES

More often than not, dose/response studies conducted by plant scientists are done improperly, making them difficult to interpret. First, the effect (response) should be plotted versus the log of the dose. Biological responses are symmetrical around the effective dose that causes a 50% effect (ED_{50}) on a log dose versus response basis (Fig. 1). Most of the time, the dose or concentration should be measured on a molar basis, rather than

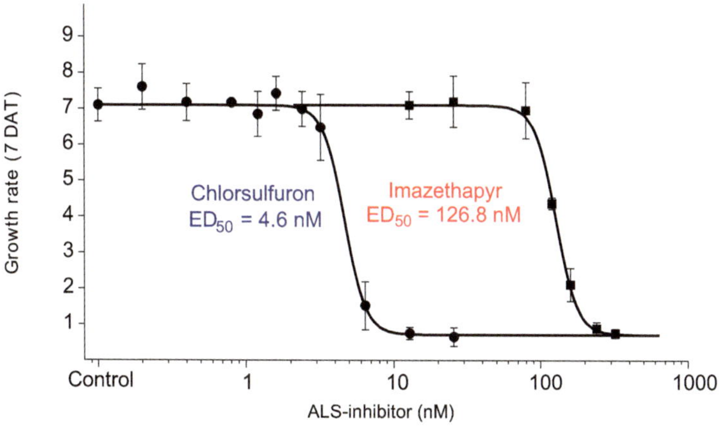

Fig. 1 Complete dose/response curves for two inhibitors of acetolactate synthase (ALS). The Y axis is growth of frond area of *Lemna pausicostata* as fold increase after 7 d treatment. DAT = d after treatment. Error bars are ± 1 SE. From the work of Michel *et al.* (2004).

by weight of the compound per unit volume. To compare toxicity of compounds, molarity is best used, since target sites interact with individual molecules, and not the mass of a molecule. These concepts are generally accepted by toxicologists. Thus, the doses administered are best when equally spaced on a logarithmic scale (e.g., 1, 3.3, 10, 33, 100 units). Furthermore, the range of doses should encompass the lowest dose for an observable effect, as well as the lowest dose for a maximal effect. Such a curve can be used to calculate the effective dose for 50% effect (ED_{50}) or effective concentration for 50% effect (EC_{50}) or for any other level of effect (Fig. 1). Sometimes this type of information is expressed as the I_{50} (the concentration causing a 50% inhibition) or any other inhibitory level (e.g., I_{80}). Statistical models and non-linear regression procedures are available to analyze dose/response data (Streibig *et al.*, 1993; Michel *et al.*, 1999; Ritz and Streibig, 2005). In this chapter, we will largely avoid the mathematics of dose/response analysis because of the general audience of this book.

If the response is quantal, that is a yes or no response (e.g., alive or dead; germinated or not germinated), the per cent response should be plotted on a probit scale, either by transformation to probits or on a transformed scale, to compensate for variation within the population (Fig. 2) (Finney, 1971). This transformation removes the sigmoid response of a normally distributed population with respect to response to a range of doses of a chemical, changing it to a linear function.

Interpretation of interactions between different compounds with respect to a particular response cannot be done without accurate determinations of the dose/response relationships of the two compounds alone and in combination. Synergisms between putative allelochemicals have often been claimed, yet proper dose/response experiments to prove this have seldom been carried out. For example, Rasmussen and Einhellig (1977) claimed synergism between ferulic and *p*-coumaric acids in their effects on germination of radish, yet they did not conduct dose/response studies. When dose/response studies were conducted with these and similar common cinnamic acid derivatives, either alone or in equimolar concentrations, the compounds were found to have either additive or antagonistic effects (Duke *et al.*, 1983, Fig. 2).

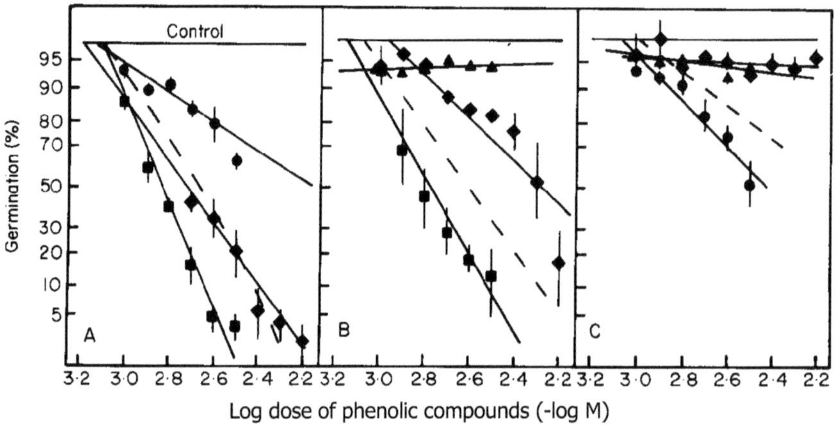

Fig. 2 Probit adjusted germination responses of lettuce seeds to log concentrations of different phenolic acids singly or in equimolar concentrations. Error bars are ± 1 SE. In A, B and C, the solid line without data points is the control germination level. A. Ferulic (●) and *p*-coumaric (■) acids alone and in combination (♦). B. Caffeic (▲) and *p*-coumaric (■) acids alone and in combination (♦). C. Caffeic (▲) and ferulic (●) acids alone and in combination (♦). The dotted lines are the predicted dose/response curves with no antagonistic or synergistic interactions. From Duke *et al.* (1983) with permission of Oxford University Press.

More elaborate methods of determining antagonistic or synergistic interactions between compounds have been outlined by Green and Streibig (1993). As allelopathy is often the result of several compounds acting simultaneously, studying interactions between allelochemicals is important to understand such a joint action process. For example, based on dose/response relationships of pure compounds, Streibig *et al.* (1999) assessed the joint action of sorgoleone analogs isolated from *Sorghum bicolor* (L.) Moench., applying an elaborate joint action model as proposed by Green and Streibig (1993). Sorgoleone and structural analogs in binary mixtures were found not to interact and were thus additive. Using the same experimental design, Kudsk *et al.* (2004) studied the characteristics of a joint action of wheat allelochemicals. As previously found by Duke *et al.* (1983), they confirmed an additive or slightly antagonistic action of mixtures of phenolic acids associated with wheat. Effects of a second group of allelochemicals associated with wheat, the benzoxazolinones benzoxazolin-2(3*H*)-one (BOA) and 6-methoxy-benzoxazolin-2(3*H*)-one (MBOA), were found to be additive as well. In contrast, mixtures of compounds of both classes of phytotoxins were usually antagonistic. Thus, synergism seems to be an unlikely phenomenon, in both allelopathy of wheat and sorghum, suggesting that it might be a rare phenomenon in allelopathy, just as it is seldom found with synthetic phytotoxins (herbicides).

DOSE/RESPONSE AND MODE OF ACTION

Knowing dose/response relationships is essential in studying the mode of action of phytotoxins. If a phytotoxin is used at a dose well beyond that required for the full effect, its mode of action might be different from that when applied at the proper dose as a herbicide or when acting as an allelochemical in nature. A single compound might affect several molecular targets with different dose/response relationships for each, but, normally, only one of these targets is affected at a realistic field dose. An exception might be the allelochemical sorgoleone, which targets both photosystem II (PS II) of photosynthesis (Einhellig *et al.*, 1993; Streibig *et al.*, 1999) and hydroxyphenylpyruvate dioxygenase (Meazza *et al.*, 2002) at very low doses.

The same dose can affect different parameters of the same organism quite differently. For example, the herbicide, acifluorfen-methyl affects cellular leakage more profoundly than chlorophyll content (Fig. 3), if the measurements are taken early enough. Results of such an experiment can suggest that one parameter is more closely associated with the primary effect of the toxicant than the other. In the case of acifluorfen-methyl, the loss of membrane integrity is related to the loss of chlorophyll, but both are secondary effects of oxidative stress caused by accumulation of the photodynamic compound protoporphyrin IX. However, the more rapid effect on membrane integrity indicates that this effect is more closely linked to the primary effect of the herbicide than the chlorophyll loss.

With a similar herbicide, acifluorfen, almost identical dose/response curves were produced with herbicide dose versus protoporphyrin IX content (Fig. 4a) and cellular leakage (Fig. 4b). In this case, the protoporphyin IX is the photodynamic compound accumulated as a result of inhibition of the enzyme protoporphyrinogen oxidase by acifluorfen.

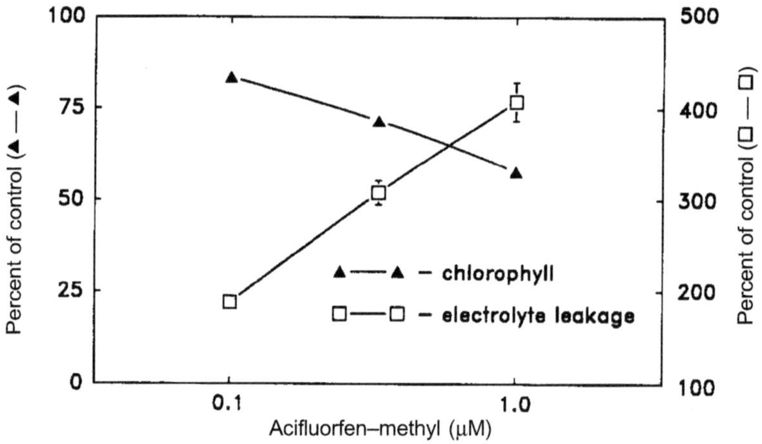

Fig. 3 Effects of different concentrations (on a log scale) of acifluorfen-methyl on chlorophyll content (measured after 24 h treatment) and cellular leakage (as measured by electrolyte leakage after 6 h treatment) of *Lemna pausicostata*. Error bars are ±1 SE. From Matsumoto and Duke (1990) with permission of the American Chemical Society.

The protoporphyrinogen IX that accumulates as a result of inhibition of this enzyme autoxidizes quickly to protoporphyrin IX (Dayan and Duke, 1997). Protoporphyrin IX causes the membrane damage that results in cellular leakage through its photodynamic properties. Thus, comparing dose/response relationships for different effects can suggest causal relationships, provided measurements of the parameters are made during the period when there is an increasing effect with dose over the same dose range.

The time period when the measurements are taken is very important, as the dose/response relationships change with time. For example, if the measured response is death, the minimum dose for 100% mortality will be larger at an earlier time than at a later time, up to the dose from which recovery occurs with some individuals. Similarly, the I_{50} values for other effects often go down with time. For example, the dose of the natural phytotoxin cornexistin needed to cause 50% of the maximal effect on cellular leakage and growth decreased with each day of treatment of duckweed (*Lemna pausicostata* Hegelm.), from 2 to 6 d (Fig. 5). Exposure (dose x time) to a compound generally increases with time unless metabolic degradation defenses are induced by the compound. Thus, the dose, the effects and the time after treatment are all parameters that should be considered, yet this is rarely the case in published studies.

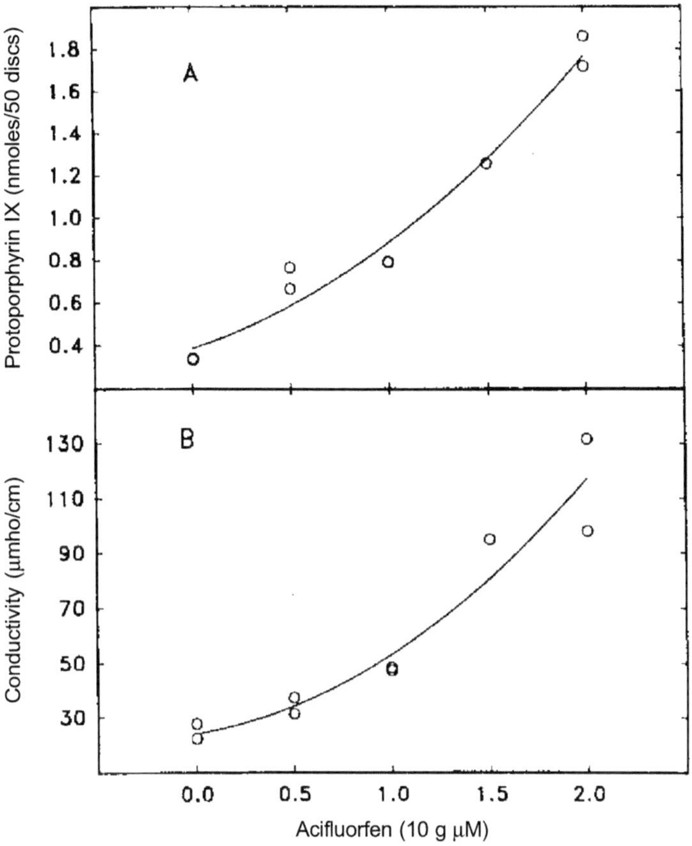

Fig. 4 Effects of different concentrations of acifluorfen applied to cucumber cotyledon discs on protoporphyrin IX (A) and cellular leakage, as measured by electrolyte released to a bathing solution (B). From Becerril and Duke (1989) with permission of the American Society of Plant Biologists.

The slopes of dose/response curves of compounds with similar modes of action are normally similar, and those with different modes of action are often different. Essentially, similar slopes are usually observed for active compounds with structural differences leading only to variation in the dose reaching the site of action or to differences in binding efficiency at the active site. For example, both chlorsulfuron and imazethapyr target the same enzyme, acetolactate synthase, and in the same system have the same dose/response slopes (Fig. 1). Dose/Response relationships of analogs of active compounds such as analogous isothiocyanates (Petersen *et al.*, 2001), sorgoleone analogs (Streibig *et al.*, 1999), or the two benzoxazolinones BOA and MBOA (Belz *et al.* 2005) proved to have a similar shape as well. In a

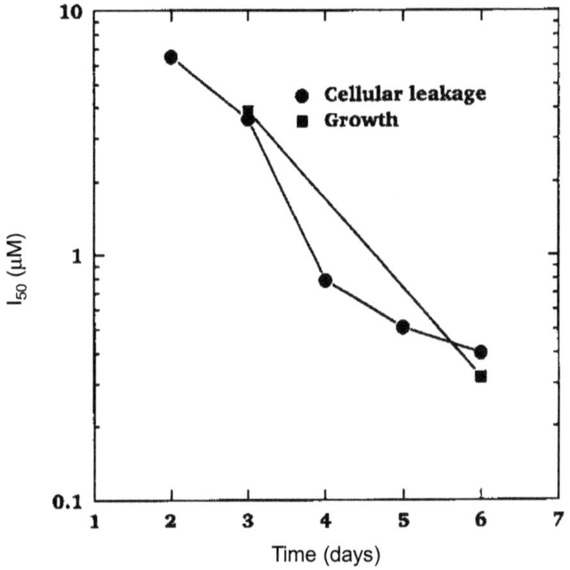

Fig. 5 The effects of duration of treatment with cornexistin on the I_{50} values for different effects on *Lemna pausicostata*. From Amagasa *et al.* (1994) with permission of Elsevier Press

survey of several phytotoxins with very different modes of action (Michel *et al.*, 2004), quite different dose/response slopes were generally found between herbicides with different modes of action, but not in every case. Furthermore, Michel *et al.* (2004) found slightly different slopes for some compounds among those that inhibit PS II. As discussed by Finney (1971), there are such exceptions to compounds with the same mode of action having the same dose/response slopes. Streibig *et al.* (1999) also observed differences in slopes with some PS II inhibitors that all bind to the Q_B binding niche on the D_1 protein of the PS II complex in a target-site specific dose/response assay using isolated thylakoid membranes. The dose/response slopes of sorgoleone analogs were similar, but differed significantly from the slopes obtained for the synthetic PS II inhibitors diuron and bentazon. They concluded that interactions with different amino acids at the Q_B binding site are responsible for observed disparity in slopes. Hence, although the same target site and, thus, mode of action was affected, the often-adopted principle of compounds with similar modes of action having similar slopes did not apply. The same phenomenon was observed by Follak and Hurle (2003) investigating the effects of two synthetic PS II inhibitors that bind to the Q_B niche. However, although dose/response relationships obtained for the inhibition of chlorophyll fluorescence in intact leaves of *Helianthus annuus* L. showed significant differences in slopes (Fig. 6a), the assessment of effects on plant growth at

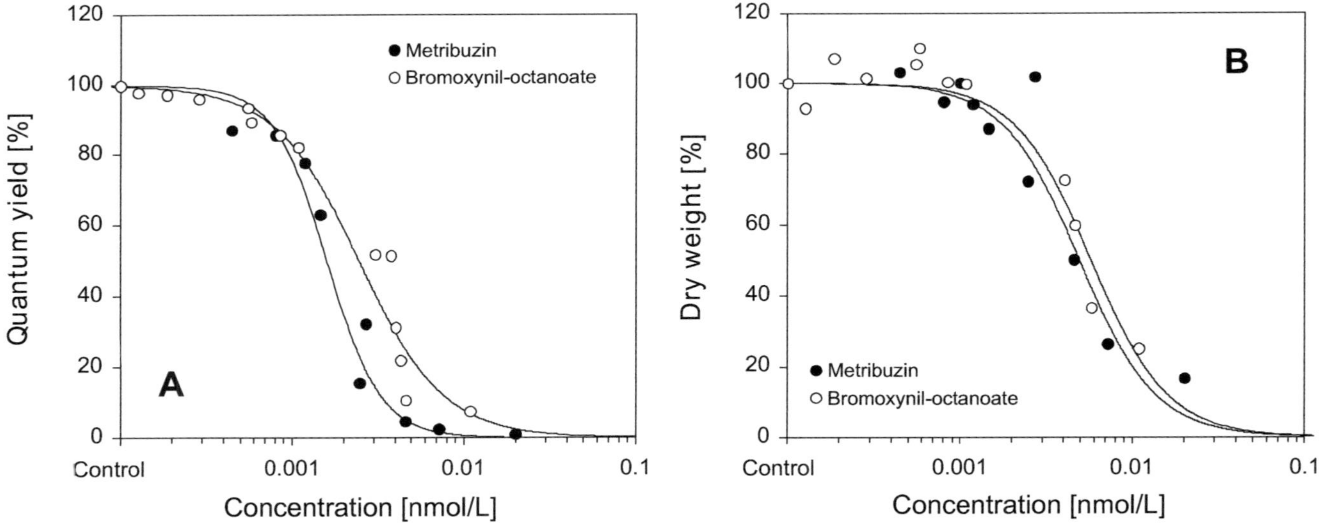

Fig. 6 The effects of response parameters measured on dose/response relationships of two PS II inhibitors affecting *Helianthus annuus*. (A) Quantum yield of electron transport at PS II of exposed leaves 1 h after exposure for 24 h. (B) Dry weight of exposed plants 16 d after exposure for 24 h. From Follak and Hurle (2003) with permission from Elsevier Press.

the end of the bioassay resulted in parallel curves for both compounds (Fig. 6b).

In this case, 24 h might have been insufficient time for the full uptake of every dose of the herbicides. Nevertheless, putative differences in the molecular mechanism of action at the target site that influence the slopes of compounds in target assays, may diminish at the whole plant level, where other physiological aspects of the mode of action may be more relevant for the shape of a dose/response curve. Although further studies are needed on these issues, results demonstrate that differences in slopes between compounds may depend on the response parameter measured. Effects closely linked to the phytotoxin's mode of action may differ from the full effect on whole-plant level. Drawing conclusions about mode(s) of action cannot be done from slopes of dose/response experiments alone. In addition, the slope of a dose/response curve is variable, depending on a variety of factors even with the same compound. In whole plant studies for example, time does not only lower I_{50} values of a compound as demonstrated in Fig. 5, but also steepens the dose/response relationship. The increase in slope is due to a recovery of plants at low levels of injury and often an intensification of effects at high levels of injury (Seefeldt *et al.*, 1995) (Fig. 7a). Furthermore, if the same active compound is applied to different target species at different rates, many factors, such as germination and seedling development, can also cause changes in the slope. In our own studies, this was the case for *Matricaria chamomilla* L. tested for effects of BOA among 10 species (Fig. 7b), as well as for *Ageratum conyzoides* L. among six species affected by the sesquiterpene lactone parthenin. In both studies, a significantly different, shallower slope occurred only for these two slow growing species. Thus, we concluded that differences in slopes were attributable to delayed development of toxicity with slow growing species.

An example of genetic variation affecting dose/response slopes was observed in studying the effect of the ALS inhibitor propoxycarbazone-sodium on sensitive and target site-resistant biotypes of three *Amaranthus* species (Wagner, 2004). In all three species, DNA-sequencing proved a specific point mutation on the ALS gene known to decrease or inhibit herbicide binding at the target site of resistant biotypes (Fedtke and Duke, 2004). In dose/response assays, such differences in herbicide binding at the target-site usually only results in a parallel displacement of dose/response curves to higher doses for the resistant biotype. This was the case for *Amaranthus blitoides* S. Watts. The exchange of proline_{197} (amino acid number as standardized by the *Arabidopsis thaliana* (L.) Heynh. sequence) by serine in the amino acid sequence of the ALS gene of the resistant biotype did not significantly change the dose/response slope for *in vitro* enzyme activity (Fig. 8a). A displacement of dose/response curves to

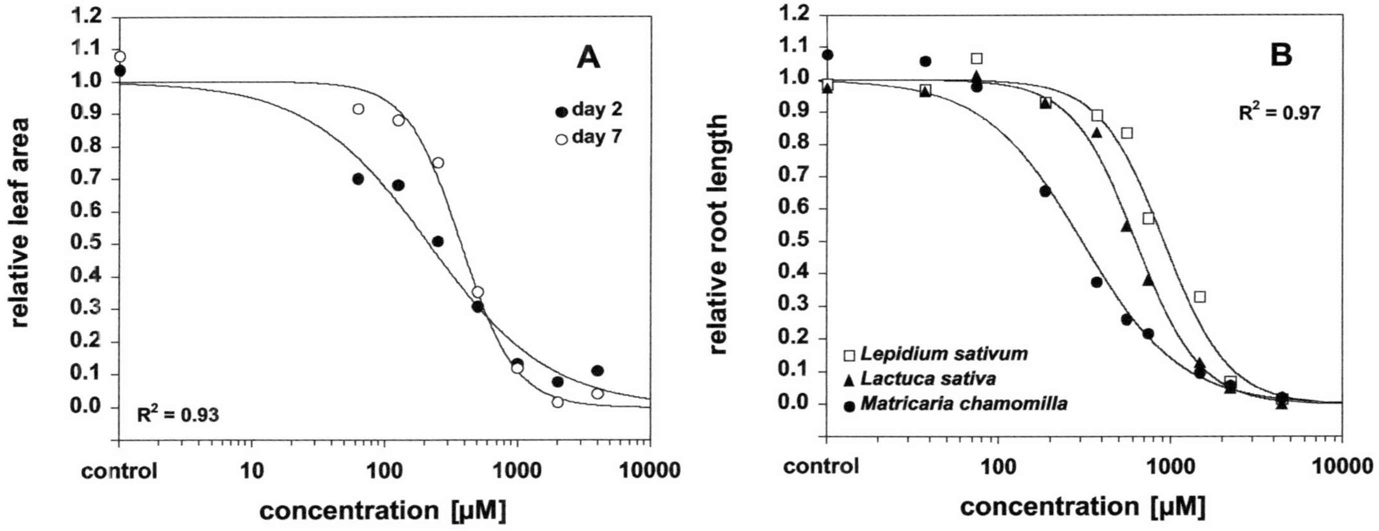

Fig. 7 Variations in slope of normalized dose/response relationships of the same active compound. (A) Effects of duration of treatment with scopoletin on *Lemna pausicostata* [from data discussed in Belz *et al.* (2005)]. (B) Effects of BOA on different species [from data discussed in Belz and Hurle (2004)].

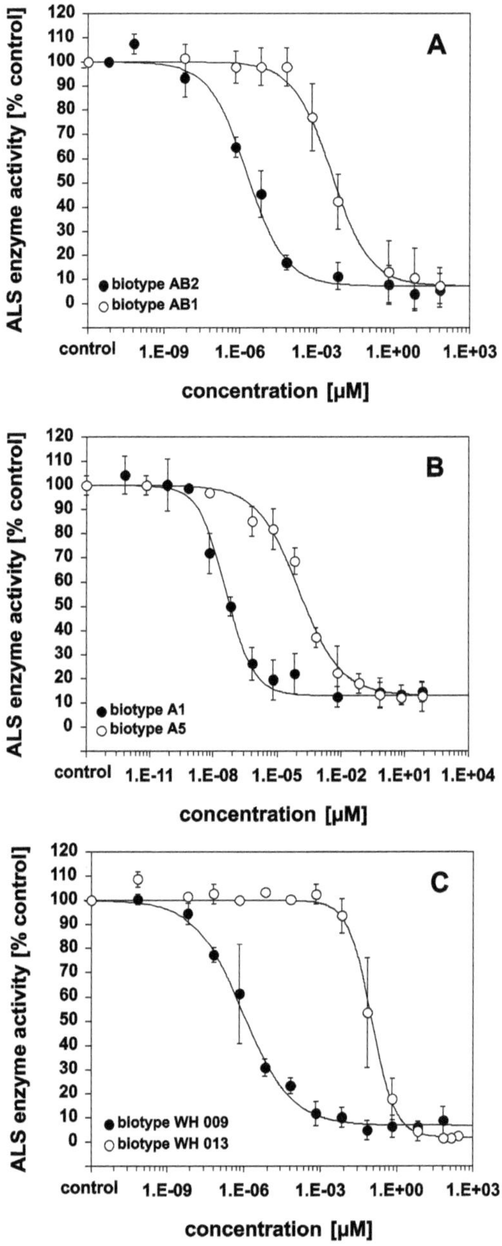

Fig. 8 Dose/response relationships for effects of propoxycarbazone-sodium concentration on ALS-enzyme activity of sensitive (●) and target-site resistant (○) biotypes of *Amaranthus blitoides* (A), *A. retroflexus* (B), and *A. rudis* (C). Error bars are ±1 SD. From Wagner (2004).

higher doses for the resistant biotypes was also observed with the other *Amaranthus* species; however, significantly different slopes appeared (Fig. 8b, c). In *Amaranthus retroflexus* L. an exchange of $proline_{197}$ by leucine in the amino acid sequence of the ALS gene of the resistant biotype led to a significantly shallower slope of the dose/response relationship. In *Amaranthus rudis* Sauer, an exchange of $tryptophan_{574}$ by leucine resulted in a significantly steeper slope. It is currently unknown what caused observed differences in slopes; however, with certain point mutations on the ALS gene, apparently not only the herbicide binding at the target site is affected. Due to the slow-binding nature of ALS inhibitors, their inhibition increases with time (Fedtke and Duke, 2004). Thus, it could be speculated that a change in protein structure of the ALS enzyme also influences the development of toxicity over time, which might ultimately change the dose/response slopes. Nevertheless, results show that even with the same compound or affected metabolism, small genetic variations of the target site among biotypes can change slopes of dose/response relationships. In contrast to the isolated enzyme level, evaluating the effects of propoxycarbazone-sodium on biomass of *A. retroflexus* biotypes at the whole-plant-level, showed a parallel displacement of dose/response curves to higher doses for the resistant biotype (Wagner, 2004). This shows once more that the *in vitro* activity of phytotoxins may be different from what is observed *in vivo*.

Besides the response parameter measured, the sampling time, and the target species or tissue, other possible sources for variation in slopes most likely not attributable to differences in the primary mode of action can be the dose range, the physiological status of the target species or tissue as well as nonconforming test conditions (Nyffeler *et al.*, 1982; Streibig *et al.*, 1995; Belz *et al.*, 2005). Still other factors that can affect dose/response slopes of even a single compound are the amount of genetic variation with respect to a factor influencing the mode of action within a plant population, factors that influence movement of the phytotoxin to the target site, and the rate of metabolic inactivation of the phytotoxin at different concentrations or under different environmental conditions. Therefore, experimental factors and test conditions must be critically evaluated before meaningful conclusions can be drawn from slopes of dose/response relations.

HORMESIS

Hormesis is the beneficial or stimulatory effect at a dose of the toxicant that is lower than the dose that causes the first detectable negative effect. This term was coined by plant pathologists who found that a natural fungicide could stimulate fungal growth at lower than fungicidal doses (Southam and Erlich, 1943). In a study in one of our laboratories, hormesis was found with the natural fungicides 4- and 7-hydroxycoumarin, 5-methoxypsoralen, and 1-methyl-2-[3′,4′-(methylenedioxy) phenyl-4-quinolone on *Phomopsis viticola* (Oliva *et al.*, 2003). In this study, hormesis was not observed with other natural fungicides on this fungus, nor with any of these natural fungicides on other fungal species.

If hormesis occurs, a complete dose/response curve will be "non-linear" or a "U curve", depending on the nomenclature and dose/response plotting method used. Molisch (1937) meant the term allelopathy to cover both stimulatory and inhibitory effects of natural compound interactions between organisms. Rice (1974) originally considered allelochemicals to have only negative effects, however, he later reconsidered (Rice, 1984), because "most, if not all, organic compounds that are inhibitory at some concentrations are stimulatory to the same processes in very small concentrations". However, in allelopathy literature the same compound has seldom been demonstrated to have both stimulatory and negative effects on the same organism, depending on the dose. When hormesis takes place, the maximum stimulatory effect is generally 130 to 160% of the control (Calabrese, 2002), although lower levels might occur, but are more difficult to prove. Calabrese and Baldwin (2001) provide a survey of several cases of hormetic responses of plants to phytotoxins.

Hormesis is commonly found, depending on the measured endpoint, when complete dose/response curves are generated for responses of plants to phytotoxins and other compounds that are not considered phytotoxins, but are toxic at high doses (Calabrese and Baldwin, 2001). However, in a study of complete dose/response relationships of *L. pausicostata* to 26 different herbicides with 19 different modes of action, Michel *et al.* (2004) found hormesis with only a few herbicides (unpublished data). An example of one of these is shown in Fig. 9.

A phytotoxin that is not hormetic with one species can be hormetic with another. For example, the herbicide glyphosate was not hormetic with *L. pausicostata* (unpublished data of S.O. Duke), but is hormetic with a variety of terrestrial plants (Wagner *et al.*, 2003; Schabenberger *et al.*, 1999) (unpublished research of E. Velini and Fig. 10). No hormesis was found with a glyphosate-resistant variety of soybean, suggesting the mechanism of hormesis is related to the mechanism of action of the herbicide (E. Velini, unpublished data). A pronounced hormetic effect was found with metsulfuron-methyl effects on growth of *Hordeum vulgare* L. (Cedergreen *et al.*, 2004) (Fig. 11), but the mechanism of this effect is unknown. Since glyphosate inhibits the shikimate pathway (Duke, 1988), reduced lignin

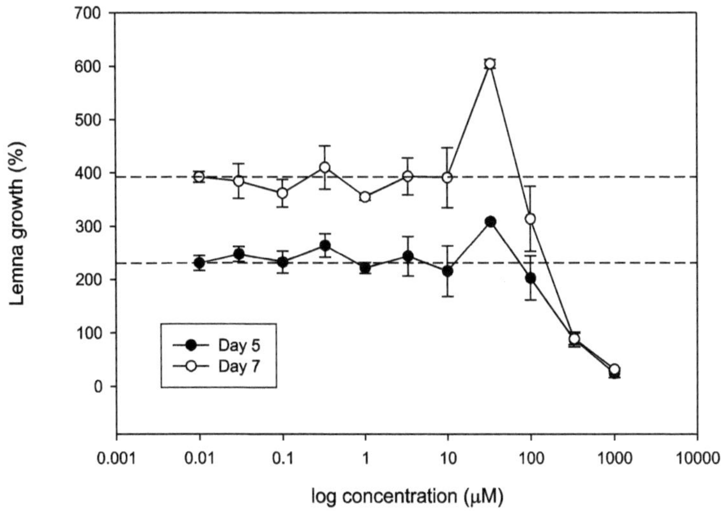

Fig. 9 Effects of the herbicide naptalam on *Lemna pausicostata* growth at 5 and 7 d after exposure. The dotted lines are the untreated means at the two times after treatment. Error bars are ±1 SE. (S.O. Duke, unpublished).

formation, resulting in increased cellular expansion which could occur at low doses.

Differences in hormesis are not only found between phytotoxins with different modes of action, but also within the same category of compounds. Dose/response relations of *Amaranthus hybridus* L. generated for six different isothiocyanates (Petersen *et al.*, 2001), phytotoxic metabolites of glucosinolates from Brassiceae species, proved significant hormesis only for allyl isothiocyanate (unpublished data). The most pronounced beneficial effects were observed on root growth, while shoot growth was barely stimulated. The highest hormetic effects on root growth occurred at 7.2 μM (216% of control) and these effects disappeared at concentrations exceeding 45.1 μM (Fig. 12a). As the biological activity of the tested alkenyl- and aryl-isothiocyanates relies upon the same reactive -NCS group which interferes with nucleophilic partners (Drobnica *et al.*, 1977), it seems unlikely that differences in the types of physiological effects were responsible for the observed disparity in hormetic effects. More often it appears that hormesis is masked because of a poorly defined stimulatory dose range or the usually higher variability of measurements near the upper limit of the response curve (Streibig *et al.*, 1995). This may partly explain the fact that stimulation at low concentrations of allelochemicals is rarely described in allelopathy research. Consequently, almost nothing is yet known about underlying mechanisms of hormetic effects or specific dose levels for stimulation.

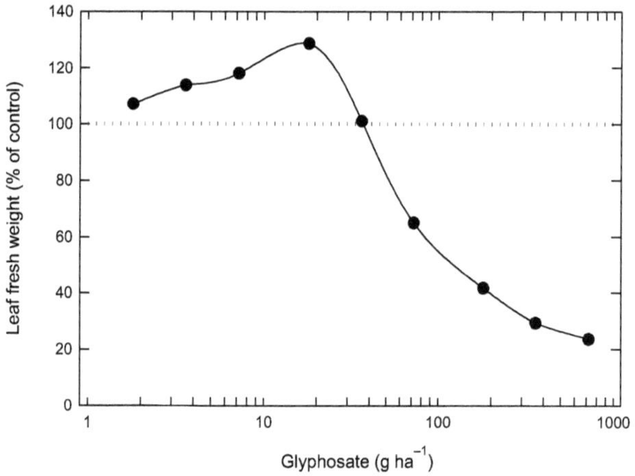

Fig. 10 Effects of glyphosate on soybean leaf fresh weight 14 d after application to non-transgenic soybean). The dotted line is the control value. The greatest fresh weight value is significantly greater than the control at the 5% probability level. Maize gave similar results. (E. Velini, unpublished data).

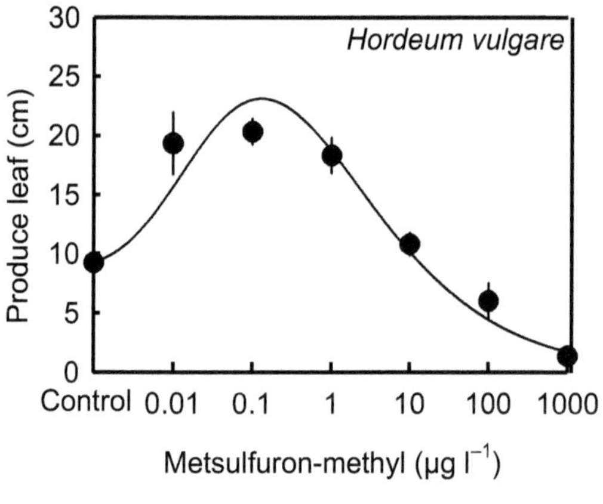

Fig. 11 An approximately 100% increase in leaf length of barley plants (*Hordeum vulgare*) treated with the herbicide metsulfuron-methyl. Data from the Cedergreen *et al.* (2004) study. Compliments of Nina Cedergreen.

Both types of allelopathy, stimulatory and inhibitory, were described for BOA (Friebe *et al.*, 1997; Belz *et al.*, 2005), the phytotoxic metabolite of the benzoxazinoid 2,4-dihydroxy-2*H*-1,4-benzoxazin-3(4*H*)-one (DIBOA), which is involved in allelopathy of graminaceous species. Friebe *et al.* (1997) showed that the plasma membrane H^+-ATPase may be involved in the biological activity of BOA since the activity of this enzyme changed from stimulatory to inhibitory as the dose of BOA increased. However, stimulation and inhibition by the same active compound does not necessarily need the same site of action (An *et al.*, 1993). Another example for hormesis in pure allelochemical studies is the sesquiterpene lactone parthenin, the principal component of capitate-sessile trichomes of certain *Parthenium hysterophorus* L. populations (Reinhardt *et al.*, 2004). Low concentrations of parthenin stimulated root growth in several species including crop plants such as *Triticum aestivum* L. (Batish *et al.*, 1997) or *Lactuca sativa* L. where maximum stimulation occurred at 191 μM (191% of control) (Fig. 12b; unpublished data). However, the limited doses for stimulation differed between test plant species. For example, the dose that caused the first detectable negative effect on *L. sativa* (746 μM) was 1.6 times higher than the dose giving an equivalent effect for *Echinochloa crus-galli* or *Eragrostis curvula* (unpublished data). As hormesis occurred at low concentrations of aqueous extracts of plant residues of *P. hysterophorus* (Pandey *et al.*, 1993; Pandey, 1994), subinhibitory levels of parthenin within extracts may be partly responsible for observed stimulation. Based on this, Pandey (1994) even suggested that extracts of *P. hysterophorus* may be explored as a possible stimulant for crop seedling growth. Besides the fact that levels of chlorophyll and carotenoids were enhanced at low extract concentrations (Pandey, 1994), potential mechanisms of hormesis by parthenin are yet unknown.

The case studies presented here demonstrate that either type of response, stimulation or inhibition, may be found with the same synthetic or natural toxicant, depending on its concentration. Calabrese and Baldwin (2003) and Stebbing (2003) indicate that hormesis is a very common phenomenon, yet, relatively few studies have claimed this phenomenon in plants, in contrast to the large body of knowledge associated with inhibitory effects. Complete dose/response curves that cover the stimulatory dose-range offer an opportunity to prove if hormesis occurs and to characterize dose levels where beneficial effects of toxic compounds recede in favor of harmful effects. Mathematical models are available to describe and analyze such stimulation-inhibition properties of toxicants (Brain and Cousens, 1989; An *et al.*, 1993; Schabenberger *et al.*, 1999; Liu *et al.*, 2003), and the application of appropriate models can considerably improve the conclusions drawn from such experiments.

Nevertheless, dose/response experiments must be conducted with care so that the control plants are also treated in every way that the treated plants are, with the exception of the active compound, as low levels of surfactants, additives, solvents or other materials with which a herbicide or phytotoxin might be formulated for treatment could be stimulatory as well. Evolutionary pressure might select against allelochemicals that provide a hormetic effect, since there would be a disadvantage to a plant

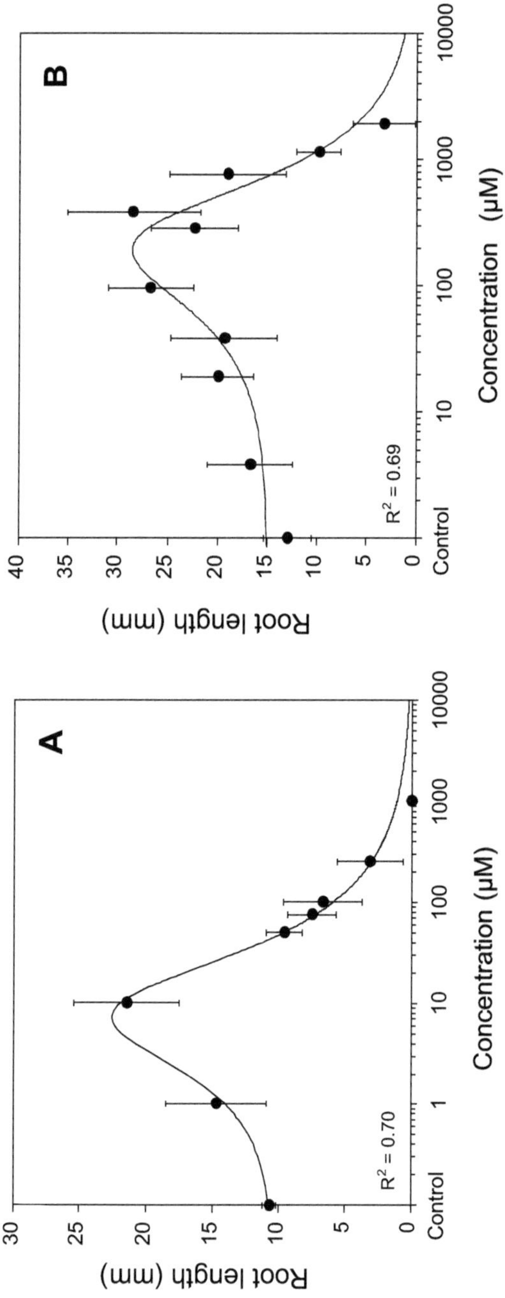

Fig. 12 Stimulatory effects of low doses of allyl isothiocyanate on root growth of *Amaranthus hybridus* (A) [from data discussed in Petersen *et al.* (2001)] and parthenin on root growth of *Lactuca sativa* (B). Error bars are ±1 SD.

species producing compounds that would enhance the growth of competitors. We know too little about hormetic effects of allelochemicals to make any conclusions about this hypothesis.

DOSING WITH PLANT DENSITIES

Two experimental dose/response approaches capitalize on varying densities of living plants to modify the dose of an active toxicant. Based on two bifocal perspectives, either the plant density of the target species as the "sink" for phytotoxins is altered, or the density of an allelopathic donor as the "source" of phytotoxins is a variable.

The principle of dosing with varying densities of target species implies that a finite dose of a toxicant is proportionally shared or diluted among an increasing number of target species. Thus, inhibitory effects result at low target plant densities and diminish or may become stimulatory at high densities where the portion of the finite dose available for each target plant is smaller. Such density-dependent phytotoxicity was first explored studying herbicides (Hoffman and Lavy, 1978; Andersen, 1981) and later introduced to allelopathy research by Weidenhamer *et al.* (1987, 1989). As these density-dependent phytotoxic effects are contradictory to yield-density responses in the case of resource competition, this experimental design allows us to distinguish and prove allelopathy in situations where competition is operative as well. Allelopathy applications mainly use the approach of dosing with varying densities of target species to evaluate the presence of phytotoxins in soils beneath a stand of suspected allelopathic plants or the release of inhibitors to the soil from plant residues.

Experiments are conducted in controlled environments or natural settings and usually establish monocultures of target species in soils where allelopathic plants have previously grown or plant residues are applied. Target plants are grown at different densities in a given soil volume and as plants start to interfere, predicted yield-density responses involve a reduction in individual plant growth, while the total yield per area remains constant across a wide range of densities (Kira *et al.*, 1953; Weidenhamer *et al.*, 1989). In the range where total yield is constant, the yield-density responses follow the universal "-1 law of constant final yield" (Kira's law), such that the relationship between log individual plant yield and log density is linear with a slope of -1. If a phytotoxin is present, density-dependent phytotoxic effects cause a decrease or a reversal of the slope of the relationship of log yield and log density and, thus, significant deviations from the law of constant yield are a clear indication for a chemical interference. Weidenhamer *et al.* (1989) for example proved the validity of this experimental design by demonstrating deviations from expected yield-density relationships when *Lycopericon esculentum* Mill. was grown in soil collected under a black walnut tree (*Juglans nigra* L.). It is well known that the plant-generated allelochemical juglone accumulates in the soil beneath the leaf canopy of walnut trees, and the presence of phytotoxins in the walnut soil was clearly indicated in the work of Weidenhamer *et al.* (1989) by a decrease in phytotoxicity with increasing plant density and a significant decrease in the slope of the log yield-log density line. We conducted an equivalent experiment with soil collected in autumn under a dense cover crop stand of winter turnip rape [*Brassica rapa* (L.) var. *rapa* spp. *oleifera* (DC.) Metzg.] (Petersen *et al.*, 2001; unpublished data), in order to evaluate the hypothesis that allelopathy might contribute to the weed suppressive ability of this cover crop. The total yield of above ground biomass of *A. hybridus* growing in turnip rape soil did not significantly change beyond target plant densities of three plants/pot (Fig. 13a). Evaluating the relationship between log-transformed data of mean plant weight and density in the range of densities where total yield was constant revealed a linear association with a slope that was not significantly different from -1 (Fig. 13b).

Thus, the results did not support the hypothesis of deviations from the expected decrease in dry weight with increasing plant density and, consequently, provided no indications for the presence of toxic substances in the soil in which turnip rape has been grown. The observed weed suppressive ability of living cover crop stands of turnip rape is therefore most likely attributable to resource competition as the main mechanism of plant-plant interaction.

The second approach that doses with densities is different, since it implies that varying doses of a toxicant are supplied by living donor plants proportionally to their density. Thus, here the source of a toxicant is variable and the sink, as the number of target species is constant, while previously a finite source was distributed to a varying sink. Inhibitory effects occur in this dose/response design at high donor plant densities where the amount of toxicants released is highest, while stimulatory effects may occur at low densities (Fig. 14).

Phytotoxicity increases with plant density which is consistent with effects of resource competition. Therefore, this experimental design allows for an allelopathic interpretation of findings only if resource competition is excluded. This restricts its application to model experiments in a controlled environment and forces experiments to be conducted under nutrient-free conditions (Wu *et al.*, 2000; Belz and Hurle, 2004). In contrast to the previous approach, this experimental design is yet mainly applied to study interactions between living plants grown on a plant-by-plant basis in mixed culture and exploits a continuous supply of allelochemicals by

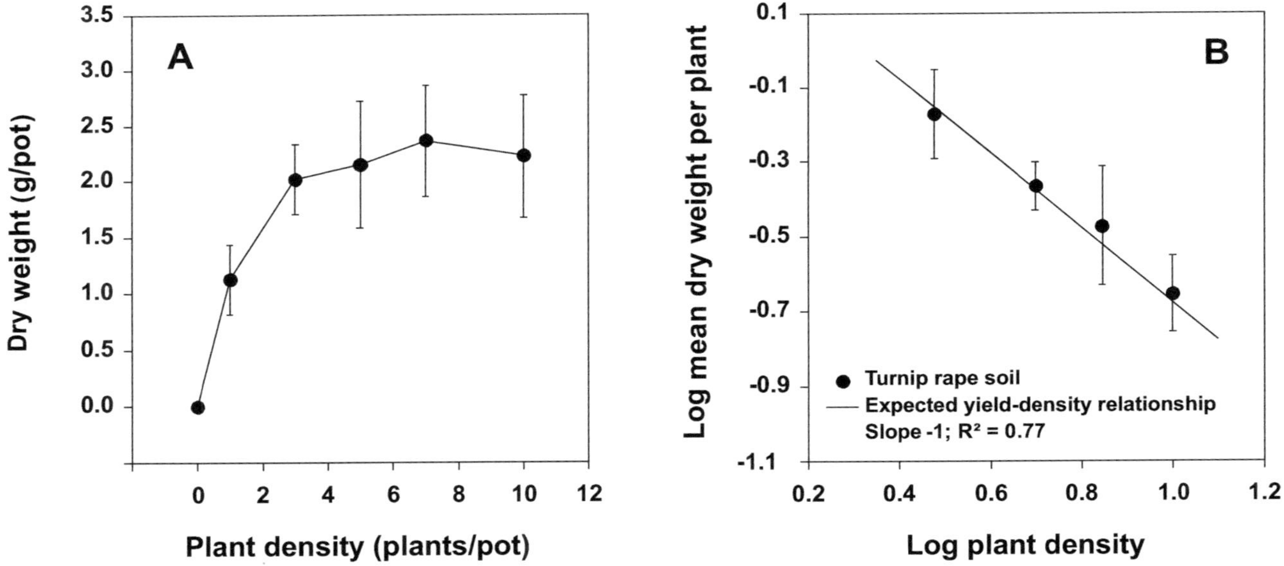

Fig. 13 Yield-density relationships for *Amaranthus hybridus* grown in turnip rape soil for 25 d under greenhouse conditions. (A) Response of total shoot dry weight to increasing plant densities. (B) Relationship of mean dry weight per plant and plant density in the range where total shoot dry weight is constant (3-10 plants/ pot) (R.G. Belz, unpublished). Error bars are ±1 SD.

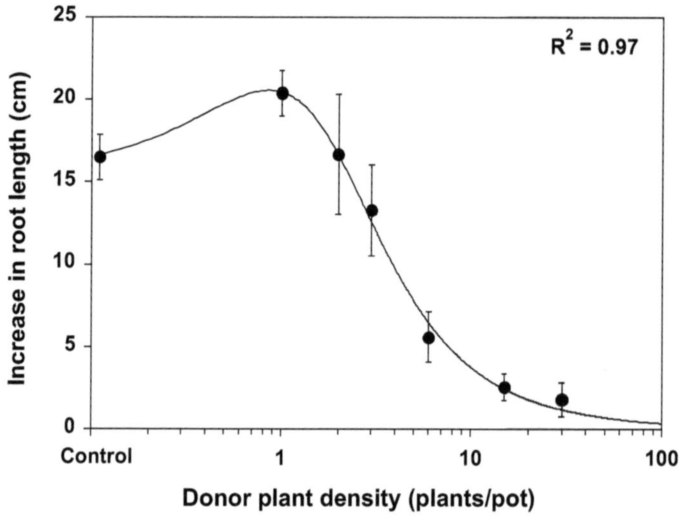

Fig. 14 Density-dependent phytotoxicity of root exudates of *Triticum spelta* L. cv. Frankenkorn on root growth of *Sinapis alba* after 7 d of mixed hydroponic culture. Error bars are ±1 SD. From data discussed in Belz and Hurle (2005).

living donor plants. Wu *et al.* (2000) demonstrated at first that root exudation of allelochemicals by a *T. aestivum* cultivar was sufficient to cause density-dependent phytotoxicity on *Lolium rigidum* Gaud. in mixed culture. Based on the obtained donor density (dose)/response relationship they selected an appropriate density to screen different cultivars for allelopathic potential using the same agar bioassay. Belz and Hurle (2004) developed a hydroponic bioassay and studied density-dependent phytotoxicity of root exudates of grain crops on *Sinapis alba* L. in a range of dose/response applications (Belz *et al.*, 2005). For example, evaluating the time course for the inhibition of *S. alba* by *Secale cereale* L. cv. Amilo illustrated that time affected the performance of exuded compounds and the shape of the dose/response relationship. The development of symptoms takes time and increased poisoning with time of exposure is reflected in whole-plant dose/response studies by a steepening of the curve with increasing time of exposure (Seefeldt *et al.*, 1995) (Fig. 15a). Although, in the present case response curves differed significantly merely in upper limits, the dependence of responses on the time after exposure is critical for comparing relationships and the choice of the right end point. Therefore, the most complete description of a donor density (dose)/ response relationship is obtained by evaluating time courses and if these are furthermore conjoined with effects on the growth pattern of the target species at different donor plant densities, a complete description of such

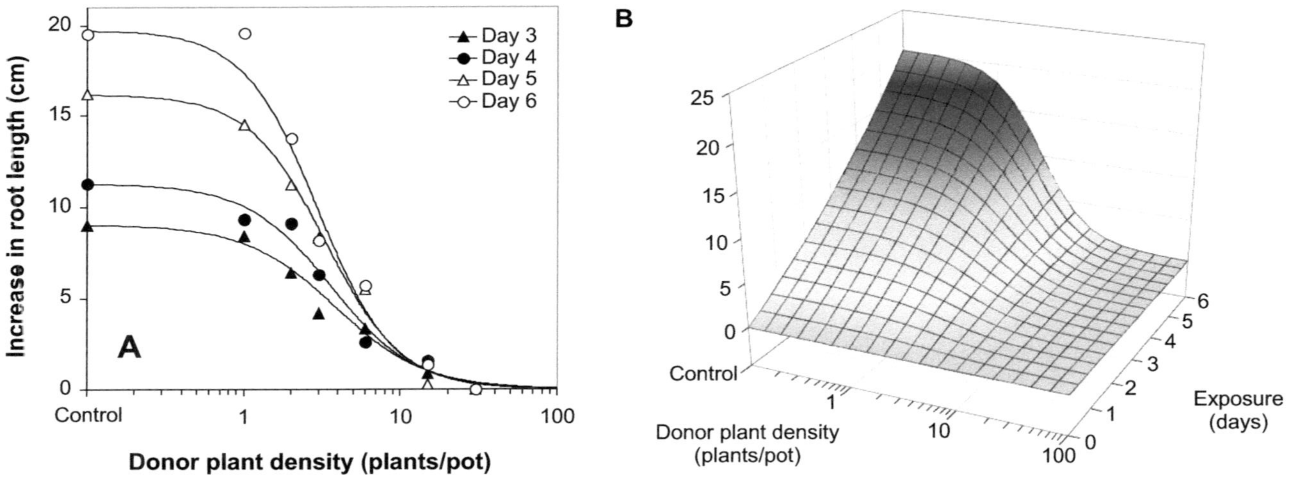

Fig. 15 Chronological development of the donor plant density (dose)/response relationship for the inhibition of *Sinapis alba* by root exudates of *Secale cereale* cv. Amilo. after 7 d of mixed hydroponic culture. (A) Density/response relationships; (B) generalized density/response growth surface with density/response relations running parallel to the density axis and growth curves running parallel to the time axis. From data discussed in Belz et al. (2005).

dynamic responses is obtained (Streibig *et al.*, 1993; Belz *et al.*, 2005) (Fig. 15b).

Belz and Hurle (2005) used their dose/response bioassay further to screen for allelopathy in a selection of grain crop cultivars and characterized the potential of each of the 146 Triticeae cultivars by the ED_{50} value of their dose/response relationships. Chemical analysis of root exudates showed that the dose of benzoxazinoids at different densities followed the expected pattern of increasing amounts with increasing donor plant density, a necessary condition for any allelochemical causally involved in the observed phytotoxicity effects. Correlating the estimated ED_{50} values with the amounts of benzoxazinoids finally proved that the ability to exude these compounds considerably contributed to the allelopathic potential in bioassay.

The principle of dosing with densities can be exploited with two different experimental approaches. The examples presented here indicate the value of such density-dependent dose/response applications for new approaches and more complex studies of the mechanisms of allelopathy phenomena under a wide range of experimental conditions. Mathematical regression models are available to describe and analyze density-dependent phytotoxicity in either approach (Finney, 1978; Streibig, 1988; Brain and Cousens, 1989; Sinkkonen, 2001, 2003), which may help to simplify the complexity of such chemical interactions. Due to the still rather limited use of such bioassay designs, the range of possible applications has not yet been determined. Nevertheless, both approaches feature certain advantages and limitations and the method of choice will depend on the particular objectives under study.

CONCLUSION

In this chapter, we have attempted to demonstrate the utility, desirability, and sometimes necessity of proper dose/response studies in allelopathy research. The type of dose/response analysis will depend on the question that is asked. The information from such studies is required as part of the proof of a compound's role as an allelochemical in nature. Furthermore, proper mode of action research cannot be conducted without first determining the dose/response relationships. Proof of interactions (antagonism or synergism) of allelochemicals cannot be made without appropriate dose/response studies. This is very important in allelopathy research, in that synergism is often invoked without proper proof to predict or explain stronger effects than expected from results with individual allelochemicals. Detailed dose/response information is essential for determination of hormesis in most cases. In allelopathy studies with allelopathic plants, dose/response studies can be made using donor plant density to provide different doses, provided one can assume that density does not significantly affect biosynthesis and/or secretion/ exudation of the allelochemical(s) produced by the donor plant. In summary dose/response relationships must be considered in many different types of allelopathy studies. We have provided some guidelines and examples for such considerations.

References

Amagasa, T., Paul, R.N., Heitholt, J.J., Duke, S.O. (1994) Physiological effects of cornexistin on *Lemna pauscicostata*. Pestic. Biochem. Physiol. 49: 37-52.

An, M., Johnson, I.R., Lovett, J.V. (1993) Mathematical modeling of allelopathy: biological response to allelochemicals and its interpretation. J. Chem. Ecol. 19: 2379-2388.

Andersen, R.N. (1981) Increasing herbicide tolerance of soybeans (*Glycine max*) by increasing seeding rates. Weed Sci. 29: 336-338.

Batish, D.R., Kohli, K.H., Singh, H.P., Saxena, D.B. (1997) Studies on herbicidal activity of parthenin, a constituent of *Parthenium hysterophorus*, towards billgoat weed (*Ageratum conyzoides*). Curr. Sci. 73: 369-371.

Becerril, J.M., Duke, S.O. (1989) Protoporphyrin IX content correlates with activity of photobleaching herbicides. Plant Physiol. 90: 1175-1181.

Belz, R.G., Hurle, K. (2004) A novel laboratory screening bioassay for crop seedling allelopathy. J. Chem Ecol. 30: 175-198.

Belz, R.G., Duke, S.O., Hurle, K. (2005) Dose-response - a challenge for allelopathy? Nonlinearity Biol. Toxicol. Med. 3: 173-211.

Belz, R.G., Hurle, K. (2005) Differential exudation of two benzoxazinoids - one of the determining factors of seedling allelopathy in Triticeae species. J. Agric. Food Chem. 53: 250-161.

Brain, P., Cousens, R. (1989) An equation to describe dose responses where there is stimulation of growth at low doses. Weed Res. 29: 93-96.

Calabrese, E.J. (2002) Hormesis: changing view of the dose-response, a personal account of the history and current status. Mutat. Res. 511: 181-189.

Calabrese, E.J., Baldwin, L.A. (2001) Hormesis: A generalizable and unifying hypothesis. Crit. Rev. Toxicol. 31: 353-424.

Calabrese, E.J., Baldwin, L.A. (2003) Toxicology rethinks its central belief. Nature 421: 691-692.

Cedergreen, N., Streibig, J.C., Spliid, N.H. (2004) Sensitivity of aquatic plants to the herbicide metsulfuron-methyl. Ecotoxicol. Environ. Safety. 57: 153-161.

Dayan, F.E., Duke, S.O. (1997) Phytotoxicity of protoporphyrinogen oxidase inhibitors: Phenomenology, mode of action and mechanisms of resistance. *In* Herbicide Activity: Toxicology, Biochemistry and Molecular Biology. R.M. Roe, J.D. Burton, and R.J. Kuhr, (eds.). I.O.S. Press, Amsterdam. The Netherlands pp. 11-135.

Drobnica, L., Kristian, K., Augustin, J. (1977) The chemistry of the -NCS group. *In* The Chemistry of Cyanates and their Thio Derivates, Part 2., S. Patai, (ed.). John Wiley & Sons, NY, USA pp. 1003-1221.

Duke, S.O. (1988) Glyphosate. *In* Herbicides-Chemistry, Degradation and Mode of Action P.C. Kearney, D.D. Kaufman, Vol. III. (eds.) Marcel Dekker, Inc., NY, USA. pp. 1-70.

Duke, S.O., Williams, R.D., Markhart, A.H. III (1983) Interaction of moisture stress and three phenolic compounds on lettuce seed germination. Ann. Bot. 52: 923-926.

Einhellig, F.A., Rasmussen, J.A ., Hejl, A.M., Souza, I.F. (1993) Effects of root exudates sorgoleone on photosynthesis. J. Chem. Ecol. 19: 369-375.

Fedtke, C., Duke, S.O. (2004) Herbicides. *In* Plant Toxicology, B., Hock and E.F. Elstner (eds.) 4th ed., Marcel Dekker, NY, USA. pp. 247-330.

Finney, D.J. (1971) Probit Analysis, 3rd ed. Cambridge Univ. Press, London, UK .

Finney, D.J. (1978) Statistical method of biological assay, 3rd ed., Charles Griffin, London, UK.

Follak, S., Hurle, K. (2003) Effect of airborne bromoxynil-octanoate and metribuzin on non-target plants. Environ. Pollut. 126: 139-146.

Friebe, A., Roth, U., Kück, P., Schnabl, H., Schulz, M. (1997) Effects of 2,4-dihydroxy-1,4-benz-oxazin-3-ones on the activity of plasma membrane H^+-ATPase. Phytochemistry 44: 979-983.

Green, J.M., Streibig, J.C. (1993) Herbicide mixtures. *In* Herbicide Bioassays, J.C. Streibig and P. Kudsk. (eds.). CRC Press, Boca Raton, FL, USA. pp. 117-135.

Hoffman, D.W., Lavy, T.L. (1978) Plant competition for atrazine. Weed Sci. 26: 94-99.

Kira, T., Ogawa, H., Sakazaki, N. (1953). Intraspecific competition among higher plants I. Competition yield-density interrelationship in regularly dispersed populations. J. Inst. Polytech., Osaka City University, Japan. 4D: 1-16.

Kudsk, P., Chunhong, J., Mathiassen, S.K. (2004). Joint action of wheat allelochemicals. Proceedings 2nd European Allelopathy Symposium, Putawy, Poland. pp. 83-84.

Liu, D.L., An, M., Johnson, I.R., Lovett J.V. (2003) Mathematical modeling of allelopathy. III. A model for curve-fitting allelochemical dose responses. Nonlinearity Biol. Toxicol. Med. 1: 37-50.

Matsumoto, H., Duke, S.O. (1990) Acifluorfen-methyl effects on porphyrin synthesis in intact *Lemna pausicostata* Hegelm. 6746 plants. J. Agric. Food Chem. 38: 2066-2071.

Meazza, G., Scheffler, B.E., Tellez, M.R., Rimando, A.M., Romagni, J.G., Duke, S.O., Nanayakkara, D., Khan, I.A., Abourashed, E.A., Dayan, F.E. (2002). The inhibitory activity of natural products on plant *p*-hydroxyphenylpyruvate dioxygenase. Phytochemistry 60: 281-288.

Michel, A., Petersen, J., Dogan, M.N., Ernst, V. (1999) Instruction for design and evaluation of quantitative dose-response relationships - shown on the efficacy of herbicides. Gesunde Pflanze 51: 10-19.

Michel, A., Johnson, R.D., Duke, S.O., Scheffler, B.E. (2004) Dose-response relationships between herbicides with different modes of action and growth of *Lemna pausicostata* - an improved ecotoxicological method. Environ. Toxicol. Chem. 23: 1074-1079.

Molisch, H. (1937) Der Einfluss einer Pflanze auf die andere-Allelopathie. Fischer, Jena, Germany.

Nyffeler, A., Gerber, H.R., Hurle, K., Pestemer, W., Schmidt, R.R. (1982). Collaborative studies of dose-response curves obtained with different bioassay methods for soil-applied herbicides. Weed Res. 22: 213-222.

Oliva, A., Meepagala, K.M., Wedge, D.E., Harries, D., Hale, A.L., Aliotta, G., Duke, S.O. (2003) Natural fungicides from *Ruta graveolens* L. leaves, including a new quinolone alkaloid. J. Agric. Food Chem. 51: 890-896.

Pandey, D.K., Kauraw, L.P., Bhan V.M. (1993) Inhibitory effect of parthenium (*Parthenium hysterophorus* L.) residue on growth of water hyacinth (*Eichornia crassipes* Mart Solms.). II. Relative effect of flower, leaf, stem, and root residue. J. Chem. Ecol. 19: 2663-2671.

Pandey, D.K. (1994) Inhibition of Salvinia (*Salvinia molesta* Mitchell) by parthenium (*Parthenium hysterophorus* L.). II. Relative effect of flower, leaf, stem, and root residue on salvinia and paddy. J. Chem. Ecol. 20: 3123-3131

Petersen, J., Belz, R., Walker, F., Hurle, K. (2001) Weed suppression by release of isothiocyanates from turnip-rape mulch. Agron. J. 93: 37-43.

Rasmussen, J.A., Einhellig, F.A. (1977) Synergistic effects of *p*-coumaric and ferulic acids on germination and growth of grain sorghum. J. Chem. Ecol. 3: 197-205.

Reinhardt, C., Kraus, S., Walker, F., Foxcroft, L., Robbertse, P., Hurle, K. (2004) The allelochemical parthenin is sequestered at high level in capitate-sessile trichomes on the leaf surface of *Parthenium hysterophours*. J. Plant Dis. Prot. 19: 253-261.

Rice, E. (1974) Allelopathy, Academic Press, NY, USA.

Rice, E. (1984) Allelopathy, 2nd Ed. Academic Press, Orlando, FL, USA.

Ritz, C., Streibig, J.C. (2005) Bioassay analysis using R.J. Statistical Software 12: www.jstatsoft.org. pp. 22.

Schabenberger, O., Tharp, B.E., Kells, J.J., Penner, D. (1999) Statistical tests for hormesis and effective dosages in herbicide dose response. Agron. J. 91: 713-721.

Seefeldt, S.S., Jensen, J.E., Fuerst, E.P. (1995) Log-logistic analysis of herbicide dose-response relationships. Weed Tech. 9: 218-227.

Sinkkonen, A. (2001) Density-dependent chemical interference - an extension of the biological response model. J. Chem. Ecol. 27: 1513-1522.

Sinkkonen, A. (2003) A model describing chemical interference caused by decomposing residues at different densities of growing plants. Plant Soil. 250: 315-322.

Southam, C.M., Ehrlich, J. (1943) Effects of extract of western red-cedar heartwood on certain wood decaying fungi in culture, Phytopathology 33: 517-524.

Stebbing, A.R.D. (2003) A mechanism for hormesis - A problem in the wrong direction. Crit. Rev. Toxicol. 33: 463-467.

Streibig, J.C. (1988) Herbicide bioassay. Weed Res. 28: 479-484.

Streibig, J.C., Rudemo, M., Jensen, J.E. (1993) Dose-response curves and statistical models. *In* Herbicide Bioassays, J.C. Streibig and P. Kudsk (eds.). CRC Press, Boca Raton, FL, USA. pp. 29-55.

Streibig, J.C., Walker, A., Blair, A.M., Anderson-Taylor, G., Eagle, D.J., Friedländer, H., Hacker, E., Iwanzik, W., Kudsk, P., Labhart, C., Luscombe, B.M., Madafiglio, G., Nel, P. C., Pestemer, W., Rahman, A., Retzlaff, G., Rola, J., Stefanovic, L., Straathof, H.J.M., Thies, E.P. (1995) Variability of bioassays with metsulfuron-methyl in soil. Weed Res. 35: 215-224.

Streibig, J.C., Dayan, F.E., Rimando, A.M., Duke, S.O. (1999). Joint action of natural and synthetic photosystem II inhibitors. Pestic. Sci. 55: 137-146.

Wagner, J. (2004) Wirkortspezifische ALS-Inhibitor-Resistenz bei *Amaranthus* spp. Dissertation University of Hohenheim, Verlag Grauer, Beuren, Stuttgart, Germany.

Wagner, R., Kogan, M., Parada, A.M. (2003) Phytotoxic activity of root absorbed glyphosate in corn seedlings (*Zea mays* L.). Weed Biol. Manage. 3: 228-232.

Weidenhamer, J.D, Mortan, T.C., Romeo, J.T. (1987) Solution volume and seed number: often overlooked factors in allelopathic bioassays. J. Chem. Ecol. 13: 1481-1491.

Weidenhamer, J.D., Hartnett, D.C., Romeo, J.T. (1989) Density-dependent phytotoxicity: distinguishing resource competition and allelopathic interference in plants. J. Chem. Ecol. 26: 613-624.

Wu, H., Pratley, J., Lemerle, D., Haig, T. (2000) Laboratory screening for allelopathic potential of wheat (*Triticum aestivum*) accessions against annual ryegrass. Aust. J. Agric. Res. 51: 259-266.

2

Can Data Derived from Field and Laboratory Bioassays Establish the Existence of Allelopathic Interactions in Nature?

Udo Blum

Department of Botany, North Carolina State University
Raleigh, North Carolina 27695-7612, USA

Characterizing actual processes (mechanisms) associated with allelopathic interactions under field conditions is very difficult, some would say impossible. This has led to the use of model systems, simplified representations of real systems, to characterize such processes. The resulting data and insights from such model systems are subsequently expressed in a variety of forms (e.g., word models, picture models, physical models, and mathematical models (empirical or mechanistic)) to communicate information, suggest new questions, solve problems, integrate concepts, and identify counter-intuitive behavior. Although all of these types of models are used frequently by researchers studying allelopathic interactions, the most commonly used are empirical mathematical models derived from field and/or laboratory bioassays. A discussion of the underlying assumptions, limitations, and uses of the latter type of models to establish the existence of allelopathic interactions in nature are presented.

Keywords: *allelopathic interactions, experimental approaches, field and laboratory bioassays, hypothesis testing, mathematical models, scientific method*

INTRODUCTION

For the science of allelopathy, like all other sciences, there is a prescribed methodology by which problems are to be solved, the scientific method. Once a problem has been identified, this method requires that alternative hypotheses, tentative explanations, be generated which can be experimentally tested. Occurrence (acceptance) or non-occurrence (rejection) of predictions deduced for each hypothesis are then determined by means of experiments. Finally, science progresses not by trying to confirm hypotheses but by attempting to falsify them since it is usually possible to find at least some confirmatory evidence for any hypothesis, but one solid piece of negative data refutes a hypothesis completely. Unfortunately, manuscripts reporting negative results are much harder to publish than confirmatory manuscripts because negative results are often mistrusted by the scientific community.

Developing and testing mathematical models derived from field and/ or laboratory bioassays is one form of hypothesis testing. Utilizing such models in hypothesis testing requires a clear understanding of their underlying assumptions and limitations. For example, models derived from laboratory experiments, which have a low spatial and temporal scale, are best for describing behavior and testing hypotheses associated with individual plants, while models derived from field experiments, which have a much larger spatial and temporal scale, are best for describing behavior and testing hypotheses associated with populations, communities or ecosystems. Simply put, models at different scales yield answers to different types of questions. Given this context, the question asked in the title might be restated as follows: At what scale should allelopathic interactions be demonstrated to establish their existence in nature? In this chapter, I plan to answer this question and the original question of the title by utilizing examples from research literature describing dicotyledonous weed emergence in no-till wheat-ecosystems.

DISCUSSION

Alternative Hypotheses

Observations and the literature provide us with ample hypotheses for how dicotyledonous weed emergence might be regulated in no-till wheat systems. Wheat cover crop residues might control weed emergence by several mechanisms: a) effects of the physical barrier of the residue and associated changes in the microclimate of the soil (Fester and Peterson, 1979; Worsham, 1989); b) immobilization of nutrients, particularly nitrogen (Broadbent and Tyler, 1962; Kimber, 1973); c) inhibition by allelopathic compounds (Patrick, 1971; Willard and Penner, 1976; Lynch, 1977; Tang and Waiss, 1978; Lieble and Worsham, 1983; Shilling *et al.*, 1985, 1986; Niemeyer *et al.*, 1989; Nair *et al.*, 1990); and d) reduced disturbance of soil and/or increased weed-seed predation in no-till systems, with no direct effect of the wheat residues (Brust and House, 1988; Worsham, 1989; Brust, 1994; Blum *et al.*, 1997;).

Experimental Approaches

The complexity of the problem, i.e., what regulates dicotyledonous weed emergence in no-till wheat agro-ecosystems, should be very evident from the alterative hypotheses proposed above. Designing field experiments and including all the appropriate controls (assuming we know all the controls that must be included) to solve such a complex problem is presently not feasible, although various aspects or subsets of the problem have been studied using field bioassays. For example, experiments have been designed to determine the influence of the physical barriers of cover crop residues and associated changes in microclimate, immobilization of nutrients, allelopathic compounds, reduced disturbance, and/or weed-seed predation (Patrick, 1971; Putnam, 1978; Putnam *et al.*, 1983; DeFrank and Putnam, 1978; Purvis *et al.*, 1985; Shilling *et al.*, 1986; Blum *et al.*, 1991, 1997, 2002;). The major difficulties with field bioassays are: a) "teasing out" direct or indirect linkages between environmental factors (e.g., colinearity); b) determining the actual forces acting on plants at a given point in time (e.g., due to the dynamics of the environment and colinearity); and c) time delays between cause and effect.

Another approach has been to study specific processes of interest utilizing laboratory model systems, frequently simply referred to as bioassays (Lieble and Worsham, 1983; Lehman and Blum, 1997; Staman *et al.*, 2001). Laboratory bioassays (e.g., wheat residue dose response studies) have problems similar to those of field bioassays, but on a smaller scale, i.e., they are smaller "black boxes" (Blum, 1999). However, the physical, chemical and biological characteristics of laboratory bioassays can be more readily manipulated to hold some characteristics constant while others are being modified. The benefits of this are obvious for characterizing cause and effect relationships in such bioassays, but these benefits come at a cost. The major difficulties with laboratory bioassays are that these bioassays: a) are carried out in environments that may never occur in nature (e.g., breaking colinearities); b) exclude feedback mechanisms that operate in the field; and c) have a low level of scale.

Data from both field and laboratory bioassays, in spite of their spatial, temporal, and environmental differences, are analyzed in a similar manner and the resulting relationships of interest are described by word models, picture models, physical models, and/or mathematical models (empirical or mechanistic). Although all of these models play an important role in our understanding of allelopathic interactions, the discussion here will be limited to mathematical relationships or what may be referred to as mathematical models.

Mathematical Models

A mathematical model is simply an equation (or set of equations) that contains variables that are used to describe a relationship by mathematical expressions. The equation (or set of equations) represents quantitatively the assumptions or hypotheses that have been made about a real system. The variables within the equation or equations are meant to be analogous to actual entities within the system under study. Mathematical equations of a model do not provide it biological significance, but simply express and interpret the assumptions or hypotheses in a quantitative manner, enabling their consequences to be deduced and showing where to look for occurrence (acceptance) or non-occurrence (rejection) of the hypotheses being tested (Thornley, 1976). Modeling, therefore, is essentially a form of quantitative hypotheses testing.

Three models are provided here as examples. First, a field model that suggests that control of pigweed seedlings increased with increasing soil pH and total phenolic acid content of a sandy loam soil, and increased and then decreased with increasing cover crop residues and soil moisture (% pigweed control = –145 + (0.28 × total soil phenolic acids) + (15.80 × pH) + (3.52 × cover crop residues) + (6.57 × water) – (0.05 × cover crop residues^2) – (0.21 × water^2), $r^2 = 0.54$, $p < 0.0001$ (where µg total phenolic acids/g soil are in ferulic acid equivalents, cover crop residues are in g/0.0625 m^2, and water is in %) (Worsham and Blum, 1992). Second, a very simple laboratory dose response model that describes how wheat residues mixed into soil may influence pigweed seedling emergence: Pigweed seedling emergence = 66.87 – (1.49 * level of wheat residue), $r^2 = 0.80$, $p < 0.0001$ (where pigweed seedling emergence is the number of seedlings on day 6 after seeding and the level of wheat residue is in mg/g soil) (Lehman and Blum, 1997). Third, a more general model of An *et al.* (1993) which describes biological responses to varying concentrations of allelochemicals: %P = 100% for control + S – I. (where the biological response (P), which may be positive, neutral or negative, is determined by the stimulatory (S) and inhibitory (I) responses at a given concentration of an allelochemical). The first two models simply describes experimental data in equation form, i.e., they are empirical models. They are generated after the data have been collected to test a hypotheses. The third model is more mechanistic in nature in that it assumes that biological responses to allelochemicals can be described by enzyme kinetics. This type of model is generally constructed before doing experiments and is tested experimentally afterwards.

Some Misconceptions about Mathematical Models

Unfortunately there are a number of misconceptions about models. For example, a) models can substitute for our lack of understanding of a system; b) there is a unique model for a given system; c) greater model complexity leads to better models; d) mathematical models provide new facts; e) a major goal of model-building is prediction, and f) mathematical models can be validated (Reynolds, 1979). In fact: a) models are only as good as our understanding of the system; b) there are many models for any given system; c) generalizations about complexity are difficult to make, but simple models often provide better predictive capabilities and transparency than more complex models; d) models cannot produce new facts, that is the role of observation and experimentation; e) although prediction is one goal of model building, most models have other goals (e.g., to communicate information, suggest new questions, solve problems, integrate concepts, and identify counter-intuitive behavior); and f) validation depends on one's criteria - for example, theoretical (mechanistic) models are really tentative hypotheses that must be tested over and over again and are thus either accepted or rejected at a level of prediction that is acceptable to a researcher, even though the underlying assumptions of the model may turn out to be incorrect; while empirical models redescribe data and thus their predictive capabilities should be judged under a narrow set of experimental conditions.

Can Empirical Models be used to Establish the Existence of Allelopathic Interactions in Nature?

Both field and laboratory bioassays represent empirical models and as such their predictive power is restricted to the range of experimental conditions of their data, i.e., they contain no information beyond the original data. In addition, since such models only redescribe data they cannot provide knowledge about causal mechanisms (Thornley, 1976). Causal mechanisms operating in the field can thus only be inferred or hypothesized. In other words, in an absolute sense, field and laboratory bioassays can never prove that allelopathy is the causal mechanism. Willis (1985) sums it up perfectly after having listed things needed to establish that allelopathy is operative (i.e., evidence of a pattern of inhibition, toxin production and release by the aggressor plant, toxin transport and accumulation in environment, sensitivity to and contact with toxin by afflicted plant, and exclusion of other potential factors) by stating:

"It should be made clear that confirmation of the above points does not prove that allelopathy is operative, only that allelopathy offers the most reasonable explanation of the observed pattern."

In my opinion, that is all field and laboratory bioassays can ever provide, a most reasonable explanation of or circumstantial evidence for an observed pattern (e.g., reasons for reduced dicotyledonous weed emergence in no-till wheat agro-ecosystems).

CONCLUSIONS

To use circumstantial evidence derived from field and laboratory bioassays to make a case for allelopathy in any ecosystems requires both system-oriented and reductionist approaches. The system-oriented approach describes the behavior of the system and suggests testable hypotheses for determining cause and effect. The reductionist approach is a way of testing the potential of the proposed (hypothesized) cause and effect relationships. In theory, good predictive empirical field and laboratory models should be useful in providing circumstantial evidence for or against the presence of allelopathic interactions. However, recall that predictions deduced for hypotheses of field and laboratory bioassays are either accepted or rejected. The acceptance or rejection of such predictions, unfortunately, does not in itself prove or disprove the presence of allelopathic interactions. Thus, the role of allelopathy in wheat no-till systems or for that matter any other ecosystem, will always be arguable.

So what regulates dicotyledonous weed emergence in no-till wheat ecosystems? There is evidence in literature for all of the alternative hypotheses proposed (see Alternative Hypotheses section). Published results suggest that the answer to this question varies with wheat cultivars, weed species, environment (e.g., soil pH, amount of cover crop residue, allelopathic agents in soil solution, precipitation events), management practices (e.g., time of desiccation of wheat), and the time of year. The bottom line? Let us stop worrying about whether allelopathy exists and just make sure we are doing good science.

Acknowledgments

The author wishes to thank J.R. Troyer, Professor Emeritus of Botany, and T.R. Wentworth, Professor of Botany, for reviewing this manuscript and for their valuable and helpful suggestions.

References

An, M., Johnson, I.R., Lovett, J.V. (1993) Mathematical modeling of allelopathy: Biological Response to allelochemicals and its interpretation. J. Chem. Ecol. 19: 2379-2388.

Blum U., Wentworth, T.R., Klein, K., Worsham, A.D., King, L.D., Gerig T.M., Lyu, S.-W. (1991) Phenolic acid content of soils from wheat no-till, wheat-conventional till, and fallow-conventional till soybean cropping systems. J. Chem. Ecol. 17: 1045-1068.

Blum, U., King, L.D., Gerig, T.M., Lehman, M.E., Worsham, A.D. (1997) Effects of clover and small grain cover crops and tillage techniques on seedling emergence of some dicotyledonous weed species. American J. Alternative Agric. 12: 146-161.

Blum, U. (1999) Designing laboratory plant debris-soil bioassays: Some reflections. *In* Principles and Practices in Plant Ecology. Allelochemical Interactions, Inderjit, K.M.M. Dakshini, and C.L. Foy, (eds.). CRC Press, NY, USA. pp. 17-23.

Blum U., King, L.D., Brownie, C. (2002) Effects of wheat residues on dicotyledonous weed emergence in a simulated no-till system. Allelopathy J. 9: 159-176.

Broadbent, F.E., Tyler, K.B. (1962) Laboratory and greenhouse investigations of nitrogen mobilization. Soil Sci. Soc. Proceedings. 27: 459-462.

Brust, G.E., House, G.H. (1988) Weed seed destruction by arthropods and rodents in low-input soybean agroecosystems. Amer. J. Alternative Agric. 3: 19-25.

Brust, G.E. (1994) Seed-predators reduced broadleaf weed growth and competitive ability. Agric. Ecosyst. Environ. 48: 27-34.

DeFrank, J., Putnam, A.R. (1978) Weed and crop responses to allelopathic crop residues, North Central Weed Control Conference. Proceedings. 33: 44.

Fenster, C.R., Peterson, G.A. (1979) Effects of No-Tillage Fallow as Compared to Conventional Tillage in a Wheat-Fallow System. University of Nebraska, Lincoln: Agricultural Experiment Station. Research Bulletin 289: 1-28.

Kimber, R.W.L. (1973) Phytotoxicity from plant residues. III. The relative effects of toxins and nitrogen immobilization on the germination and growth of wheat. Plant and Soil. 38: 543-555.

Lehman, M. E., Blum, U. (1997) Cover crop debris effects on weed emergence as modified by environmental factors. Allelopathy J. 4: 69-88.

Liebl, R.A., Worsham, A.D. (1983) Inhibition of morningglory (*Ipomoea lacunosa* L.) and certain other weed species by phytotoxic components of wheat (*Triticum aestivum* L.) straw. J. Chem. Ecol. 9: 1027-1043.

Lynch, J.M. (1977) Phytotoxicity of acetic acid produced in the anaerobic decomposition of wheat straw. J. Applied Bact. 42: 81-87.

Nair, G.N., Whitenack, C.J., Putnam, A.R. (1990) 2, 2'-oxo-1,1'-azobenzene: A microbially transformed allelochemical from 2,3-benzoazolione: I. J. Chem. Ecol. 16: 353-364.

Niemeyer, H.M., Pesel, E.S., Capaja, S.V., Bravo, H.R., Franke, S., Francke, W. (1989) Changes in hydroxamic acid levels of wheat plants induced by aphid feeding. Phytochemistry. 28: 447-449.

Patrick, Z.A. (1971) Phytotoxic substances associated with the decomposition in the soil of plant residue. Soil Sci. 111: 13-18.

Purvis, C.E., Jessop, R.S., Lovett, J.V. (1985) Selective regulation of germination of annual weeds by crop residues. Weed Res. 25: 415-421.

Putnam, A.R., DeFrank, J., Barnes, J.P. (1983) Exploitation of allelopathy for weed control in annual and perennial cropping systems. J. Chem. Ecol. 9: 1001-1010.

Reynolds, J.F. (1979) Some misconception of mathematical modeling. What's New in Plant Physiology. 10: 41-43.

Shilling, D.G., Liebl, R.A., Worsham, A.D. (1985) Rye (*Secale cereale* L.) and wheat (*Triticum aestivum* L.) mulch: The suppression of certain broadleaved weeds and the isolation and identification of phytotoxins. *In* The Chemistry of Allelopathy: Biochemical Interactions Among Plants, A.C. Thompson, (ed). ACS Symposium Series 268. American Chem. Society. Washington, D. C. USA. pp. 243-271.

Shilling, D.G., Worsham, A.D., Danehower, D.A. (1986) Influence of mulch, tillage, and diphenamid on weed control, yield, and quality in no-till flue-cured tobacco (Nicotiana tabacum L.). Weed Sci. 34, 738-744.

Staman, K., Blum, U., Louws, F., Robertson, D. (2001) Can simultaneous inhibition of seedling growth and stimulation of rhizosphere bacterial populations provide evidence for phytotoxin transfer from plant residues in the bulk soil to the rhizosphere of sensitive species? J. Chem. Ecol. 27: 807-829.

Tang, C.S., Waiss, A.C. (1978) Short-chain fatty acids as growth inhibitors in decomposing wheat straw. J. Chem. Ecol. 4: 225-232.

Thornley, J. (1976) Mathematical Models in Plant Physiology. Academic Press, NY, USA.

Willard, J.I., Penner, D. (1976) Benzoxazinones: Cyclic hydroxamic acids found in plants. Residue Rev. 64: 67-76.

Willis, R.J. (1985) The historical bases of the concept of allelopathy. J. History of Biol. 18: 71-102.

Worsham, A.D. (1989) Current and potential techniques using allelopathy as an aid in weed management. *In* Phytochemical Ecology: Allelochemicals, Mycotoxins, and Insect Pheromones and Allomones, C.H. Chou, G.R. Waller (eds). Monograph Series No. 9. Academia Sinica, Taipei, Republic of China. pp. 275-291.

Worsham, A.D., Blum, U. (1992) Allelopathic cover crops to reduce herbicide inputs in cropping systems. *In* the Proceedings of the First International Weed Control Congress 17-21 February, 1992. 2: 577-579.

3

Plant-box Method: A Specific Bioassay to Evaluate Allelopathy through Root Exudates

Yoshiharu Fujii[1], Dolorosa Pariasca[1], Tomoko Shibuya[1], Tamaki Yasuda[1], Brian Kahn[2] and George R. Waller[2]

[1]Allelopathy Laboratory, Chemical Ecology Unit, National Institute for Agro-Environmental Sciences, Tsukuba, Ibaraki 305-8604 Japan

[2]Department of Biochemistry and Molecular Biology, Oklahoma State University, Stillwater, Oklahoma 74078 USA

A specific method for assessing allelopathy was developed using intact plants growing in an agar medium without nutrients. There are several routes of action of allelopathy and this method deals with transferring allelochemicals through root exudates. A heat-resistant, transparent plant-box apparatus was developed to contain non-sterile plant tissue culture because of its ability to withstand autoclaving and easy observation of the root zone. The candidate donor plant was grown in sand for one to two mon to adapt it for the agar medium; then transplanted into a root zone-separating tube positioned inside the plant-box. A special low-temperature gelatinizing agar that solidifies at 30-31°C was used in order to avoid heat shock. An acceptor plant (lettuce) was seeded on the surface of the agar and incubated under certain conditions for 5 d. Allelopathic activity was calculated from the first regression curves between the radicle and hypocotyl length and the distance from the root zone. Using this method, a total of 336 plant species belonging to 48 families were evaluated. Among these plants, velvetbean (*Mucuna pruriens*) and oat species (*Avena* sp.) were found to have the strongest inhibitory activity.

Keywords: *agar medium, Avena sp., bioassay, leaf leacheates, Mucuna pruriens (L.) DC., oat, plant-box, root exudates, velvetbean*

INTRODUCTION

Several bioassay methods have been presented in allelopathic studies (i.e., Anderson, 1985; Leather and Einhellig, 1986; Shilling and Yoshikawa, 1987; Fujii *et al.*, 1990a, b) and most of them are used for evaluating the allelopathic potential of the species concerned. Determination of activity that follows extraction, purification and identification of bioactive compounds is done to guide the researcher. Bioassay selection, in most cases, depends on the availability of species and sensitivity of the method, with little regard for standardization. Therefore, most of these reports were not always correlated with the plant-to-plant interference observed in the field.

It is quite difficult to separate the effects of allelopathy from those due to competition for light, nutrients, water and other biological and physical factors. Thus, proving the existence of allelopathy in the field is very complicated and there are many compounds reported as allelochemicals *in situ* without their activity being fully proven. Once a natural product with inhibitory activity to plants has been extracted, purified, identified, and once its field experiment results imply allelopathy, it tends to be generally accepted as an allelochemical. Due to these problems, scientists doubt the importance of allelopathy in plant, soil and microbial interactions.

In order to erase these doubts, we have made preliminary reports on trials that separate allelopathy from other factors of competition (Fujii *et al.*, 1991b,c,d; Yasuda *et al.*, 1991); however, some of them are labor intensive. Thus, a new bioassay method was developed producing results that may be interpreted in relation to field observations. By using this method, we have examined the allelopathic activity of crop, vegetables and weeds grown in Japan.

MATERIALS AND METHODS

Plant Preparation

River sand was passed through a 3 mm-sieve and washed with water. Black vinyl plastic pots measuring 6 × 6 (width) × 7 cm (height) were then filled. Four to eight seeds of each plant were seeded in each pot. Nutrient solution (Hoagland and Arnon, 1950) with pH adjusted to 6.8 by adding 0.1 M NaOH or HCl was applied according to the need of plants. For Graminaceous species, ferrous iron was increased by Fe-EDTA up to four times of the original formula. Plants were grown for several months.

Seeds of test plants mostly obtained from the Japanese Gene Bank of the Ministry of Agriculture, Forestry and Fisheries, Tsukuba. Vegetable seeds and pasture legume seeds were gifts from Kaneko Seeds Co., Takii Seeds Co., Yukijirushi Seeds Co., Sakata no Tane Co. and Hokuren Coop Co.

Root Zone-separating Tube Preparation

A vinyl acetate tube for tap water pipe with 32 and 25 mm, outer and inner diameter, respectively, was cut into 65 mm length. Then, a window was opened leaving a 90° angle wall (Fig. 1) and was covered with polyester gauze (Toray Tetron, Japan, #C-119 Skylark) using an adhesive agent (Ethron). A 2-mm plate was attached and glued by the same agent at its bottom. The constituents of this adhesive agent were cyclohexane (40%), methy ethyl ketone (35%) and acetone (25%). The tube was oven dried at 60°C to eliminate all solvents and washed with distilled water.

Fig. 1 Root zone-separating tube for plant-box method

Selection of Gel-supports

To select the best gel-support for plant-box method, gel-making agents such as agar (Wako Chemical, Tokyo, Japan; Regent Grade (GR), Nacalai Tesque, Kyoto, Japan; Reagent Grade (RG), and Low Temp. Agar, and Difco, USA; Bacto-Agar), agarose, gellun-gum (Wako Chemical, Tokyo, Japan; Special Grade), gelatin (Wako Chemical, Tokyo, Japan; Special Grade, and Difco, USA; Bacto-Gel), cellulose, dextran (Pharmacia, Sweden, SephadexG-10, 100, 200), and glass beads were tested at their best concentration.

Calculation of Allelopathic Activity

With the plant-box apparatus's transparent characteristic, the effect of donor plant against lettuce, acceptor plant, (expressed by their radicle lengths) became highly visible. Radicle lengths of lettuce seedlings in close proximity to the donor plant tended to be shorter for allelopathic plants.

As shown in Fig. 4, distances from the root zone-separating tube and the radicle length were plotted and the linear regression curve was fitted with a computer calculation using Microsoft Excel or equivalent statistical analysis (Steel and Torrie, 1980). From this calculated curve, the radicle length of seedlings inside root zone (zero distance) was compared to the length of control. This length shows the inhibitory effect of allelochemicals released from the root of the donor plant. The decline of each regression equation shows the migration speed of allelochemicals in agar medium. The change of migration speed may be due to their solubility in water, molecular weight and the shape of molecules. Prior to the bioassay, migration speed of known model compounds was calculated by the same approach. Riboflavin and blue dextran with molecular weights of 268 and 2500, respectively, were used. The agar was removed with a corkscrew (6 mm diameter) from a corner of the box and then filled up with the solution of each model compound. The migration speed was then measured by their color (yellow or blue) in the agar as they spread throughout the plant-box.

Selection of Optimum Agar Concentration

The earlier experiment was conducted to examine the appropriate gelling medium for good seed germination and seedling growth of lettuce. In addition, an optimum concentration of agar was evaluated from various levels ranging from 0.5, 1.0, 1.5, 2.0, to 2.5%. The effect of incubation time on the growth of lettuce was also evaluated. Five lettuce seeds were placed on the gelatinized surface of agar-containing Petri dishes. Lengths of both radicle and hypocotyl of lettuce seedlings were measured every day after imbibition onset for 3 d. In addition, the germination percentage of seeds was recorded.

Test for Migration Speed of Compounds

Migration speed using model compounds with visible color was also calculated. Riboflavin (0.2 mg/ml) and blue dextran (2 mg/ml) with molecular weights of 376 and 2000, respectively, were used. By a corkscrew of 6 mm diameter, agar at the corner of the box was removed and the hole was filled up with the solution of each model compounds. Then migration speed was measured optically. The results showed that migration speed of both compounds is about 1 cm per day. At this rate, the leachates could likely reach the box side walls within 5 d, and thus the incubation period was set at 5 d.

Agar Preparation

Low-temperature gelatinizing agar (Nacalai Tesque, Kyoto, Japan), concentration of 0.5 to 0.75% (W/V) and gelatinizing temperature of

30-32°C was autoclaved at 115°C for 15 min and kept at 40°C in a water bath until used.

Transplanting of Test Plants into Plant-box

Donor plants grown in sand culture with estimated root dry weight of 100 to 300 mg were washed with distilled water and placed into the root zone-separating tubes of the plant-box (Magenta Box, Magenta Corporation, Chicago, USA, 60 × 60 (width) × 100 mm (height)). Five guide-dots at 10 mm distance were marked on two upper sides of the box at about 65 mm from the base for precise positioning of seeds (Fig. 2). The tubes containing the test plants were fixed at the corner of the box through a cellophane tape attached to the upper part of the plant (Fig. 2). Agar was poured into the box up to the level indicated with dots. The boxes were then cooled immediately in ice water and allowed to stand for 15 min until the agar gelatinized. Seeding of lettuce, the acceptor plant species, followed after 1 h.

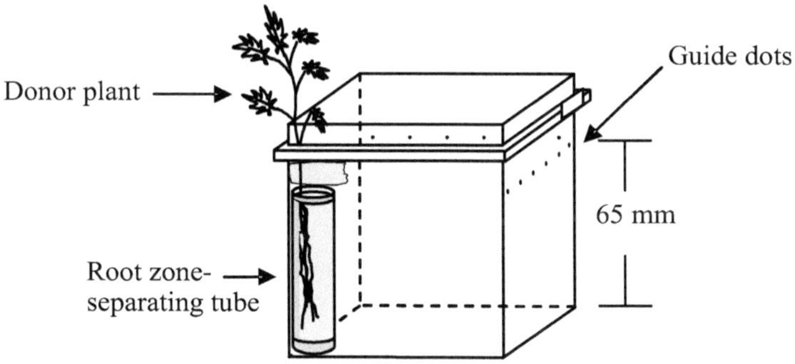

Fig. 2 Diagram of a plant-box (lateral view)

Seeding Process

Lettuce seeds (*Lactuca sativa*, cv. Great Lakes 366) were used because of their high sensitivity, simultaneous and rapid germination, reliable germination percentage and homogeneity. Seeding was done (narrow tip downward) at distances of 1 cm apart as shown in Fig. 3.

Plant Growth Conditions

Plant-boxes were fully covered (except the upper surface) with black plastic film in order to avoid root phototropism, and the top surface with polyethylene film to avoid possible bacterial contamination. They were set in the incubator and their daily conditions were checked. The incubation

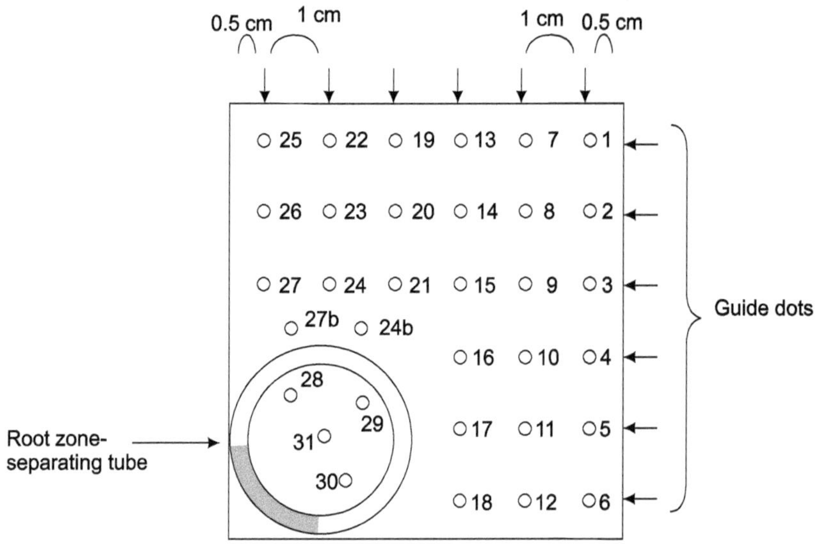

Fig. 3 Top view of plant-box (seeding diagram) with the guide-dots followed, a lattice distribution of seeds has resulted

temperature was at 25/20°C (12/12) and the agar was kept moist by adding appropriate amount of distilled water everyday. The radicle and hypocotyl lengths of lettuce seedlings were measured after 5 d of incubation. Dry weights and heights of donor plant were also calculated. Usually, radicle length was affected more than the hypocotyl length (Shilling and Yoshikawa, 1987), thus, radicle growth inhibition was compared to the control.

PROTOCOL FOR PLANT-BOX METHOD

I. Dissolution of agar
   (1) Add 7.5 g of low-temperature melting agar powder (0.75% w/v) to 1,000 ml of distilled water.
   (2) Autoclave for 15 min at 115°C.
   (3) Cool down to 40°C in a water bath until used.

II. Transplanting of donor plants and seeding of acceptor species
   (1) Wash the roots gently and thoroughly with distilled water, then wrap them in water-soaked tissue paper to prevent wilting.
   (2) Place the plant into the root zone-separating tube, and fix to its position using cellophane tape (Fig. 2).
   (3) Label the boxes indicating the plant name and sowing date (or collection date when uncertain).

(4) Pour the agar solution slowly (to avoid bubbles) until the 6.5-cm level marked with dots on the box lateral sides (Fig. 2).
(5) Cool down the box immediately by dipping it in ice-chilled water (30 min) and let it stand at room temperature for another few minutes before seeding with the acceptor species.
(6) Sow the seeds on the agar surface (narrow tip downward) following the indicated distances shown in the seeding diagram (Fig. 3).
(7) Cover the box surface with polyethylene film as soon as the seeding process is done.
(8) Place the box in a black vinyl receptacle and make sure all lateral sides are well covered.
(9) Incubate the boxes in an incubator for 5 d and check the daily condition of the samples (temperature and agar) and add water as required everyday.
(10) Measure the radicles and hypocotyl lengths of lettuce seedlings over a graph paper, following the numbering indicated in the seeding diagram (Fig. 3).
(11) Measure the height of the donor plant.
(12) Uproot the donor plant from the agar medium, wash the roots thoroughly to remove agar, then dry it gently with paper towels.
(13) Separate the roots from the over-ground part of the plant.
(14) Take their corresponding fresh weights.
(15) Place them in a paper bag and then in an air-forced oven dryer at 60°C for 1 to 2 d.
(16) Take their corresponding dry weights after drying.

RESULTS AND DISCUSSIONS

Tang and Zhang's Teflon Ring Separation Method

Tang and Zhang (1986) separated the root zone of germinating mung beans (donor) from lettuce (acceptor) plant using teflon tubing; however a modification of this device into a glass dish was developed in our laboratory. Using this method, strong growth inhibition in acceptor plant species (lettuce) was caused by some plants, such as *Mucuna pruriens, Canavalia ensiformis, Vicia villosa* and *Avena sativa* (Table 1). The original concentration of agar used was 1.5% (w/v); seemingly high that it caused inhibition on acceptor root growth. Moreover, a much-increased concentration of agar within the area restricted by the separation device

Table 1 Effect of root exudate on the growth of lettuce by teflon ring separation method

Donor plant	*Radicle length [%]*[*1]	*Hypocotyl length [%]*[*1]
Leguminosae		
Astragalus sinicus (Chinese milk vetch)	88	111
Canavalia ensifolmis (Sword bean)	9	51
Cassia tora (Oriental senna)	58	87
Glycine max cv. Fukuyutaka (Soybean)	41	128
Lathyrus sativus (Grass pea)	33	59
Lupinus albus (Lupine)	62	97
Mucuna pruriens var. *utilis cv. ana* (Velvetbean)	10	36
Mucuna pruriens var. *utilis* cv. *cinza*	11	43
Trifolium incarnatum (Crimson clover)	60	165
Trifolium pratense (Red clover)	19	57
Trifolium repens r. *giganteum* (Ladino clover)	62	118
Vicia sativa (Common vetch)	16	128
Vicia villosa (Hairy vetch)	29	65
Gramineae		
Avena sativa (Oat)	18	71
Eleusine coracana (Finger millet)	89	108
Secale cereale (Rye)	42	79
Secalotricum ryedax (Triticale)	28	84
Sorghum saccharatum (Sweet sorghum)	84	142
Zea mays var. indentata (Dent corn)	67	150
Other family		
Cucumis sativus (Cucumber)	94	114
Kalanchoe pinnata (Never die)	56	113
Phytolacca americana (Pokeweed)	56	114
Tagetes patula (French marigold)	93	105

*1 [%] means the percentage of the length of lettuce in agar inside the wall compared to that of outside; no contact with root of donor plant was made.

was observed as a result of donor plant transpiration. With transpiration, donor plant water uptake increased, resulting in an increased agar concentration. Such an increase resulted in an inhibited root growth of the acceptor plant which is difficult to separate from the real effects of allelopathy.

Establishment of the Plant-box Method

In the course of "Tang and Zhang Teflon Ring Separation Method", it was found that lettuce seedlings in close proximity to the donor plant showed inhibited root growth; the closer they were, the more inhibition was observed. In order to make these relationships clear, the root zone of the donor plant was separated by cellulose dialyzing membrane tube and the relationship between root distance (acceptor from donor plant) and elongation were measured. With cellulose dialyzing membrane as a separation device, these distances needed to be measured in each experiment. Additionally, preparation of the dialyzing membrane tubes required a lot of time. To solve the problem, the root zone separation technique was developed and dialyzing membrane tube was changed into vinyl acetate tube attached with polyester or nylon gauze (Fig. 1).

Selection of Gel-supports

Several gel-making agents were tested as a medium. Table 2 shows the results with lettuce as the test plant. Agar and gellan-gum showed suitable growth and their transparency was satisfactory. However, to make a gel out of gellan-gum, addition of Ca^{++} or K^{+} as cation is required and this is likely to create chemical interaction with the allelochemicals released from the roots. As such agar was chosen. Its optimum concentration for plant growth was determined and result showed that 0.75% (W/V) concentration was the most satisfactory. This concentration is the lowest to make a stiff gel by this agar.

Table 2 Selection of gel-supports for the plant-box method using lettuce as test plant

Gel materials	*Concentration (%)*	*Radicle length (mm)*	*Hypocotyl length (mm)*
Agar (Wako, Rg.)	0.5	33 b[*1]	20 bc
Agar (Nakarai, Rg.)	0.5	35 b	20 bc
Agar (Difco, Bacto-Agar)	0.5	28 c	22 ab
Agar (Wako, Rg.) + Kaolinite	0.5 + 0.2	30 c	21 ab
Agar (Wako, Rg.) + Silica-gel	0.5 + 2.0	43 a	21 ab
Gelatin (Wako)	5.0	5 fg	3 h
Gelatin (Difco, Bacto-Gel)	5.0	2 g	1 i
Dextran (SephadexG-200)	im[*2]	18 d	11 e
Dextran (SephadexG-100)	im	11 e	9 f
Dextran (SephadexG-10)	im	3 fg	6 g

[*1]Letters on the right hand of data shows the results of Duncan's Multiple Range Test. The same letter means not significant (P > 0.01). Replication number is 6.

[*2]Immersion with water to make the full expansion gel, excess water was decanted carefully.

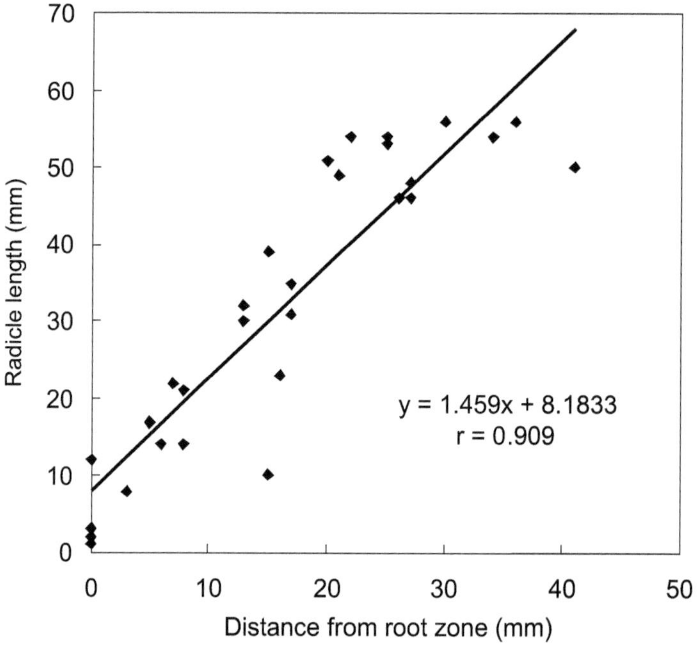

Fig. 4 Calculation of allelopathic activity by plant-box method

Assessment of Allelopathic Activity by the Plant-box Method

Table 3 shows the allelopathic activity between species and cultivars against lettuce as the acceptor plant. In Table 3 for Leguminous species, among 70 species tested, velvetbean showed the strongest inhibitory activity. We isolated L-3,4-dihydroxypenylalanine (L-DOPA) as a major allelochemical from velvetbean (Fujii *et al.*, 1990). Among its cultivars, "hassjo", semi-dwarf type from Japan was the strongest. *Vicia* species such as *Vicia faba,* a winter legume crop cultivated for food and cover crop in Japan and the USA and *Vicia villosa,* known as hairy vetch or woolly pod vetch, also used as a cover crop in the USA, was found allelopathic. We isolated cyanamide as a new allelochemical from *Vicia* species (Kamo *et al.*, 2003). Other promising legumes are *Medicago* spp, *Leucaena leucosephala, Canavalia ensiformis, Melilotus* spp, *Puerarila lobata* and *Vigna* spp.

In Table 3 for the allelopathic activity of Gramineae species, this family showed lesser inhibitory effects than the leguminous species. *Avena* spp like Oat (*Avena sativa*) known for its allelopathic activity since ancient times (Schreiner and Reed, 1907) showed the strongest inhibition with wild oats, *A. sterillis, A. murphy* and *A. barbata* showing the highest activity. Other Gramineae species that seem promising are *Setaria italica, Panicum miliaceum, Anthoxanthum odoratum, Triticum* spp, *Sorghum* spp and

Table 3 Assessment of allelopathic activity by the plant-box method.

Scientific name (English name, and/or Japanese name)	*Radicle length (%)*[*1]	*n*[*2]
Leguminosae		
Arachis hypogaea (Peanut)	49	7
Astragalus sinicus (Chinese milk vetch)	59	7
Cajanus cajan (Pigeon pea)	40 *	10
Calopogonium mucunoides (Calopogonio)	49 *	8
Canavalia ensiformis (Jack bean)	28 ***	11
Cassia tora (Sickle senna)	34 **	2
Cicer arietinum (Chickpea)	44 *	6
Crotaralia spectabilis (Sunn hemp)	40 **	6
Glycine max (Soybean, cv. Tachi-nagaha)	65	3
Lablab purpurea (=Dolicos lablab) (Lablab bean)	41 *	5
Latyrus sativus (Grass pea)	41 *	7
Lupinus albus (White lupine)	60	10
Medicago sativa cv. *dupy* (Alfalfa)	34 **	6
Melilotus albus (White sweet clover)	23 ***	4
Mucuna pruriens var. *utilis* (Velvetbean, av. of 5 cv)	7 *****	18
Mucuna pruriens var. *utilis* cv. *Ana*	11 ****	8
Mucuna pruriens var. *utilis* cv. *Florida*	12 ****	3
Mucuna pruriens var. *utilis* cv. *Hassjo*	4 *****	3
Pisum sativum (Pea, average of 4 cultivar)	42 *	16
Pueraria lobata (Kudzu)	28 ***	4
Sesbania cannabina cv. *Densuke* (Sesbania)	61	7
Trifolium incarnatum (Crimson clover)	36 **	4
Trifolium pratense (Red clover)	72	5
Trifolium repens (White clover)	53	6
Vicia angustifolia var. *segetalis* (Karasunoendou)	55	10
Vicia faba (Broad bean, average of 4 cultivar)	19 ****	14
Vicia faba (Broad bean, cv. *Tokushima-zairai*)	2 *****	3
Vicia faba (Broad bean, cv. *Shimizu-wase*)	11 ****	3
Vicia faba (Broad bean, cv. *Otafuku*)	54	3
Vicia sativa (Common vetch)	25 ***	5
Vicia villosa var. *villosa* (Hairy vetch)	20 ****	21
Vicia villosa var. *dasycarpa* (Woolly pod vetch)	19 ****	9
Vigna angularis (Adzuki bean, av. of modern cv.)	84	12
Vigna unguiculata (Cowpea)	33 **	10
Vigna unguiculata subsp. *sesquipedalis* (Asparagus pea)	49 *	16
Gramineae		
Anthoxanthum odoratum (Sweet vernalgrass)	28 ***	7
Avena sativa (Oat, average of 12 cultivar)	30 ***	21

(Table 3 Contd)

(Table 3 Contd)

Eleusine coracana (Finger millet, Shikoku-bie)	34 **	5
Festuca arundinacea (Tall fescue)	45 *	8
Hordeum vulgare (Barley, average of 7 cultivars)	38 **	11
Imperata cylindrica var. *koenigii* (Cogon grass)	12 ****	3
Lolium multiflorum (Italian ryegrass)	74	11
Lolium perenne (Perennial ryegrass)	64	7
Panicum coloratum var. *kabulabula* (Kabulabula grass)	38 **	3
Panicum maximum (Guinea grass)	40 **	4
Panicum miliaceum (Millet, Japanese native cv.)	14 ****	7
Phalaris arundinacea (Reed canary grass)	76	4
Phleum pratense (Timothy)	96	3
Poa pratensis (Kentucky bluegrass)	60	12
Secale cereale (Rye, average of 5 cultivars)	29 ***	12
Sorghum dochna (Sorghum)	47 *	8
Sorghum sudanense (Sudan grass)	45 *	6
Secalotricum ryedax (Triticale)	23 ***	3
Triticum aestivum (Wheat, cv. *Nourin 61*)	34 **	2
Triticum polonicum (Polish wheat)	13 ****	3
Zea mays (Corn, cv. Pioneer Dent)	66	3
Zoysia japonica (Japanese lawn grass, *No-shiba*)	58	3
Other Family		
Abutilon theophrasti (Chinese jute)	9 *****	5
Allium cepa (Onion)	65	4
Allium fistulosum (Welsh onion)	83	6
Amaranthus spinosus (Spiny amaranthus)	36 **	3
Amaranthus tricolor (Keitou)	39 **	5
Artemisia absinthium (Wormwood, for medicinal use)	38 **	3
Artemisia princeps (Yomogi, for medicinal use)	21 ***	2
Bidens frondosa (Beggar's tick, Stickright)	74	4
Brassica campestris (Field mustard, average of 5 cultivars)	50 *	14
Brassica napus (Rape, average of 3 cultivars)	41 *	9
Brassica oleracea var. *italica* (Broccoli)	29 ***	6
Capsicum annuum (Red pepper, average of 4 cv.)	39 **	6
Chenopodium album (Common lambsquarter)	71	4
Chenopodium ambrosioides (Aritasou)	47 *	2
Cucumis melo (Melon)	61	4
Cucumis sativus (Cucumber)	58	6
Cucurbita pepo (Pumpkin)	62	3
Helianthus annuus (Sunflower)	78	5
Hyssopus officinalis (Hyssop)	61	4
Lagenaria siceraria var. *hispida* (Hyoutan)	42 *	2

(Table 3 Contd)

(Table 3 Contd)

Lavandula angustifolia (Lavender)	32 **	3
Linum usitatissimum (Flax)	20 ****	5
Lycopercicum esculentum (Tomato, average of 6 cv)	46 *	15
Melissa officinalis (Lemon balm)	99	3
Mentha arvensis (Japanese mint)	72	3
Mentha piperita (Peppermint)	58	4
Mentha pulegium (Pennyroyal mint)	83	3
Momordica charantia (Balsam pear, *Tsuru-reishi*)	58	3
Nepeta cataria (Catnip)	83	3
Nigella damascena (Devil-in-a-Bush)	46 *	5
Perilla frutescens var. *crispa* (Beefsteak Plant)	60	4
Perilla frutescens (Perilla, Egoma)	60	7
Phytolacca americana (Pokeweed)	62	3
Polygonum tinctorium (Chinese indigo, Ai)	94	3
Portulaca oleracea (Purslane)	24 ***	3
Raphanus sativus (Radish)	53	2
Ricinus communis (Castor bean)	56	6
Ruta graveolens (Common rue)	18 ****	3
Solanum melongena var. *esculentum* (Egg plant)	51	2
Solidago altissima (Golden rod)	79	5
Spinacia oleracea (Spinach)	93	3
Symphytum officinale (Comfrey)	21 ***	3
Symphytum peragrinum (Russian comfrey)	11 ****	2
Tagetes patula (French marigold)	68	6
Tropaeolum majus (Common nasturtium)	37 **	4
Verbascum thapsum (Mullein)	38 **	4

*1 Radicle length (%) means the allelopathic activity at the surface of the root zone separation tube compared to that of control. Values approaching 100 indicates no inhibitory activity.

*2 *n* means the number of set of replications in different stages of growth, time of seasons, and these data are a summary of the results of five years (1990-1995) in duration.

*3 Lettuce (*Lactuca sativa* cv. *Great Lakes 366*) was used as acceptor plant.

*4 Allelopathic activity was grouped into next symbols for convenience: *****(<10%),
****(11~20%), ***(21~30%), **(31~40%), *(41~50%); values above 50% were not discussed.

Hordeum spp. Most of these species have been reported as allelopathic plants (Rice, 1984).

As for the allelopathic activity of Compositae species, *Artemisia absinthium* and *A. princeps*, both medicinal herbs, were found remarkably allelopathic, but lesser compared to Leguminous and Gramineous family. Lettuce (*Lactuca sativa*), the acceptor plant used in this bioassay, belongs to the same family. This could be the likely reason of its reduced sensitivity. Additionally, the Compositae family, contain hydrophobic chemicals and these compounds could not migrate under the plant-box condition where the medium is water based.

In the Cruciferae family, it shows that the Brassica family, known to contain sulfur compounds, is allelopathic. Some of these compounds such as isothiocyanates are well known to inhibit the growth of soil-born fungi and bacteria. These compounds are also inhibitory to plants. Using this family as a cover crop to get rid of soil-born diseases is important in sustainable agriculture. Broccoli and cauliflower are the strongest in this group. How to use these vegetables for weed suppression, poses a challenge.

In the Solonaceae family, some red and sweet peppers, and some native types of tomato are allelopathic. It is known that red pepper contains capsicin, which causes irritation to many animals and humans, however it is not known if pepper is allelopathic in the field.

The results from the Labiatae family, well known for the production of essential oils, showed that some plants known as herbs, such as lavender (*Lavandula angustifolia*), oregano (*Origanum vulgare*), clary (*Salvia sclarea*) and savory (*Satureja hortensis*) are inhibitory.

The Amaranthaceae family is well known for its plant smothering effect in the fields of onion, carrot and itself (Fisher and Quijano, 1985), but showed only slight inhibition in our test. It is therefore not clear if its weed suppression capability is due to allelopathy through root exudates. Further testing using acceptor plants other than lettuce perhaps could elucidate its contribution to allelopathy if there is any. Some *Amaranthus* spp are promising ground cover crops that suppress weeds in tropical countries.

In the Cucurbitaceae family, there are several important vegetables but only *Lagenaria siceraria* var. *hispida* and *Cucurbita maxima* were slightly allelopathic. These plants with large leaf systems usually inhibit the growth of weeds once established since they cover the surface of the land completely. Combination of competition for light and allelochemical factors including volatile compounds may be responsible for the complete suppression of weeds.

Most of the Polygonaceae family and Chenopodiaceae family showed little activity. Only *Chenopodium ambrosioides*, a traditional medicine in Japan, showed slight inhibition. *Chenopodium album* was reported to have chemicals inhibitory to wheat, radish and soybeans.

In Liliaceae, *Lycoris radiata* showed allelopathic activity. This plant was introduced into Japan more than 1,500 years ago from China and distributed by farmers into their footpaths of rice paddy fields and graveyards. Consequently the Japanese people call this plant "flowers of the dead". Rats and moles shy away from this plant because it produces toxic alkaloids in its tubers. These chemicals may play an important role as allelochemicals for defense.

Most of the Caryophyllaceae family showed no allelopathic activity. As for other families, Chinese jute or velvet leaf (*Abtilon theophrasti*), already reported very allelopathic, (Rice, 1984b), likewise showed strong allelopathic inhibition in our test, but the active chemicals remain unknown. *Ruta graveolens* (Rue) and *Symphytum* spp. (Comfrey) were also found allelopathic and so was the ground cover plant, *Portulaca oleracea*, or common purslane, and some members of this group producing beautiful flowers.

CONCLUSION

In conclusion, plants belonging to species such as *Avena, Mucuna, Abtilon, Erigeron, Brassica* and *Vicia* that have been reported allelopathic, have likewise shown strong inhibitory activity by the plant-box method. This method used lettuce as an acceptor plant, but it could be substituted with other plants such as weeds. If weeds are used, more replications and pre-incubation of seedlings should be done.

There are many bioassays reported during the past 30 y (Leather and Einhellig, 1986) but most of them are only tested for germination using extracts, exudates or leachates from plants. There are some reports that aim at standardization (Lehle and Putnam, 1982; Fujii *et al.*, 1990a, b) but few bioassays use intact plants (Anderson, 1985) as our method does. This method makes it possible to assess the direct allelopathic activity of a donor plant and its direct effect to an acceptor plant.

Presently, more tests using more plant species and different cultivars are being evaluated. Our hope is to find plants that are highly allelopathic and to identify and assess the presence of possible allelochemicals in the field where the interaction takes place. By a combination of this method and field tests (Fujii *et al.*, 1990a, b, 1991a-i; Yasuda *et al.*, 1991) our team could separate the effects of allelopathy from those due to competition for light, nutrients, and water and possibly answer the questions proposed by Harper (1977).

Acknowledgments

We appreciate the advice of Dr. Akio Ichikawa suggesting the use of the plant-box for bioassay of allelopathy. This research was supported partially from a Bio-Cosmos Project, No. BC 05-1-001, National Institute of Agro-Environmental Sciences, Ministry of Agriculture, Forestry and Fisheries, Japan. Part of this work was done under the "US-Japan Science and Technology Joint Agreement on Cooperation in Research and

Development", by the USDA and MAFF as project No. 91-8A, "Basic Research on Bioactive Substances in Plant Ecosystems" between Y. Fujii and G.R. Waller.

References

Anderson, L.W.J. (1985) Use of bioassays for allelochemicals in aquatic plants. *In* The Chemistry of Allelopathy, Biochemical Interactions Among Plants, A.C. Thompson, (ed.) ACS Symposium Series 268, American Chemical Society, Washington, DC, USA. pp. 351-370.

Chou, C-H., Waller, G.R. (eds.) (1983) Allelochemicals and Pheromones, Institute of Botany, Academia Sinica Monograph Series, No. 5, Academia Sinica, Taipei, Republic of China.

Chou, C-H., Waller, G.R. (eds.) (1989) Phytochemical Ecology; Allelochemicals, Mycotoxins, and Insect Pheromones and Allomones, Institute of Botany, Academia Sinica Monograph Series, No. 9, Academia Sinica, Taipei, Republic of China.

Fisher, N.H., Quijano, L. (1985) Allelopathic agents from common weeds, *Amaranthus palmeri, Ambrosia artemisifolia,* and related weeds. *In* The Chemistry of Allelopathy, Biochemical Interactions among Plants, A.C. Thompson, (ed.) ACS Symposium Series 268, American Chemical Society, Washington, DC, USA. pp. 133-147.

Fujii, Y., Shibuya, T., Yasuda, T. (1990a) Method for screeningallelopathic activities by using the logistic function (Richards' function) fitted to lettuce seed germination and growth curves. Weed Res., Japan. 35: 353-361.

Fujii, Y., Shibuya, T., Yasuda, T. (1990b) Survey of Japanese weeds and crops for the detection of water-extractable allelopathic chemicals using Richards' function fitted to lettuce germination test. Weed Res., Japan. 35: 362-370.

Fujii, Y., Shibuya, T., Yasuda, T. (1991a) L-3,4- Dihydroxyphenylalanine as an allelochemical candidate from *Mucuna pruriens* (L.) DC. var. *utilis*. Agr. Biol. Chem. 55: 617-618.

Fujii, Y., Shibuya, T., Yasuda, T. (1991b) Discrimination of allelopathy of tomato plant by stairstep experiment and rotary greenhouse experiment. Jap. J. Soil Sci. Plant Nutri. 62: 150-155.

Fujii, Y., Shibuya, T., Yasuda, T. (1991c) Discrimination of allelopathy of velvetbean (*Mucuna pruriens*) with stairstep experiment and rotary greenhouse experiments. Jap. J. Soil Sci. Plant Nutrition. 62: 258-264.

Fujii, Y., Shibuya, T., Yasuda, T. (1991d) Discrimination of allelopathy of upland rice, taro, and oat by substitutive experiment and its modified experiments. Jap. J. Soil Sci. Plant Nutrition. 62: 357-362.

Fujii, Y., Shibuya, T., Yasuda, T. (1991e) Intercropping of velvetbean (*Mucuna pruriens*) by substitutive experiments: Suggestion of companion plants with corn and kidney bean. Jap. J. Soil Sci. Plant Nutri. 62: 363-370.

Fujii, Y., Shibuya, T. (1991f) A new bioassay for allelopathy with agar medium I. Assessment of allelopathy from litter leachate by sandwich method. Weed Res., Japan. 36(s): 150-151.

Fujii, Y., Shibuya, T. (1991g) A new bioassay for allelopathy with agar medium II. Mixed culture of allelopathic candidates with acceptor plants in agar medium. Weed Res., Japan. 36(s): 152-153.

Fujii, Y., Furukawa, M., Sugawara, M., Hayakawa, Y., Shibuya, T., Yasuda, T. (1991h) Survey of Japanese medicinal plants for the detection of allelopathic properties. Weed Res., Japan. 36: 36-42.

Fujii, Y., Shibuya, T., Usami, Y. (1991i) Allelopathic effect of *Mucuna pruriens* on the appearance of weeds. Weed Res., Japan 36: 43-49.

Fujii, Y., (1993) The allelopathic effect of some rice varieties. Technical Bulletin for Food and Fertilizer Center 134: 1-6.

Fujii, Y., (1994) Screening of allelopathic candidates by new specific discrimination and assessment methods for allelopathy, and the identification of L-DOPA as the allelopathic substance form the most promising velvetbean (*Mucuna pruriens*). The Bulletin of the National Institute of Agro-Environmental Sciences 10: 115-218.

Harper, J.L. (1977) Mechanism of Interaction between Species. *In* Population Biology of Plants. Academic Press, London, UK. pp. 347-381.

Hoagland, D.R., Arnon, D.I. (1950) The water-culture method for growing plants without soil. California Agricultural Experiment Station Circular No. 347.

Inderjit, S., Dakshini, K.M.M., Einhellig, F.A. (eds), (1995) Allelopathy, Organism, Processes, and Applications. ACS Symposium Series 582, American Chemical Society, Washington, D.C. USA.

Iqbal, Z., Hiradate, S., Noda, A., Isojima, S., Fujii, Y. (2002) Allelopathy of buckwheat: Assessment of allelopathic potential of extract of aerial parts of buckwheat and identification of fagomine and other related alkaloids as allelochemicals. Weed Biology and Manage. 2: 110-115.

Iqbal, Z., Hiradate, S., Noda, A., Fujii, Y. (2003) Allelopathic activity of buckwheat: isolation and characterization of phenolics. Weed Sci., 51: 657-662.

Iqbal, Z., Hiradate, S., Araya, H., Fujii, Y. (2004) Plant growth inhibitory activity of *Ophiopogon japonicus* Ker-Gawler and role of phenolic acids and their analogues: a comparative study. Plant Growth Regul., 43: 245-250.

Kamo, T., Hiradate, S., Fujii, Y. (2003) First isolation of natural cyanamide as a possible allelochemical from hairy vetch *Vicia villosa.* J. Chem. Ecol. 29: 275-283.

Leather, G.R., Einhellig, F.A. (1986) Bioassays in the study of allelopathy. *In* The Science of Allelopathy. A.R. Putnam., C.S. Tang. (eds.) John Wiley & Sons. NY, USA. pp 133-145.

Lehle, F.R., Putnam, A.R. (1982) Plant Physiol. 69: 1212-1216.

Mallik, M.A.B., Puchala, R., Grosz, F.A. (1994) A growth inhibitory factor from lambsquarters (*Chenopodium album*). Journal of Chem. Ecol. 20: 957-967.

Narwal, S.S. (1994) Allelopathy in Crop Production. Scientific Publishers, Jodhpur, India.

Putnam, A.R., Tang, C.S. (ed.). (1986) The Science of Allelopathy, John Wiley & Sons, NY, USA.

Rice, E.L. (1984) Allelopathy, 2nd ed. Academic Press. London, UK.

Rice, E.L. (1984b) Manipulated ecosystems: roles of allelopathy in agriculture, *In* Allelopathy, 2nd ed. Academic Press. London, UK. pp. 8-12.

Rice, E.L. (1995) Biological Control of Weeds and Plant Diseases, Advances in Applied Allelopathy University of Oklahoma Press, Norman, OK, USA.

Rizvi, S.J.H., Rizvi, V. (eds). (1992) Allelopathy, Basic and Applied Aspects. Chapman & Hall, London, UK..

Schreiner, O., Reed, H.S. (1907) The production of deleterious excretions by roots. Bull. Torrey Bot. Club 34: 279-303.

Shilling, D.G., Yoshikawa, F. (1987) A rapid seedling bioassay for the study of allelopathy. *In* Allelochemicals, Role in Agriculture and Forestry, G.R. Waller, (ed.) ACS Symposium Series 330, American Chemical Society. pp. 334-342.

Steel, R.G.D., Torrie, J.H. (1980) Linear regression. *In* Principles and Procedures of Statistics, a Biometrical Approach, 2nd ed. McGraw-Hill Book Company, NY, USA. pp. 239-271.

Stevens, K.L. (1986) Polyacetylenes as allelochemicals. *In* The Science of Allelopathy, A.R. Putnam, and C.S. Tang. (ed.) John Wiley & Sons, NY, USA. pp. 219-228.

Tang, C.S., Zhang, B. (1986) Qualitative and quantitative determination of the allelochemical sphere of germinating mung bean. *In* The Science of Allelopathy, A.R. Putnam and C.S. Tang (ed). John Wiley & Sons. NY, USA. pp. 229-242.

Thompson, A.C. (ed.), (1985) The Chemistry of Allelopathy, Biochemical Interactions Among Plants. ACS Symposium Series 268, American Chemical Society, Washington, DC, USA.

Waller, G.R., Nowacki, E.K. (ed.) (1978) Alkaloid Biology and Metabolism in Plants, Prenum Press, New York, USA.

Waller, G.R. (ed.), (1987) Allelochemicals, Role in Agriculture and Forestry, ACS Symposium Series 330, American Chemical Society, Washington, DC, USA.

Waller, G.R., Yamasaki, K. (eds.), (1996a) Saponins Used in Traditional and Modern Medicine, Advances in Experimental Medicine and Biology Vol. 404, Prenum Press, NY, USA.

Waller, G.R., Yamasaki, K. (eds.), (1996b) Saponins Used in Food and Agriculture, Advances in Experimental Medicine and Biology Vol. 405, Prenum Press, NY, USA.

Yasuda, T., Shibuya, T., Fujii, Y. (1991) Discrimination of allelopathy of common lambsquarters by stairstep experiments. Jap. J. Soil Sci. Plant Nutri. 62: 252-257.

SECTION 2

New Allelochemicals (Pharmaceuticals, Degradation, Promotion, Ion Dissolution)

4

Isolation, Structural Elucidation and Synthesis of Biologically Active Allelochemicals for Potential Use as Pharmaceuticals

Stephen J. Cutler[1], Rosa M. Varela[3], Miguel Palma[3], Francisco A. Macias[3] and Horace G. Cutler[2]

Department of Medicinal Chemistry, School of Pharmacy, University of Mississippi, University, MS 38677, USA

[2]Natural Products Discovery Group, Southern School of Pharmacy, Mercer University, 3001 Mercer University Drive, Atlanta, GA 30341, USA

[3]Department of Chemistry and Analytical Chemistry, Faculty of Sciences, University of Cadiz, Poligono Rio San Pedro S/N, 11510 Puerto Real, Spain.

Our group has isolated many biologically active natural products that possess effects as agrochemicals, including chaetoglobosin K, moniliformin, cyclopenin, cytochalasin B, and H. Not only do many of these have potential use as agrochemicals, they also possess structural templates that may serve as leads in the development of pharmaceuticals. One such group of agents is the flavones. These allelochemicals are widely distributed throughout nature and are responsible for a lot of of the coloring found in the plant kingdom. Flavanoids have been shown to possess a broad range of properties including antibacterial, antifungal, antiviral and anticancer. One compound in particular, flavone acetic acid (FAA), has been shown to be very effective against slow growing solid tumors, i.e., pancreatic adenocarcinoma, colon 38 adenocarcinoma and mammary adenocarcinoma. Based on the biological effects of FAA, we have synthesized several analogs and demonstrated their usefulness against B16 melanoma and PC3, a prostate tumor. In order to better understand the molecular mechanism of action of these agents, we report their effects on preventing the inhibition of gap junction-mediated communication between cells. From this, we have found that allelochemicals serve as excellent templates in the development of new pharmaceuticals.

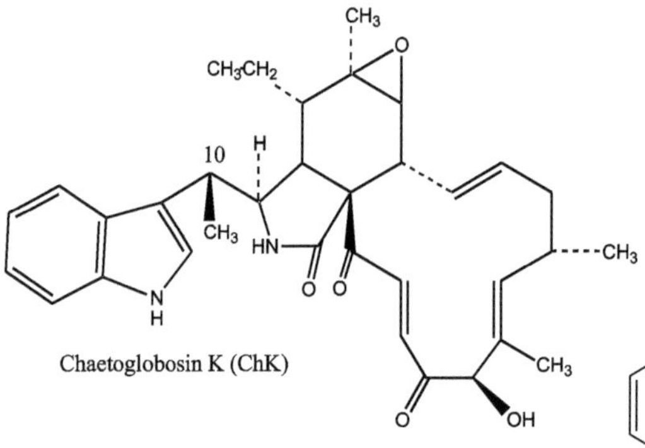

Chaetoglobosin K (ChK)

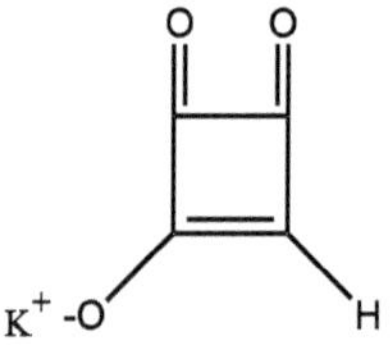

Moniliformin

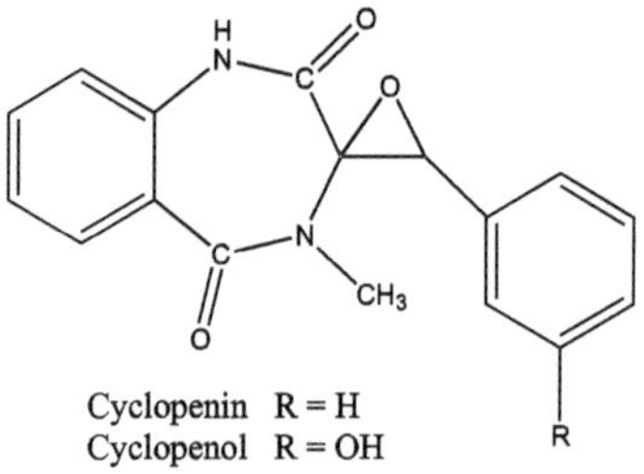

Cyclopenin R = H
Cyclopenol R = OH

Keywords: *etiolated wheat coleoptile, flavone acetic acid, natural products*

INTRODUCTION

The flavone ring (Fig. 1) structures are ubiquitous throughout nature. These compounds are found in fruits, vegetables, grains, nuts, stems, leaves and flowers (Harbourne, 1967; Harbourne, *et al.*, 1975) and are an integral part of normal plant growth and development. Additionally, they constitute the active ingredients of many "folk" medicines and are part of the arsenal of herbalists.

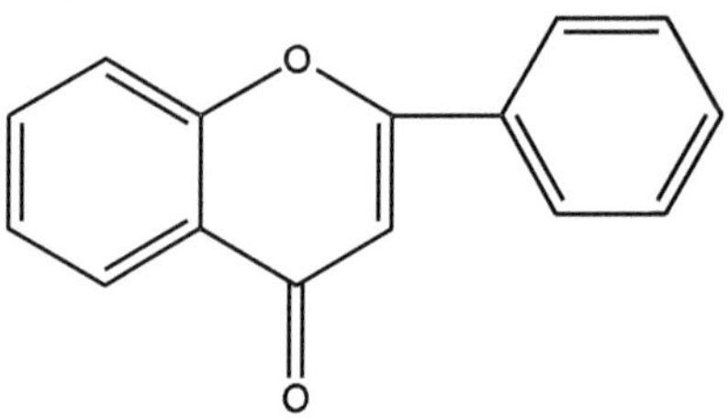

Fig. 1 Flavone nucleus

Flavanoids have been shown to possess a broad range of activities including control of carcinomas, viruses, fungi and bacteria (Ishitsuka *et al.*, 1982; Hirano *et al.*, 1995).

Hundreds of flavones have been evaluated for antitumor activity at the National Cancer Institutes (NCI) drug development program (Armand *et al.*, 1988). For a number of years, the mouse leukemia L1210 and P388 were the primary screening tumor models at NCI. In 1990, the National Institutes of Health restructured the primary bioassay to include other tumor lines. This was in part due to the primary screen of leukemia cell lines not detecting activity against solid tumors. It is well established that agents which are highly active against P388, are often inactive against other tumors such as solid neoplasms. Of 1,500 agents active against P388, less than 2% were active against Lewis Lung Carcinoma, Colon Adenocarcinoma, and CX-1 —a human colon tumor. Although hundreds of agents effective against leukemia have been identified using the P388 and L1210 models, few agents have been identified which are active against solid tumors of various organs. Also, agents which are inactive against P388 can be active against other neoplasms. Examples include taxol, bleomycin, prednisone, busulfan, hexamethylmelamine and Baker's Antifol (Frei, 1985). One such agent which has demonstrated effects against solid tumors with limited uses against soft leukemias is flavone acetic acid 1 (Fig. 2) (Plowman *et al.*, 1986; Bissery *et al.*, 1988). This agent was synthesized by the Lyonnaise Industrielle Pharmaceutic (LIPHA, Lyon, France) (Briet *et al.*, 1983a,b) and was investigated as a broad spectrum antineoplastic agent against pancreatic adenocarcinoma (No. 2) and colon 38 adenocarcinoma, which is often relatively resistant to cytotoxic agents. Other tumors on which flavone acetic acid has been shown to possess activity include reticulum cell sarcoma and mammary adenocarcinoma. Furthermore, it is more effective on these slow growing solid tumors than on rapidly proliferating leukemias such as L1210 and P388. In addition, flavone acetic acid possessed none of the usual properties of cytotoxic agents such as nausea, vomiting and myelosuppresion (Kerr *et al.*, 1986). This may, in part, be due to its limited effects against the rapidly proliferating leukemia models. The mechanism of action has not been demonstrated, but some reports suggest that flavone acetic acid may augment natural killer (NK) cell activity in normal and tumor-bearing mice and in human patients (Hornung *et al.*, 1988; Wiltrout *et al.*, 1988; Dalessandro and Borsellino *et al.*, 1996; Kitson *et al.*, 1998). These findings suggest a lead for further chemical studies to develop a potential chemotherapeutic drug which may also function as a biological response modifier (BRM) (Aitken *et al.*, 1997, 1998). However, because of the emphasis on the pharmaceutical properties of FAA, the agrochemical properties have been largely ignored.

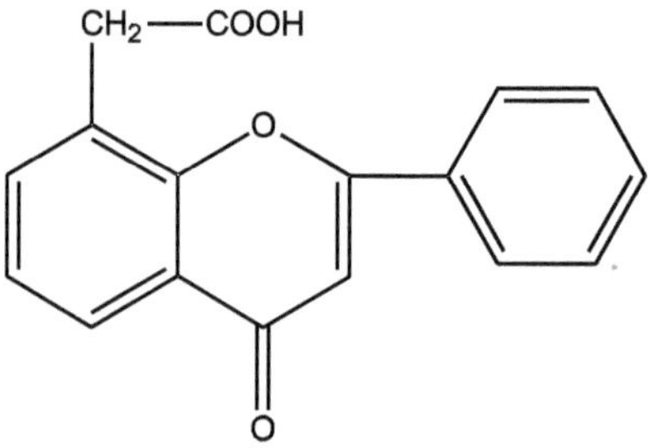

1

Fig. 2 Flavone acetic acid

Due to the unique spectrum of activity of flavone acetic acid, the search for other bioactive congeners which may exert a more potent and selective activity against neoplastic cells derived from solid tumors and, additionally, may possess agrochemical properties, is a tantalizing project.

MATERIALS AND METHODS

Various flavones (Fig. 3) were synthesized as previously described (Cutler *et al.*, 1990, 1993). In essence, this procedure utilizes various substituted furan *o*-aminonitriles as dienes in Diels-Alder cycloaddition reactions to prepare *o*-hydroxyacetophenones (Scheme 1).

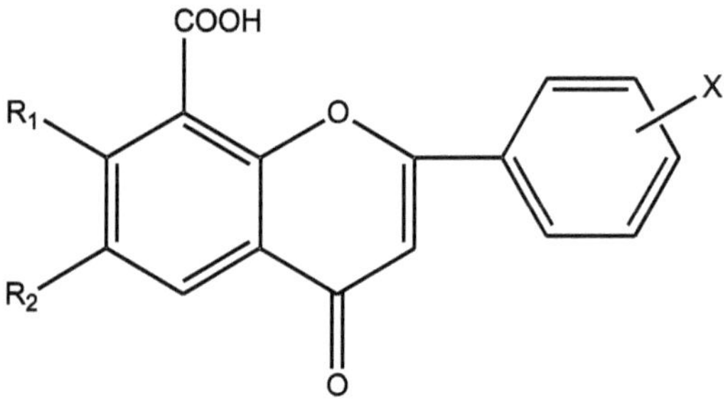

2

Fig. 3 Flavone derivatives

This reaction occurs under mild conditions (refluxing dioxane) which are in sharp contrast to the drastic conditions which are normally employed in Diels-Alder reactions with 2, 5-disubstituted furans. Work-up of the adduct with aqueous hydrochloric acid yields the substituted *o*-hydroxyacetophenone. This provides a useful precursor for the synthesis

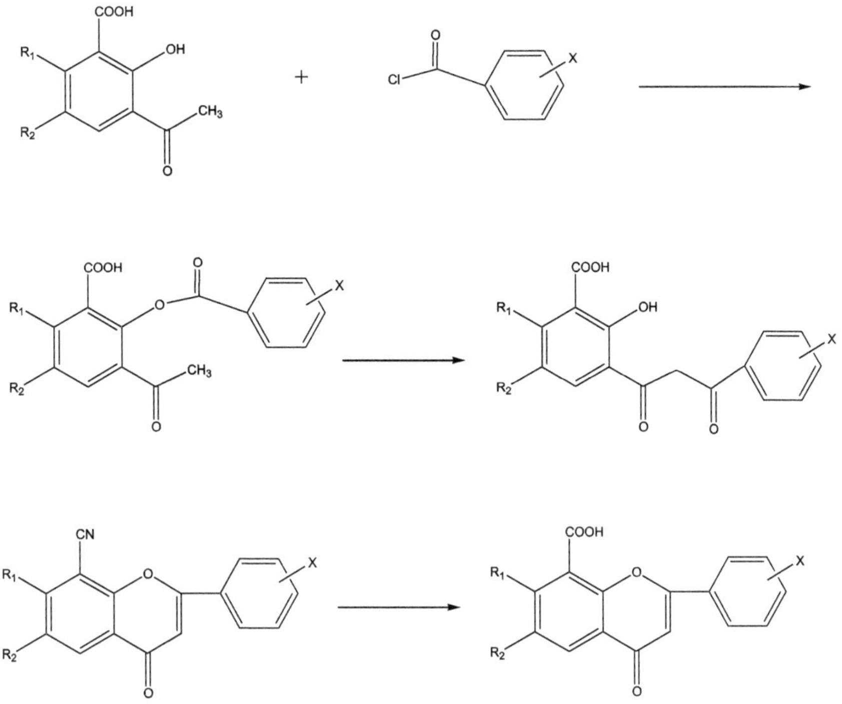

Scheme 1

of uniquely substituted flavones which are structurally related to flavone acetic acid (Scheme 1). Using this route, several highly substituted flavones (Fig. 4) were synthesized and evaluated for their potential antitumor properties. Selected analog were submitted to the NCI for evaluation against P388 leukemia. However, most analog were evaluated against solid tumors such as B16 melanoma and PC3, a prostate tumor cell line. Also, the compounds were screened in the wheat coleoptile bioassay, which is a good indicator of biological activity in both plant and zoological systems (Jacyno and Cutler, 1993; Cutler and Cutler, 1999).

B16 Melanoma

The effects of the flavones against B16 melanoma were performed in mice (Cutler *et al.*, 1993). All B16 melanoma studies were performed on female B6D2F$_1$ mice (17-21 g) which had been inoculated ip with 0.5 ml of a 1:10 brei of B16 melanoma prepared from subcutaneous tumors passaged in syngeneic C57BL/6 mice. The animals were randomized to treatment groups of seven mice each and were treated ip on days one to eight. Drugs were formulated as aqueous solutions by preparation of sodium salts

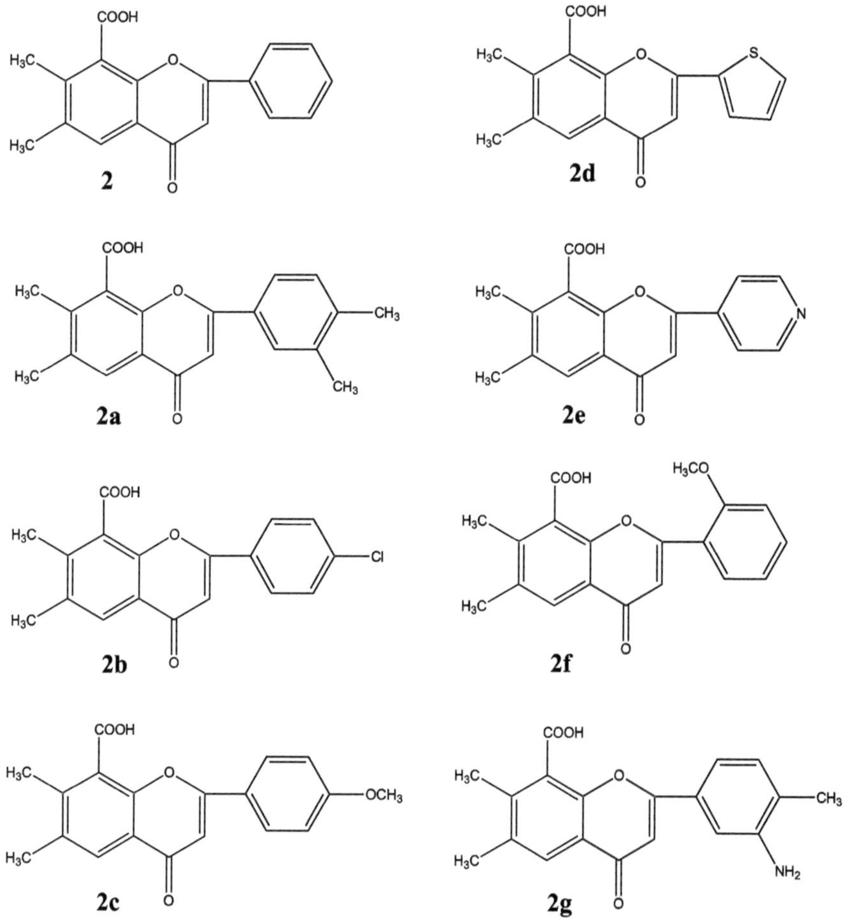

Fig. 4 Selected flavones synthesized and evaluated in this study

achieved by addition of a slight molar excess of sodium hydroxide. The mice were monitored for survival for 60 d. An increase in lifespan of > 40% is considered to reflect significant antitumor activity.

PC3

PC3 cells (ATCC, Manassas, VA) were cultured in F12K media and supplemented with 20 mM L-Glutamine, 100 μg/ml penicillin/streptomycin/Amphotericin B (Mediatech, Herndon, VA) and 10% heat-activated fetal bovine serum (Atlanta Biologicals, Inc. Norcross, GA). Cells were cultured at 37°C with 5% CO_2 unless otherwise specified. PC3 cells were passed into Falcon 24 well tissue culture plates (Fisher Scientific) 24 h prior to the addition of test flavones. Drugs were formulated as aqueous solutions by preparation of sodium salts achieved by addition of a slight molar excess of sodium hydroxide. At determined time periods the flavones were added in quadruplicate in conjunction with cell only and vehicle controls. The cells were incubated at 37°C with 5% CO_2 until assayed by the Neutral Red based toxicological assay. The protocol was adapted from Sigma Chemicals, Inc (Saint Louis, MS). At the specified time point 100 μl of Neutral Red (Sigma Chemicals, Saint Louis, MS) diluted 1:1 in 1 M HEPES (pH 6.8) was added to each well (1/10th the volume). The cultures were then incubated for 3 h at 37°C with 5% CO_2. The dye and media were removed and the cells washed twice, once with PBS + 1 mM $CaCl_2$, 1 mM $MgCl_2$ and the other in PBS only. 500 μl of solubilization buffer (1% acetic acid, 50% EtOH) was then added and the cultures incubated at room temperature for 10 min while shaking. Distilled water (500 μl) was then added to each well and the absorbance taken at 540 nm using a Perkin Elmer UV/VIS spectrophotometer. Data are calculated as percent of control with mean ± STD.

Etiolated Wheat Coleoptiles

The etiolated wheat coleoptile bioassay was used as the primary agrochemical bioassay (Hancock *et al.*, 1964). Wheat seed, *Triticum aestivum* L. cv. Wakeland, was sown on moist vermiculite in plastic trays, sealed with aluminum foil and placed in the dark at 22+1°C. Four-day-old seedlings were removed and prepared under a green safelight. The roots and caryopses were removed from each seedling and the apical portions were fed into a van der Weij guillotine. The first 2 mm were cut off and discarded, and then the next 4 mm were removed from each coleoptile and saved for the bioassay. These were floated on distilled water, and then added to test tubes containing a molar concentration of the compound to be assayed. A 10^{-3} M stock solution of each compound was formulated by weighing the necessary amount, adding 25 μl of acetone and bringing the total volume to 10 ml in phosphate-citrate buffer at pH 5.6 supplemented with 2% sucrose. From this, serial dilutions were made. Then, 2 ml of each concentration were added to individual test tubes along with 10 coleoptile sections. The tubes were placed in a roller tube apparatus rotating at 0.25 rpm for 18 h. After 18 h, the sections were removed, blotted, placed on a glass plate and enhanced with a photographic enlarger to produce an image x3. All data were statistically analyzed (Kurtz *et al.*, 1965).

RESULTS AND DISCUSSION

The initial evaluation of selected 8-cyano and 8-carboxamido derivatives in the P388 leukemia assay of the National Cancer Institute (NCI) failed to demonstrate any effects. However, the 8-carboxylic acid demonstrated significant anti-leukemia properties in this assay. It exhibited a confirmed activity with a Treated/Control = 128-161% at 120-240 mg/kg. A T/C value of 125% represents significant activity. This suggests the importance of the acid moiety in these flavones for antitumor activity. Since FAA was more effective on solid tumors than on rapidly proliferating leukemias, emphasis was placed on evaluating the flavone-8-carboxylic acids against B16 melanoma and PC3 prostate carcinoma.

The effects of the flavone derivatives against B16 melanoma in mice are presented in Table 1. Several of the agents synthesized are equally as effective as FAA with one showing (the 3′-NH_2, 4′-CH_3 derivative) a higher level of activity. This demonstrates that substituents on the 2-aryl ring can markedly influence activity and dose-potency. This observation was also noted in the PC3 cell line assay which is presented in Fig. 5.

Table 1 Effects of selected flavone derivatives against B16 tumors. Comparative anti-tumor effects in mice bearing ip B16 Melanoma ILS % (increase life span) of flavone-8-carboxylic acid

Compound	*Dose (mg/kg, ip, d 1-8)*				
X =	*160*	*80*	*40*	*20*	*10*
H	Tox	46	35	23	4
FAA	Tox	Tox	71	46	25
4-CH_3	Tox	71	73	60	38
FAA	Tox	Tox	71	46	25
3'-NH_2, 4'-CH_3	100	120	13	13	0
FAA	Tox	40	20	7	7

The available data suggests that an 8-carboxylic acid moiety is essential for antitumor activity among these flavone analog since the 8-cyano and 8-carboxamido analog were inactive. Furthermore, it is not necessary to separate the acid functional group by one methylene unit from the aromatic portion of the system. It is clear that varying the substituents on the 2-aryl ring can have a pronounced effect on the biological activity of these agents. These results suggest that electron donating groups such as a methyl or methoxy increase the antineoplastic activity more than electron withdrawing groups such as a chloro substitution.

Several of the flavone derivatives significantly inhibited wheat coleoptile growth at 10^{-3} M concentrations. The results of this bioassay are presented in Table 2. While the concentrations may appear to be high, it has been our experience, after testing several hundred compounds in the coleoptile bioassay, that while some may exhibit activity only at 10^{-3} M,

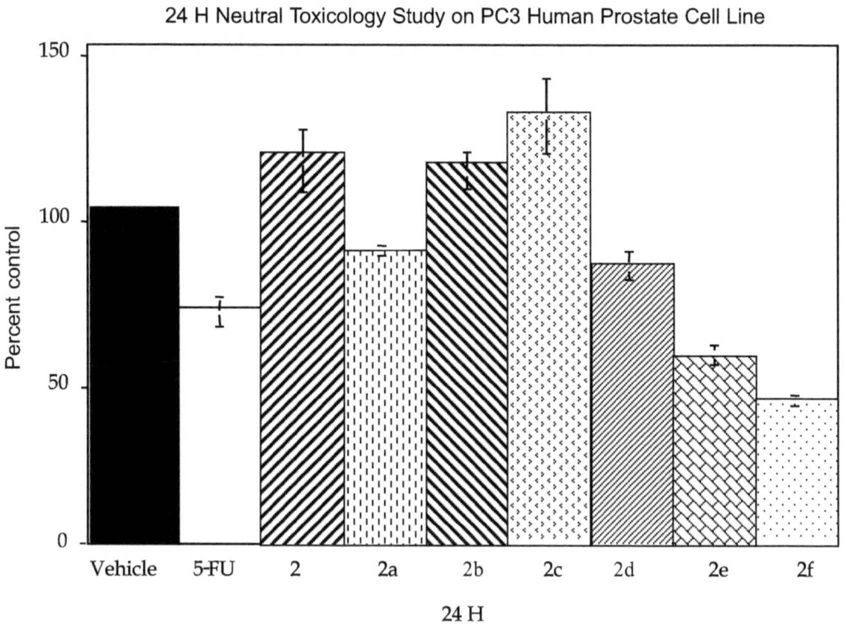

Fig. 5 Effects of selected flavones on human prostate cancer cells (PC3 cell line)

Table 2 Activity of selected substituted flavones against etiolated wheat coleoptiles, *Triticum aestivum* cv. Wakeland. Values are expressed as percent inhibition relative to controls.

Compound	*Molar Concentration*		
	10^{-3}	10^{-4}	10^{-5}
2	20*	0	0
2a	38*	0	0
2b	26*	0	0
2c	50*	0	0
2e	38*	0	0
2f	0	0	0
2g	0	0	0

*Significant inhibition ($p < 0.01$)

the response is greatly amplified when the compound is placed in the correct biological niche. Such an example is 6-pentyl-a-pyronc (Parker *et al.*, 1997). Furthermore, the coleoptile bioassay is a very non-specific bioassay but, importantly, it serves as a primary bioassay in revealing general biological activity. Using this assay, several allelochemicals have been identified (e.g., chaetoglobosin K, moniliformin, cyclopenin, cytochalasin B, and H). In the current study, it demonstrated that the flavone derivatives possess allelochemicals properties. These results lead to the probability that further synthetic manipulation of the parent molecule may lead to more potent allelopathic and zoologically important products.

CONCLUSION

While our main emphasis has been placed on the flavone structure and its manipulation for the production of pharmaceuticals, it should be reiterated that the other structures, chaetoglobosin K, moniliformin, cyclopenin and cyclopenol, have found "other" uses. While all were initially identified as allelopathic agents from fungi that had plant growth regulating properties (Cutler *et al.*, 2004), further applications of these natural products have been found to be experimentally useful in apparently unrelated fields. Chaetoglobosin K is presently being evaluated as an antineoplastic agent and has been determined to have significant effects on intercellular communications across gap junctions in cultured rat brain glial cells. In cases where insecticides close the passage of important molecules across gap junctions, for example the insecticides dieldrin and endosulfan, chaetoglobosin K has been shown to open those channels (Matesic *et al.*, 2001) resulting in cellular homeostasis. Moniliformin, a highly polar compound, that occurs naturally as either the potassium or sodium salt, has been transformed into an experimental herbicide. As pointed out earlier, its mode of action is by inhibiting mitosis at the C-metaphase plate. Hence, it may substitute for colchicine, which is a highly lipophilic compound. Additionally, moniliformin has been used as a potential agent in the liquid crystal industry in Japan (personal communications). Cyclopenin, originally found to have both plant growth regulatory and tranquilizing properties, has undergone further synthesis in our laboratories to make potentially useful bioactive compounds.

It would be a short sighted and costly mistake to consider allelopathic agents solely for use in agronomic crops. These bioactive chemical templates offer platforms for further valuable derivatives for use in what may appear to be unrelated, but highly important and lucrative disciplines.

References

Aitken, R.A., Biddy, M.C., Double, J.A., Laws, A.L., Ritchie, R.B., Wilson, D.W. (1997) Synthesis and antitumor activity of new derivatives of flavone-8-acetic acid (FAA). Part 2: Ring-substituted derivatives. Archiv Der Pharmazie. 330: 215-224.

Aitken, R.A., Biddy, M.C., Bielefeldt, F., Double, J.A., Laws, A.L., Mathieu, A.L., Ritchie, R.B., Wilson, D.W.J. (1998) Synthesis and antitumor activity of new derivatives of flavone-8-acetic acid (FAA). Part 3: 2-Heteroaryl derivatives. Archiv Der Pharmazie, 331: 405-411.

Armand, J.P., De Forni, M., Recondo, G., Cals, L., Cvitkovic, E., Munck, J.M. (1988) *In* Plant Flavonoids in Biology and Medicine II. Biochemical, Cellular, and Medicinal Properties, Alan R. Liss, Inc. 235-241.

Bissery, M.-C., Valeriote, F.A., Chabot, G.G., Crissman, J.D., Yost, C., Corbett, T.H. (1988) Flavone acetic acid (NSC 347512)-induced DNA damage in Glasgow osteogenic sarcoma *in vivo*. Cancer Res. 48, 1279-1285.

Briet, P., Berthelon, J.J., Collonges, F. (1983a) European Pat 80419 Chem. Abstract. 99: 122305t.

Briet, P., Berthelon, J.J., Collonges, F. (1983b) European Pat 80934 Chem. Abstract. 99: 175594t.

Cutler, H.G., Cutler, S.J. (1999) Agrochemicals and Pharmaceuticals: The Connection. *In* Biologically Active Natural Products: Agrochemicals and Pharmaceuticals H.G. Cutler, S.J. Cutler, (eds.), CRC Press, Boca Raton Fl, USA. pp. 1-14.

Cutler, H.G., Cutler, S.J., Matesic, D. (2004) Mode of Action of Phytotoxic Fungal Metabolites. *In* Allelopathy: Chemistry and Mode of Action of Allelochemicals, F.A., Macias, J.C.G. Galindo, J.M.G. Molinillo, H.G. Cutler (eds.), CRC Press LLC, Boca Raton Fl, USA. pp. 253-270.

Cutler, S.J., El-Kabbani, F.M., Keane, C., Fisher-Shore, S.L., Blanton, Jr., C.D. (1990) Furan *o*-Aminonitriles as Precursors to Flavone Analogues. Heterocycles. 31: 651-661.

Cutler, S.J., El-Kabbani, F.M., Fisher-Shore, S.L., McCabe, F.L., Johnson, R.K., Blanton, Jr., C.D. (1993) Synthesis of Flavone-8-Carboxylic Acid Analogues as Potential Antitumor Agents. European J. Med. Chem. 28: 407-414.

Dalessandro, N., Borsellino, N. (1996) In vivo effects of tumor necrosis factor-alpha or flavone acetic acid in combination with doxorubicin on multidrug-resistant B16 melanoma. Anti-cancer Drugs. 7: 281-287.

Frei, E. (1985) Curative cancer chemotherapy. Cancer Res. 45, 6523-6537.

Hancock, C.R., Barlow, H.W., Lacey, H.J. (1964). The East Malling coleoptile straight-growth test method. J. Exp. Bot. 15: 166-176.

Harbourne, J.B. (1967) Comparative Biochemistry of the Flavonoids, Academic Press, NY, USA.

Harbourne, J.B., Mabry, T.J., Mabry, H. (1975) The Flavonoids, Academic Press, NY, USA.

Hirano, T., Abe, K., Gotoh, M., Oka, K. (1995) Citrus flavone tangeretin inhibits leukemia HL-60 cell growth partially through induction of apoptosis with less cytotoxicity on normal lymphocytes. British J. Cancer. 72: 1380-1388.

Hornung, R.L., Young, H.A., Urba, W.J., Wiltrout, R.H. (1988) Immunomodulation of natural killer cell activity by flavone acetic acid: Occurrence via induction of interferon a/b. J. Natl. Cancer Inst. 80: 1226-1231.

Ishitsuka, H., Ohsawa, C., Ohiwa, T., Umeda, I., Suhara, Y. (1982) Antipicornavirus flavone Ro 09-0179. Antimicrobial Agents Chemotherapy. 22: 611-616.

Jacyno, J.M., Cutler, H.G. (1993). Detection of herbicidal properties: scope and limitations of the etiolated wheat coleoptile bioassay. PGRSA Quarterly. 21:15-24.

Kerr, D.J., Kaye, S.B., Graham, J., Cassidy, J., Harding, M., Setanoians, A., McGrath, J.C., Vezin, W.R., Cunningham, D., Forrest, G., Soukop, M. (1986). Phase I and pharmacokinetic study of LM985 (flavone acetic acid ester). Cancer Res. 46: 3142-3146.

Kitson, R.P., Ohashi, M., Brunson, K.W., Goldfarb, R.H. (1998). Flavone acetic acid enhances accumulation of IL-2 activated NK cells within established metastases. In Vivo. 12: 593-597.

Kurtz, T.E., Link, R.F., Tukey, J.W., Wallace, D.L. (1965) Short-cut multiple comparisons for single and double classifications. Part 1, Results. Technometrics. 7: 95-161.

Matesic, D., Blomme, I., Sunman, J.A., Cutler, S.J., Cutler, H.G., (2001) Prevention of Organochlorine-induced Inhibition of gap Junctional Communication by Chaetoglobosin K in Astricytes. Cell Biol. 17: 395-408.

Parker, S.R., Sale, P.P., Hill, R.A., Cutler, H.G. (1997). The biological activity and application of 6-pentyl-2H-pyran-2-one and its analogs. HortRes. 1-26.

Plowman, J., Narayanan, V.L., Dykes, D., Szavasi, E., Briet, P., Yoder, O.C., Paull, K.D. (1986) Flavone acetic acid: a novel agent with preclinical antitumor activity against colon adenocarcinoma 38 in mice. Cancer Treatment Report. 70: 631-635.

Wiltrout, R.H., Boyd, M.R., Back, T.C., Salup, R.R., Arthur, J.A., Hornung, R.L. (1988). Flavone-8-acetic acid augments systemic natural killer cell activity and synergizes with IL-2 for treatment of murine renal cancer. J. Immunol. 140: 3261-3265.

5

Recent Chemical Aspects of Wheat Allelopathy

Terry Haig

Australian Agricultural, Food and Wine Institute, and School of Science and Technology, Charles Sturt University, Wagga Wagga, NSW, 2678, Australia

The scope of this chapter covers in review some of the more recent developments in wheat (*Triticum* spp.) allelochemistry, mainly across the years 2000–2004. Whilst the attempt to do this is meant to be comprehensive, it is not necessarily exhaustive. Outcomes from the ongoing European FATEALLCHEM project are discussed, along with quantification of wheat allelochemicals, the properties of wheat root exudates, important correlations between concentrations of allelochemics and observed phytotoxicity, and metabolic studies on wheat allelochemicals.

Keywords: *benzoxazinoids, correlations, metabolites, phenolic acids, quantification, root exudates, scopoletin, synergism, wheat allelochemicals*

INTRODUCTION

As an important world crop, wheat has become a significant focus of allelopathic attention over the last three decades. Research on allelopathy in wheat has covered such ground as its genetic basis, effects upon weeds, pests, various other crops, diseases, autotoxicity, cultivar screening and isolation, identification and quantification of the active allelochemics (Wu *et al.*, 2001a). A wheat cultivar laboratory screening test for allelopathic potential using annual ryegrass (*Lolium rigidum* Gaud.) as the test plant (a common wheat weed in Australia) and a novel means of excluding the plant competition factor, ECAM (the equal-compartment-agar-method), was successfully employed by Wu *et al.* (2000a) to evaluate 92 cultivars.

Root length of ryegrass was inhibited over a range from 24% to 91%, thereby indicating that there were significant differences between the potentials of common wheat cultivars. This same ECAM method was then later used by Wu *et al.* (2000b) to evaluate a 50-country collection of 453 cultivars for allelopathic potential and for evidence of genetic control of allelopathy in wheat. In this collection, inhibition of ryegrass root length ranged from 10% to 91%, and other evidence pointed to the conferring of the allelopathy trait via multiple genes. The allelopathic properties of wheat were examined further against a herbicide-resistant (HR) variety of annual ryegrass, when Wu *et al.* (2003) showed that aqueous extracts from the shoots of 39 wheat cultivars demonstrated significant inhibition of germination and root growth of this resistant biotype. Germination of the HR biotype was inhibited from 3% to 100% depending upon the wheat cultivar, while root growth responded with a range from 12% stimulation to 100% inhibition. As this ryegrass biotype was resistant to acetyl CoA carboxylase inhibitors, acetolactate synthase inhibitors, photosystem II inhibitors, and tubulin formation inhibitors (Herbicide Groups A, B, C, D), the allelopathic result was seen as offering a new potential to deal with even resistant weeds without the need for artificial herbicides.

On a broader front, the importance of wheat allelopathy has been officially recognized by the Commission of the European Communities with the launch of the project "FATEALLCHEM" (Fate and Toxicity of Allelochemicals in Relation to Environment and Consumer) which involves collaboration between Danish, Italian, Spanish, Lithuanian, Austrian and Polish research institutes. The project is to investigate the prospects of exploiting wheat allelopathy to control weeds, and perhaps disease and insects also. Part of this investigation entails human and environmental risk assessment, and therefore the setting of certain key objectives (Fomsgaard, 2004a). Five objectives have so far been nominated:

(1) to identify and quantify the amount of allelochemicals and their metabolites in wheat of different origins and grown under different climates, environments, farming systems and soil types;
(2) to generate dose-response relationships of major wheat allelochemicals on the most important target weeds, pests and fungi;
(3) to assess the risk to environment and consumers of wheat allelochemicals and their metabolites;
(4) to compare the empirical results with theoretical results using rules-based (e.g. QSAR) predictions of toxicity and environmental transformation;
(5) to quantify allelochemicals in old wheat cultivars from Eastern Europe that have not been subjected to breeding programs.

Some of the outcomes from ongoing work on this project are reported in summary within this chapter.

There are a large number of studies on the allelochemics of wheat in the literature, with emphasis on such compounds as scopoletin (Baghestani *et al.*, 1999), phenolic acids (Lodhi *et al.*, 1987; Wu *et al.*, 2001b) and the benzoxazinoids (Copaja *et al.*, 1991; Perez and OrmenoNunez, 1991; Petho, 1992; Wu *et al.*, 2001c). Some studies have been conducted on the allelochemics within wheat tissues, while others have been done on the root exudates from wheat plants. An important issue appears at this point, because the necessary contact between living wheat plants and weeds is via exuded allelochemics through soil transportation, and may have very little to do with content within tissues. This was made quite evident in the work of Wu *et al.* (2001c) where, of a global selection of 58 wheat cultivar exudates, only 11 cultivar exudates were found to contain measurable amounts of the main allelochemical DIMBOA [2,4-dihydroxy-7-methoxy-2H-1,4-benzoxazin-3-(4H)-one] despite the fact that 57 of those cultivars had measurable quantities if DIMBOA within their roots. Exudation of DIMBOA appears to be an active metabolic process dependent upon genetic factors.

QUANTIFICATION OF ALLELOCHEMICALS IN WHEAT

The allelochemics in wheat known as the phenolic acids, describes a small range of water-soluble, aromatic, carboxylic group-containing phenols (such as salicylic acid), which are very broadly distributed within higher plants. Certainly, as a group of allelochemicals, the phenolics have received great attention, with a lot of supporting literature implicating them in some allelopathic activity (Einhellig, 2004). The accumulated evidence points to phenolics as playing a significant role in allelopathy, and a number of reviews on their biochemistry and ecology have been written (Blum *et al.*, 1999; Blum, 2004; Einhellig, 2004; Inderjit, 1996) but despite this their phytotoxicity is usually quite low. Pure substance chemoassay usually points to the need for minimum concentrations in the range 0.1 to 1.0 mM for significant plant toxicity (Huang *et al.*, 2003), a result which is about 100 to 1,000 times too high to be herbicidally useful. Typical structures are shown in Fig. 1.

While HPLC-DAD analysis of plant components for the phenolic acids has been the main technique of earlier days, better analyses for the individual acids have been achieved in more recent times using either HPLC or GC with a mass detector, especially a detector capable of MS/MS techniques which gives rise to great selectivity and sensitivity (Haig, 2004). Thus, the individual phenolic acid contents of wheat roots and shoots have

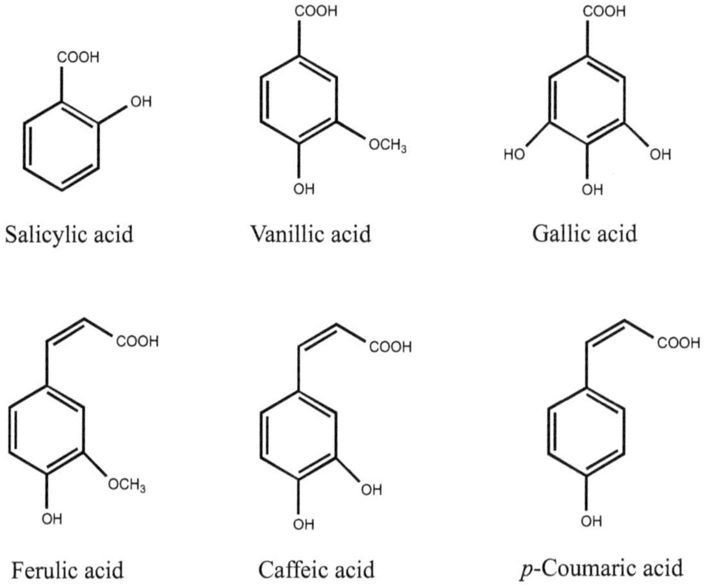

Fig. 1 Typical structures for phenolic acid allelochemics in wheat

been quantitatively determined by a GC/MS/MS technique via trimethylsilyl derivatization by Wu *et al.* (2000c; 2001b). Despite this attention, phenolic acids are not considered to be the most significant group of allelochemicals in wheat. This place is reserved for the class of compounds known as the benzoxazinoids, examples of which appear in Fig. 2.

(a) DIMBOA [2,4-dihydroxy-7-methoxy-2H-1,4-benzoxazin-3-(4H)-one]

(b) DIBOA [2,4-dihydroxy-2H-1,4-benzoxazin-3-(4H)-one]

(c) BOA Benzoxazolin-2-(3H)-one

As a preliminary view for the FATEALLCHEM project of what the allelopathic literature described for allelochemic quantification in wheat, Eljarrat and Barcelo (2001) reviewed sample handling and chromatographic analytical methods published until the year 2000. They found that CE, LC and GC methods had been used with such detectors as UV, DAD, MS and MS/MS, and inferred that an LC/MS/MS technique offered the most advantages. These authors also cautioned analysts on the instability of the hydroxamic acid forms of benzoxazinoids (such as DIBOA and DIMBOA with an OH group attached to the N atom) which can easily decompose to less phytotoxic compounds such as BOA and MBOA (6-methoxy-BOA) during sample preparation. Heat, alkali and

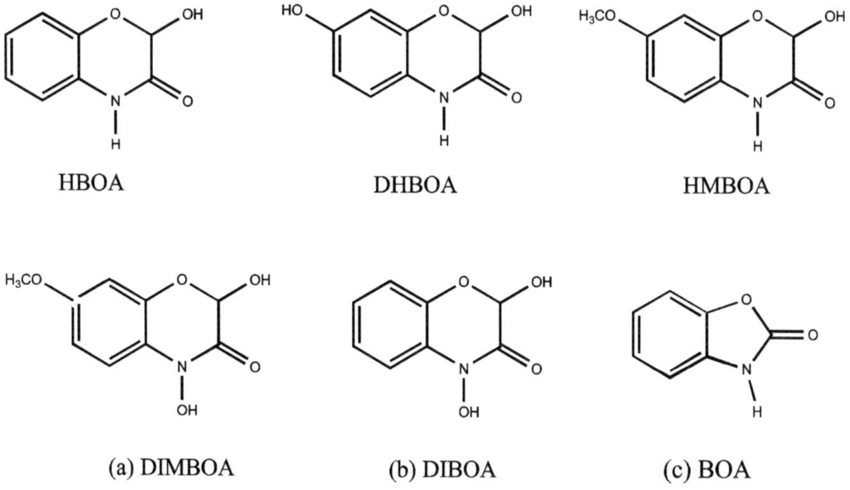

Fig. 2 Some benzoxazinoid allelochemicals present in wheat

certain enzymes speed up this decomposition. As the active benzoxazinones (6-membered heterocyclic ring) occur naturally within the wheat plant as their non-toxic conjugated 2-β-D-glucosides, any attempt to measure the separated aglucones must first allow for hydrolysis of the glycosidic linkage through either artificial acid hydrolysis or by the natural β-glucosidase enzymatic process that occurs once the wheat cells are mechanically injured. The HPLC analysis techniques using anti-hydrolysis safeguards are best suited to differentiate between aglycone and glycosidic contents for the benzoxazinones (Lyons *et al.*, 1988). FATEALLCHEM project workers (Bonnington *et al.*, 2003) followed through from the above considerations to devise an HPLC-ESI-MS/MS method of benzoxazinoid derivatives analysis from wheat tissue samples using a PLE-SPE sample clean-up procedure. HPLC column resolution, peak shape and analyte stability were enhanced from earlier work by the use of a dodecyl (C_{12}) TMS end-capped Synergi MAX-RP stationary phase. Compounds analyzed in MRM MS/MS negative ion mode were DIMBOA, DIBOA, the lactams HBOA and HMBOA, plus the two benzoxazolinone (5-membered heterocyclic ring) degradation products BOA and MBOA. The concentration range for the analytes was between 0.5 and 207 μg/g dry weight across six wheat cultivars, while individual analyte method LOD ranged between 0.1 and 1.1 μg/g dry weight. No attempt was made to measure the contents of wheat exudate. A further step again was made in the FATEALLCHEM project's concern over the reliability of benzoxazinoid analysis when Eljarrat *et al.* (2004) published the results of an interlaboratory study across four European countries involving six laboratories. The samples analyzed were comprised of seven benzoxazinoids in two standard solutions, plus one purified wheat root extract. Pooled results from all six laboratories on the analysis of the two standard solutions showed relative standard deviations for individual compounds ranging from 2% to 14%, but the precision was notably poorer with the actual root extract where standard deviations ranged from 19% to 47% between laboratories. Interestingly, the acidified solution stability of the chosen seven benzoxazinoids was also examined using standard spiked solutions stored at three different temperatures: room temperature, 4°C, and at -20°C, using triplicate LC/MS analyses. Storage times ranged over 1, 2, 3 and 7 d. Significant degradations were found to occur when solutions were stored above -20°C. At -20°C, MBOA, BOA and the glucoside of DIMBOA, remained stable, while DIBOA and HBOA showed losses of about 10%, and DIMBOA and HMBOA suffered some 20% loss. Obviously, any serious attempt to monitor such wheat allelochemicals accurately will in future require allowance for sample instability. Of all the analysis techniques tested in this study, LC/MS/MS was again recommended as the best choice.

Quantification of both benzoxazinone and phenolic acid allelochemics from the same set of derivatized wheat exudate samples using a GC/MS/MS technique proved to be a convenient method for Huang *et al.* (2003). Until this date, no wheat study had been published which deals simultaneously with production dynamics of both phenolics and benzoxazinoids within root exudates. The monitoring at six different time intervals across 15 d of wheat seedling growth in sterile, no-nutrient, 0.3% agar-water of the exudate content from the species *Triticum turgidum* ssp. Durum (Desf.) Husn. (cv. Khapli), led to a maximum in the total from five benzoxazinones at the 6-d stage equal to 0.735 ± 0.38 μmol/L agar. The corresponding maximum for the seven monitored phenolic acids occurred at the 8-d stage where the total content amounted to 1.590 ± 0.682 μmol/L agar. The timing of these two maxima was observed to coincide with the maximum level of phytotoxicity against the test plant annual ryegrass (*Lolium rigidum* Gaud.) which was being used to monitor the allelopathic effect of wheat plants.

Practical Application of Quantitative Techniques

For the reader in search of more specific analytical details concerning the use of GC/MS/MS methodology for wheat allelochemics, the recent detailed book chapter discussion on this procedure given by Haig (2004), is recommended. This practical case study examined wheat root exudates, but the techniques are equally applicable to extracts from plant parts also. Although the GC-based procedure requires an extra analyte derivatization step (trimethylsilyl derivatives) for volatility (when compared to a HPLC method), the troubling need to consider benzoxazinone instabilities during the period of analysis is of little concern because of the chemical locking of the labile functions achieved during the derivatization reactions.

A wider interest to date has been shown in the application of LC/MS/MS techniques for allelochemic quantifications in wheat, even though spontaneous benzoxazinoid degradations can occur during the analysis period. Some repression of this degradation during HPLC is achieved by working with an acidic eluant, but a long period between sample preparation and final analysis can still lead to errors. Though a number of workers have employed a HPLC/MS technique, probably the best outline of methodology comes from the dedicated publication by Bonnington *et al.* (2003), who described the practical details of wheat sample preparation; HPLC conditions; mass spectrometry conditions; calibration methods and limits of detection with notable rigor. A summary of their general approach is given here.

Sample preparation

A sample of previously cleaned and lyophilized wheat root or foliage (0.1 g) was spiked with stable internal standards (2-OMe-DIBOA and 2-OMe-HBOA) and the mixture extracted by pressurized liquid extraction (PLE) in an 11 mL stainless steel extraction cell filled with diatomaceous earth. The optimized PLE conditions used were: pressure 1500 psi; flush volume 60%; cell purge 60 sec; solvent composition MeOH (1% HOAc); temperature 150°C; three 5-min static cycles; cell preheat, 5 min with no N_2 purge. Concentration and purification of the extract solution for benzoxazinoids was performed with 500 mg RP C_{18} SPE cartridges which were subsequently eluted with 5 mL of MeOH/acidified water (60:40). Spiking experiments indicated that recovery of analytes ranged from 72% to 89%.

Chromatographic conditions

The HPLC used was a Waters 2690 Alliance fitted with quaternary pump and 120-vial sample management system and a 250 × 4.6 mm, 4 μm Synergi MAX-RP 80A C_{12} column (Phenomenex) using a 50 μL injection. The solvent line operated at 1 mL/min and consisted of elution solvent A (water with .05% HOAc) plus solvent B (MeOH) in a gradient of 0-2 min, 70% A; 2-19 min, 70-40% A; 19-21 min, 40-0% A; 21-23 min, 0% A; 23-25 min, 0-70% A; 25-38 min, 100% A. Total run time was 38 min allowing for column cleaning and regeneration. The line was also fitted with a zero-volume T solvent splitter where 0.2 mL/min was directed to MS detection and remaining 0.8 mL/min directed to waste. Online UV detection was carried out prior to the split using a Waters 996 PDA detector over the range 250-440 nm.

Mass spectrometry conditions

The MS/MS detector consisted of a Micromass Quattro triple quadrupole spectrometer with pneumatically assisted electrospray probe and Z-spray interface. Nitrogen was used as the desolvation and cone gas while the collision gas was argon at 3.6×10^{-5} mbar. MS data were acquired and processed with Masslynx V3.5 software. Using previously optimized analytical parameters based on infusions of pure reference compounds and precursor ion/product ion scans, quantifications of benzoxazinoids were achieved in MRM ESI-MS mode measuring negative ions with the following instrument settings: capillary voltage, 2.8 kV; source temp, 150°C; desolvation temp, 350°C; desolvation gas flow rate, 600 L/h; cone gas flow rate, 50 L/h; extractor 7 V; lens, 0.6 kV; multiplier 650; ion energy, 1.0 V; entrance, -2; exit energy, 1; resolution 12.0; interchannel delay, 0.02 s; total scan time, 1.01 s. Two ion transitions for each analyte were selected, a primary transition for quantification purposes and a secondary transition for identity confirmation. Positive target compound identification relied upon a standard retention time ±2% together with a ratio of ion abundances of two specific precursor/product ion transitions being within 10% of measured standards.

Two convenient chromatographic time windows were also selected having eight transitions per window to match the compounds (a) HBOA; DIBOA; HMBOA; DIMBOA (5 - 13.5 min), and (b) BOA; MBOA; MeO-DIBOA; MeO-HBOA (13.5 - 19 min). Three representative sets of compound-optimized MRM MS/MS parameters are included below to illustrate the analytical choices made.

First transition: (quantification)

Compound	*Precursor-product (m/z)*	*Cone (V)*	*Collision (eV)*	*Dwell (s)*	
DIMBOA	164–149	15	13	0.05	
DIBOA	134–78	35	18	0.2	
MBOA	164–149	15	13	0.05	
Second transition: (confirmation)					
					*ratio**
DIMBOA	164–121	15	20	0.1	4.36
DIBOA	134–91	35	20	0.2	1.87
MBOA	164–121	15	20	0.05	3.73

*ratio = MRM first transition/MRM second transition

Quantifying method

Both external standard and internal standard quantifications could be achieved. External standard methods based upon comparison with linear calibration curves from pure reference compounds yielded regression coefficients in excess of 0.99 even for analyte concentration ranges over four orders of magnitude. For analysts in possession of stable artificial benzoxazinoid analogues (such as 2-OMe-DIBOA) which can be used as internal standards, excellent results were obtainable within a 0.005-5 ng/µL range. For example, the primary transition of MBOA relative to that of 2-OMe-DIBOA gave a calibration curve regression coefficient of 0.994. Typical LOD for the entire PLE-SPE-HPLC-MS/MS method (Bonnington *et al.*, 2003) as applied to wheat foliage for analyses were as follows: HBOA (0.2); DIBOA (1.1); HMBOA (0.1); DIMBOA (0.2); BOA (0.5); MBOA (0.1 µg/g dry wt.) An estimate of method precision was made on triplicate analyses for each of the above six benzoxazinoids, with the following results: HBOA (5.14); DIBOA (15.60); HMBOA (1.74); DIMBOA (17.35); BOA (7.11); MBOA (5.95% rsd).

WHEAT ROOT EXUDATES

As mentioned in the introduction, the significance of plant root exudates as the prime source of allelopathic contact between competing plants in the field should not be overlooked. It is simply not adequate to measure wheat tissue allelochemic content in shoots, leaves or roots as a reliable indication of allelopathic performance against weeds (Perez and Ormeno-Nunez, 1991; Huang *et al.*, 2003). This is because exudation is likely to be an active metabolic process dependent upon the genetic makeup of the particular cultivar (Wu *et al.*, 2001c). In addition, there may be some need for certain environmental stimulation before particular wheat plants exude defensive allelochemics. The work of Wu *et al.* (2000d) demonstrated for example that wheat (*Triticum aestivum* L.) cultivars AUS#18060 and HY-65 yielded no detectable amounts of the active key allelochemic DIMBOA in their growth medium (0.3% agar-water) despite the fact that for AUS#18060 shoot content for DIMBOA was 694.22 µg/g dry wt. and root content was 720.15 µg/g, while for HY-65 shoot content was 389.16 µg/g and root content was 669.93 µg/g dry wt. Both cultivars did however exude phenolic acids, and both exhibited significant phytotoxicity in the bioassay on root length of test plant annual ryegrass. The work of Wu *et al.*, (2001d) on the association of phenolic acids in exudates with that within corresponding wheat shoots also demonstrated a notable lack of correlation, with the best result being only $r = 0.45$ ($P < 0.01$) between ferulic acid in shoots and exudate.

Typical levels of DIMBOA measured in the agar exudates of 11 *Triticum aestivum* cultivars grown for 17 d by Wu *et al.* (2001c) are displayed in Table 1.

Table 1 Exudate concentrations of DIMBOA in wheat cultivars

Wheat cultivar	*Agar concentration µmol/L*
Excalibur	0.145 ± 0.012
Sunco	0.055 ± 0.002
Eretria	0.063 ± 0.001
Virgilio	0.049 ± 0.011
PF8716	0.199 ± 0.008
Cranbrook	0.074 ± 0.007
Tunis 2	0.071 ± 0.005
L 1512-2721	0.375 ± 0.028
SST 6	0.041 ± 0.003
Tasman	0.162 ± 0.023
Khapli	0.331 ± 0.023

The exudate concentrations displayed in Table 1 are based upon the outputs from the sowing of 12 pre-germinated, surface-sterilized wheat seeds onto 30 mL of 0.3% no-nutrient, sterilized agar-water within a 500 mL parafilm-wrapped beaker. Controlled growth cabinet conditions were a temperature cycle of 25°C/13°C in a corresponding light/dark cycle of 13 h/11 h across 17 d. Although the growth conditions for these wheat seedlings in an agar/beaker environment were quite mild, given the known instability of DIMBOA (Eljarrat *et al.*, 2004), there is a distinct possibility that measurement after 17 d may not reflect maximum levels attained, although such figures are still consistent for internal group comparison.

The Indian Khapli cultivar from the above study (Wu *et al.*, 2001c) exhibited not only a high exudate concentration of DIMBOA but also was known to contain a high level (>400 µg/g root tissue) of total phenolic acids (Wu *et al.*, 2000c). Khapli was therefore chosen by Huang *et al.* (2003) as a suitable target cultivar for further exudate dynamics studies. Within these studies, and employing a GC/MS/MS method, the researchers aimed at monitoring changes in the concentrations of a selected group of seven phenolic acids plus a selected group of five benzoxazinones (12 analytes in total) in the exudates of 24 Khapli wheat seedlings grown over 15 d under controlled conditions in sterile 0.3% agar-water. At the same time they recorded the corresponding changes in phytotoxicity to the wheat weed annual ryegrass (*Lolium rigidum*) by root growth bioassay

Table 2 Dynamic quantifications of phenolics and benzoxazinones across 15 d of Khapli wheat seedling growth in no-nutrient, sterile agar-water medium

Compound	***2 D***	***4 D***	***6 D***	***8 D***	***11 D***	***15 D***
DIBOA	0.006 ± .001	0.083 ± .017	0.146 ± .018	0.070 ± .004	0.006 ± .001	0.010 ± .000
DIMBOA	ND	0.076 ± .010	0.241 ± .015	ND	ND	ND
HBOA	0.042 ± .002	0.189 ± .027	0.168 ± .022	0.424 ± .023	0.082 ± .016	0.129 ± .009
HMBOA	0.020 ± .001	0.092 ± .011	0.154 ± .020	0.209 ± .004	0.115 ± .013	0.103 ± .012
DHBOA	ND	0.023 ± .002	0.026 ± .000	0.019 ± .000	0.016 ± .000	0.016 ± .000
Total benzoxazinones	0.068 ± .002	0.463 ± .035	0.735 ± .380	0.722 ± .024	0.219 ± .021	0.258 ± .015
Total phenolics acids	0.215 ± .038	0.907 ± .370	0.832 ± .473	1.590 ± .682	0.415 ± .04	0.169 ± .023

ND = not detected. Data as mean (N = 3) concentrations ± SD in µmol/L agar-water

across the same 15 d period. Six monitoring growth intervals were used: 2, 4, 6, 8, 11 and 15 d. A summary of their results is shown in Table 2.

It may be seen from Table 2 that under these conditions, the more active benzoxazinone allelochemics in Khapli wheat seedling exudate maximized in the 6-8 d growth interval before decreasing, thus effectively overlapping with the phenolic acids total maximum over the broader 4-8 d interval. This joint maximization would appear to account for the concurrently observed maximum in phytotoxicity measured by the root growth of test plant annual ryegrass which occurred during the same short period. In fact, there appeared to be a smooth association between the concentrations of the combined groups of allelochemics and the observed ryegrass toxicity. This association led Huang *et al.* (2003) to attempt statistical correlation between the two, the results from which are discussed in the next section. The general decline in the later-interval concentrations of both groups of allelochemics under these mild, near-neutral, sterile conditions is of some interest, especially to those who have regard to the prospects of allelopathic control of competing weeds in a field situation. While the reason for this decline was not experimentally pursued, several reasons could be hypothetically advanced. For the benzoxazinones, their natural instability would lead to a decline over time, especially of the hydroxamic acid type. Bioactive DIMBOA for example has a half-life at 28°C and pH 6.75 of 5.3 h (Woodward *et al.*, 1978), decomposing gradually into the less active benzoxazolinone MBOA. The concentrations of individual phenolic acids fluctuated across the 15 d period, but as a group their total declined in the later period also - a process observed in the work of Kobayashi *et al.* (1996). The explanation for decline given by these workers is that the growing wheat seedlings reabsorbed the phenolics as nutrient material. Of course in the complex environment of field soil, any number of other complicating factors may also have a role in affecting allelochemical concentrations. The decline in hydroxamic acid output from Khapli wheat finds a consistent parallel in the studies of Shulz *et al.* (1994) who also found a similar decline with DIBOA from quackgrass exudates (*Agropyron repens* L.). While the hydroxamic acids are usually regarded as more bioactive than the phenolic acids, the correlation analyses carried out by Huang *et al.* (2003) [next section] showed higher levels of association between combined phenolics and hydroxamic acids than with just phenolics alone or benzoxazinones alone when comparing their observed phytotoxicities. In a similar vein, Korableva *et al.* (1969) observed that the allelochemic scopoletin (a coumarin) is more effective as a growth inhibitor when combined with the phenolic compound caffeic acid. This recent work therefore raises again the old issue about the role of phenolic acids as true allelochemicals. It does appear that the effect of phenolic acids on plant roots is to alter their cell permeability (Blum *et al.*, 1999), and that by doing this they may enhance the uptake of the more toxic hydroxamic acids, thereby producing a joint effect which is more toxic than the hydroxamic acids alone.

CORRELATIONS BETWEEN WHEAT ALLELOCHEMICAL CONCENTRATIONS AND OBSERVED PHYTOTOXICITY

Strict, scientific, molecular level, cause-effect explanations of allelopathic activity in the field are extremely difficult to achieve because of the many unknown factors influencing plant growth and inhibition. For this reason, progress in this direction has had to work with more limited laboratory models and the inferences that can be drawn from them. Causative explanations in molecular terms can nevertheless be greatly assisted by significant correlations found to link phytotoxic effects with specific compounds. Certain attempts to make such linkages do exist in the allelopathic literature (Ben-Hammouda *et al.*, 1995) (phenolics content correlation with allelopathy of sorghum against wheat, $r^2 = 0.436$); Burgos *et al.* (1999) (BOA/DIBOA content correlation with allelopathy of *Secale cereale* against *Eleusine indica*, $r^2 = 0.50$); Wu *et al.* (2002) (phenolics plus DIMBOA in wheat exudate correlation against *Lolium rigidum*, $r^2 = 0.504$). Some correlative success has also been forthcoming from linking laboratory allelopathy results with allelopathic field results in studies of rice phytotoxicity against star fruit (*Damasonium minus*) weed, $r^2 > 0.7$ (Seal, Pratley, Haig unpublished results), thus holding out hope that laboratory results on allelopathy may eventually be able to prove predictive value for field situations. It is to be expected that the best correlations in compound-based allelopathy explanations would arise from using the concentrations of putative allelochemics present in the donor plant exudate rather than from its plant parts. This is obvious from the means of transportational contact with the receiver plant. One interesting aspect of the work by Wu *et al.* (2002) is the finding that the root length of bioassay test plant annual ryegrass was significantly (but weakly) associated with the wheat exudate individual compounds *p*-hydroxybenzoic acid ($r^2 = 0.32$); vanillic acid ($r^2 = 0.27$); and *trans*-ferulic acid ($r^2 = 0.18$); but not at all with DIMBOA, the reputably main allelochemic. However, multiple (8 -variate) regression analysis of the combination of all seven phenolics plus DIMBOA was significantly associated with the allelopathic potential of wheat exudate on ryegrass ($r^2 = 0.504$; $P < 0.01$). This last result would appear to point to a joint additive (or perhaps synergistic) action by all eight compounds in their inhibition of ryegrass. Nevertheless, even at this level of analysis, only

50.4% of bioassay plant response was being accounted for. By taking further account of the other contributing hydroxamic acid present in wheat exudate (DIBOA), Huang *et al.* (2003) were able to improve the correlation using a univariate (Ln [DIMBOA + DIBOA + 7 phenolics]) linear model, to a figure of $r^2 = 0.914$ ($P < 0.005$), thereby explaining over 91% of test plant response based upon nine compounds. This result was better than a two-compound alternative using a linear model, univariate (Ln [DIMBOA + DIBOA]) calculation which generated a value for $r^2 =$ 0.678 ($P < 0.05$), thus, again lending support to the notion that a joint action with phenolic acids and hydroxamic acids was taking place. The quality of the 0.914 natural dose-response correlation can be gauged by a direct comparison with the artificially controlled chemoassay result obtained by Shulz *et al.* (1994) whose pure allelochemical (DIBOA) dose-response treatment of *Lolium perenne* yielded a correlation (on the same calculation base) where $r^2 = 0.965$. Clearly progress is being made in our understanding of the chemical basis for wheat allelopathy. Even so, additional improvement is likely if others of the known allelochemics in wheat exudate, such as scopoletin, BOA, MBOA, etc., are included in future studies.

Similar correlation work has been performed by Belz and Hurle (2001) who examined the relationship in various wheat species between phytotoxicity against *Sinapsis alba* and the concentration of putative allelochemics in wheat root exudate. These workers used ED_{50} values from dose-response curves for various wheat species and cultivars created from a range of wheat plant densities (0-30 plants per pot) and corresponding *S. alba* root length responses (as percentage control) in order to construct a phytotoxicity-level versus exudate concentration of allelochemic plot, from which they could by regression analysis, derive the level of correlation. As one example from their results for *Triticum aestivum*, the plot of DIMBOA concentration in exudate versus ED_{50} against *S. alba* for a range of cultivars yielded a value for $r^2 = 0.99$ using a non-linear model.

The exact nature of the joint interaction between the putative components of wheat exudate which produces phytotoxicity has remained puzzling to this date. One of the mysteries is the effectiveness of natural low-concentration mixtures of benzoxazinoids and/or phenolic acids for inducing toxic outcomes (Huang *et al.*, 2003). In pure chemical assay work, the concentration of individual phenolics or benzoxazinones for toxicity appears to be required at levels 100 times the concentrations being measured in plant-plant bioassays where phytotoxicity is evident. One explanation offered in the literature for this has been synergism, even though it is a rare phenomenon, and nobody has as yet been able to reproduce synergism from such chemical components at any significant level. In fact, antagonism seems to be more prevalent in tested artificial mixtures (Kudsk *et al.*, 2004). With this in mind, Huang *et al.* (2003) investigated the properties of an artificial mixture designed to mimic the composition measured for the components of the root exudate of their cv. Khapli wheat at its period of maximum phytotoxicity against ryegrass in agar-water growth medium. During their 15-d dynamic study of Khapli exudate, the sum of total phenolics plus total benzoxazinones had never exceeded 3.02 μmol/L of agar. The maximum phytotoxicity concentrations in units of μmol/L water-agar (as at day 6) used for each compound were: *p*-hydroxybenzoic acid 0.418; vanillic acid 0.03; coumaric acids 0.189; syringic acid 0.077; ferulic acids 0.118; HBOA 0.168; DIBOA 0.146; HMBOA 0.154; DIMBOA 0.241 (lactam DHBOA unavailable); total mixture concentration 1.55 μmol/L. The actual outcome from testing this artificial mix against the root growth of annual ryegrass was not an inhibition but a small stimulation! If synergism was indeed the reason for such low-concentration phytotoxicity in day 6 Khapli exudate, then this designed combination would appear to have the wrong composition. Of course the natural wheat exudate may also contain other known bioactive compounds such as scopoletin, BOA, MBOA and other as yet undiscovered compounds even in these sterile conditions, which may be contributing towards toxicity. Discovering all such possible compounds and their correct molecular ratios for any correct synergistic combination (if indeed this be the true explanation) is going to be a very difficult task. In the non-sterile conditions of soil, many additional metabolites from phenolics (e.g. vinylphenols) and benzoxazinoids (e.g. aminophenoxazinones) would also play some role, in addition to natural soil components.

METABOLIC STUDIES ON WHEAT ALLELOCHEMICS

The objectives specified by the FATEALLCHEM project (see Introduction) have been an incentive to investigate the properties and fate of metabolites originating from the allelochemicals of wheat. The most studied allelochemics in this context are the benzoxazinoids. Fomsgaard *et al.* (2004b) have published a useful review which collects most data relating to the known metabolism and fate of these compounds, together with identifying information, up until the end of 2003. Since that time, the subject has been further advanced experimentally through the work of Fomsgaard *et al.* (2004c) and by Macias *et al.* (2004).

The aim of Macias *et al.* (2004) was to characterize the complete degradation sequence of DIMBOA in wheat crop soil. DIMBOA degradation in soil is completely analogous to that of DIBOA degradation (Fig. 3) in that both materials begin their sequences as the originating

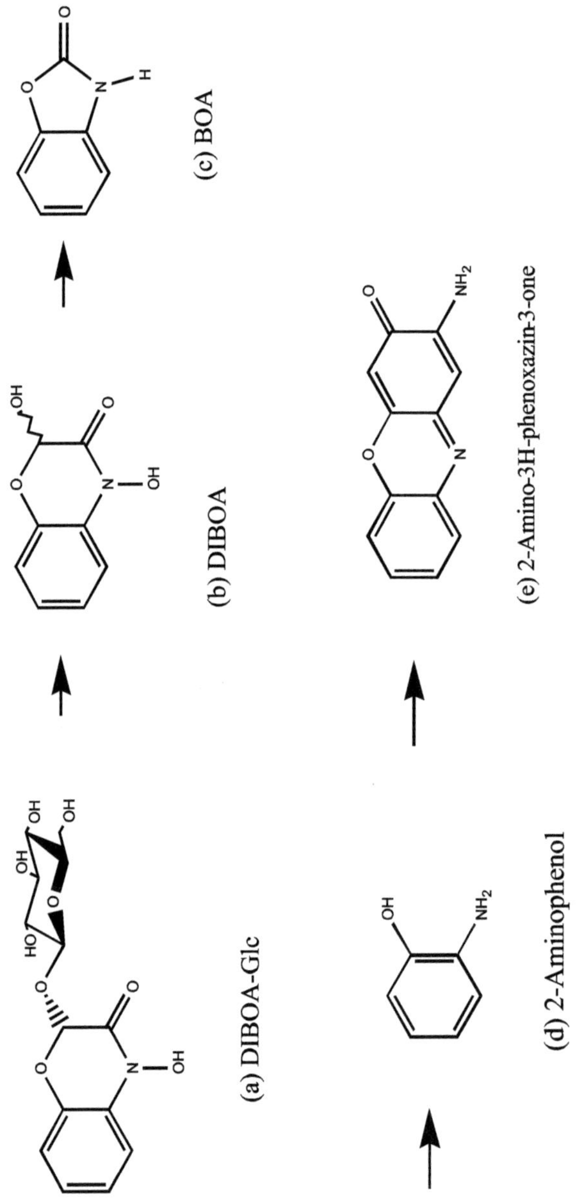

Fig. 3 Main degradation sequence of DIBOA-glucoside, (a) to (e)-[APO], in soils

2-β-D-glucosides and finish as their corresponding aminophenoxazinone end-products. Depending upon circumstances and the nature of micro-organisms available, further soil metabolic transformations can also be traced (Zikmundova *et al.*, 2002a,b) to such end-products as 2-acetylaminophenoxazin-3-one (AAPO); 2-(N-hydroxy)acetylamino-3H-phenoxazin-3-one (NHAAPO); 2-(2-hydroxyacetyl)amino-3H-phenoxazin-3-one (HAAPO); 2-amino-7-hydroxyphenoxazin-3-one (AHPO); 2-acetylamino-7-hydroxyphenoxazin-3-one (AAHPO); 7-hydroxy-2-(2-hydroxyacetyl)aminophenoxazin-3-one (HHAAPO) and 2-acetylamino-7-methoxyphenoxazin-3-one (AAMPO).

Macias *et al.* (2004) studied two soils cultivated with two separate *Triticum aestivum* cultivars (Astron and Ritmo) and measured the degradation rate of DIMBOA at various soil-dosage levels. The half-lives for DIMBOA conversion to MBOA in Astron soils at 1,000 and 5,000 mg/kg of soil were 32 ± 12 and 31 ± 1 min respectively, while the half-lives in Ritmo soils at the corresponding dosages were 31 ± 4 and 30 ± 1 min respectively. These results were compared to earlier ones obtained by similar procedures using DIBOA, which proved itself much slower in conversion to BOA, with its rates being measured in hours rather than minutes. Thus, Astron soil half-lives for DIBOA were determined as 24 ± 5 h and 62 ± 7 h at 1,000 and 5,000 mg/kg soil respectively, while in Ritmo soil the corresponding values were 28 ± 5 h and 57 ± 16 h. The group also studied the soil dynamics of conversion of MBOA into AMPO (2-amino-7-methoxy-3H-phenoxazin-3-one) (C^{13} NMR supplied) and found such benzoxazolinones more resistant to degradation. Thus MBOA half-lives in Astron and Ritmo soils at 2,000 mg/kg soil dosage were 4.3 ± 0.5 and 6.0 ± 2.0 d respectively, while corresponding BOA half-lives were 2.7 ± 0.4 and 2.2 ± 0.3 d. Within the same study, the soil stabilities of the AMPO and APO molecules were examined and both found to be undegraded after a period of three mon. The authors state that owing to the high persistence of these aminophenoxazinones, they could have great importance in the ecological role previously assigned to the benzoxazinones.

In a similar, more limited, but useful comparison study, Fomsgaard *et al.* (2004c) have investigated the soil degradation of BOA and MBOA at three soil-dosage levels: 0.0004 mg/g; 0.400 mg/g and 4.0 mg/g soil. They found that at 0.0004 mg/g levels the degradation of both BOA and MBOA was quite fast with half-lives below 1 d. Under these conditions, no APO, AAPO, AMPO, or AAMPO metabolites could be detected. At the two higher dose levels however, all four of these aminophenoxazinone derivatives could be detected in the soil for 60 d or longer. Fritz *et al.* (2004) have measured some comparative bioactivities and state generally that in ranking bioactivities APO >> BOA, and AMPO >> MBOA; APO > AAPO > AMPO >> AAMPO.

Further useful recent reviews in connection with the metabolism and mode of action of benzoxazinoids are available from the authors Sanchez-Moreiras *et al.* (2004) and Sicker *et al.* (2004), while one important earlier study by Friebe *et al.* (1997) correlating the biological actions of DIMBOA, DIBOA, MBOA, and BOA with the action of plasma membrane H^+ -ATPase is worthy of note.

Finally, the other group of allelochemics in wheat, the phenolic acids, has received a measure of attention in respect of their metabolism also, with accounts being given by recent reviews from Blum (2004) and Einhellig (2004). The interesting proposal by Kobayashi *et al.* (1996) that phenolics of the coumaric acid type may experience an enzymatic decarboxylation to 4-hydroxystyrenes which display significant bioactivity is also a consideration for any future studies on the chemical basis for wheat allelopathy.

References

Baghestani, A., Lemieux, C., Leroux, G.D., Baziramakenga, R., Simard, R.R. (1999) Determination of allelochemicals in spring cereal cultivars of different competitiveness. Weed Sci. 47: 498-504.

Belz, R., Hurle, K. (2001) Tracing the source - Do allelochemicals in root exudates of wheat correlate with cultivar-specific weed-suppressing ability? Brighton Crop Protection Conference - Weeds. 4D-4: 317-320.

Ben-Hammouda, M., Kremer, R.J., Minor, H.C., Sarwar, M. (1995) A Chemical Basis for Differential Allelopathic Potential of Sorghum Hybrids on Wheat. J. Chem. Ecol. 21: 775-786.

Blum, U. (2004) Fate of Phenolic Allelochemicals in Soils - the Role of Soil and Rhizosphere Microorganisms. *In* Allelopathy: Chemistry and Mode of Action of Allelochemicals, F.A. Macias, J.C.G. Galindo, J.M.G. Molinillo, H.G. Cutler (eds). CRC Press LLC, Boca Raton. Fl, USA. pp 57-76.

Blum, U., Schafer, S.R., Lehman, M.E. (1999) Evidence for inhibitory allelopathic interactions involving phenolic acids in field soils: Concepts vs. experimental model. Crit. Rev. Plant Sci. 18: 673-693.

Bonnington, L., Eljarrat, E., Guillamon, M., Eichhorn, P., Taberner, A., Barcelo, D. (2003) Development of a Liquid Chromatography-Electrospray-Tandem Mass Spectrometry Method for the Quantitative Determination of Benzoxazinone Derivatives in Plants. Anal. Chem. 75: 3128-3136.

Burgos, N.R., Talbert, R.E., Mattice, J.D. (1999) Cultivar and Age Differences in the Production of Allelochemicals by *Secale cereale*. Weed Sci. 47: 481-485.

Copaja, S.V., Niemeyer, H.M., Wratten, S.D. (1999) Hydroxamic acid levels in Chilean and British wheat seedlings. Ann. App. Biol. 118: 223-227.

Einhellig, F.A. (2004) Mode of allelochemical action of phenolic compounds. *In* Allelopathy: Chemistry and Mode of Action of Allelochemicals, F.A. Macias J.C.G., Galindo, J.M.G., Molinillo and H.G. Cutler (eds.), CRC Press LLC, Boca Raton. Fl, USA. pp. 217-238.

Eljarrat, E., Barcelo, D. (2001) Sample handling and analysis of allelochemical compounds in plants. Trends Anal. Chem. 20: 584-590.

Eljarrat, E., Guillamon, M., Seuma, J., Mogensen, B.B., Fomsgaard, I.S., Olivero-Bastidas, A., Macias, F.A., Stochmal, A., Oleszek, W., Shakaliene, O., Barcelo, D. (2004) First European interlaboratory study of the analysis of benzoxazinone derivatives in plants by liquid chromatography. J. Chrom. A 1047: 69-76.

Fomsgaard, I.S. (2004a) Introduction of the FATEALLCHEM Project. *In* the Proceedings of the Second European Allelopathy Symposium, Institute of Soil Science and Plant Cultivation Press Service, Pulawy, Poland, 3-5 June. p. 33.

Fomsgaard, I.S., Mortensen, A.G., Carlsen, S.C.K. (2004b) Microbial transformation products of benzoxazolinone and benzoxazinone allelochemicals - a review. Chemosphere. 54: 1025-1038.

Fomsgaard, I.S., Mortensen, A.G., Gents, M.B., Understrup, A.G. (2004c) Time-Dependent Transformation of Varying Concentration of the Hydroxamic Acid Metabolites MBOA and BOA in Soil. *In* the Proceedings of the Second European Allelopathy Symposium, Institute of Soil Science and Plant Cultivation Press Service, Pulawy, Poland, 3-5 June. p. 61.

Fritz, J., Blumel, S., Hansen, L.M., Kudsk, P., Macias, F., Oleszek, W. (2004) Structure depending activities of wheat allelochemicals on target and non-target organisms. Proceedings of the Second European Allelopathy Symposium, Institute of Soil Science and Plant Cultivation Press Service, Pulawy, Poland, 3-5 June. p. 91.

Haig, T. (2004) Application of Analytical Techniques to the Determination of Allelopathic Agents in Wheat Root Exudates. A Practical Case Study. *In* Allelopathy: Chemistry and Mode of Action of Allelochemicals, F.A. Macias, J.C.G. Galindo, J.M.G. Molinillo and H.G. Cutler (eds.), CRC Press LLC, Boca Raton. Fl, USA. pp. 149-161.

Huang, Z., Haig, T., Wu, H., An, M., Pratley, J. (2003) Correlation between phytotoxicity on annual ryegrass (*Lolium rigidum*) and production dynamics of allelochemicals within root exudates of an allelopathic wheat. J. Chem. Ecol. 29: 2263-2279.

Inderjit. (1996) Plant phenolics in allelopathy. Botanical Review. 62: 186-202.

Kobayashi, A., Kim, M.J., Kawazu, K. (1996) Uptake and exudation of phenolic compounds by wheat and antimicrobial components of the root exudate. Z. Naturforsch. 51c: 527-533.

Kudsk, P., Chunhong, J., Mathiassen, S.K. (2004) Joint Action of Wheat Allelochemicals. *In* the Proceedings of the Second European Allelopathy Symposium, Institute of Soil Science and Plant Cultivation Press Service, Pulawy, Poland, 3-5 June. p. 83.

Lodhi, M.A.K., Bilal, R., Malik, K.A. (1987) Allelopathy in agroecosystems: Wheat phytotoxicity and its possible roles in crop rotation. J. Chem. Ecol. 13: 1881-1891.

Lyons, P.C., Hipskind, J.D., Wood, K.V., Nicholson, R.L. (1988) Separation and Quantification of Cyclic Hydroxamic Acids and Related Compounds by High-Pressure Liquid Chromatography. J. Agric. Food Chem. 36: 57-60.

Macias, F.A., Oliveros-Bastidas, A., Marin, D., Castellano, D., Simonet, A.M., Molinillo J.M.G. (2004) Degradation Studies on Benzoxazinoids. Soil Degradation Dynamics of 2,4-Dihydroxy-7-methoxy-(2H)-1,4-benzoxazin-3(4H)-one (DIMBOA) and its Degradation Products, Phytotoxic Allelochemicals from Gramineae. J. Agric. Food Chem. 52: 6402-6413.

Perez, F.J., Ormeno-Nunez, J. (1991) Difference in hydroxamic acid content in roots and root exudates of wheat (*Triticum aestivum* L.) and rye (*Secale cereale* L.): Possible role in allelopathy. J. Chem. Ecol. 17: 1037-1043.

Petho, M. (1992) Occurrence and physiological role of benzoxazinones and their derivatives IV. Isolation of hydroxamic acids from wheat and rye root secretions. Plant Physiol. Agrochem. 41: 167-175.

Sanchez-Moreiras, A.M., Coba de la Pena, T., Martinez, A., Gonzalez, L., Pellisier, F., Reigosa, M.J. (2004) Mode of Action of the Hydroxamic Acid BOA (sic) and Other Related Compounds. *In* Allelopathy: Chemistry and Mode of Action of Allelochemicals, F.A. Macias, J.C.G. Galindo, J.M.G. Molinillo, H.G. Cutler (eds.). CRC Press LLC, Boca Raton. Fl, USA. pp. 239-252.

Shulz, M., Friebe, A., Kuck, P., Seipel, M., Schnabl, H. (1994) Allelopathic Effects of Living Quackgrass (*Agropyron repens* L.). Identification of Inhibitory Allelochemicals Exuded from Rhizome Borne Roots. Angew. Bot. 68: 195-200.

Sicker, D., Hao, H., Schulz, M. (2004) Benzoxazoline-2(3H)-ones - Generation, Effects and Detoxification in the Competition among Plants. *In* Allelopathy: Chemistry and mode of action of allelochemicals, F.A. Macias, J.C.G. Galindo, J.M.G. Molinillo, H.G. Cutler (eds.). CRC Press LLC, Boca Raton Fl, USA. pp. 77-102.

Woodward, M.D., Corcuera, L.J., Helgeson, J.P., Upper, C.D. (1978) Decomposition of 2,4-dihydroxy-7-methoxy-(2H)-1,4-benzoxazin-3(4H)-one in aqueous solutions. Plant Physiol. 61: 796-802.

Wu, H., Pratley, J., Lemerle, D., Haig, T. (2000a) Laboratory screening for allelopathic potential of wheat (*Triticum aestivum*) accessions against annual ryegrass (*Lolium rigidum*). Aust. J. Agric. Res. 51: 259-266.

Wu, H., Pratley, J., Lemerle, D., Haig, T. (2000b) Evaluation of seedling allelopathy in 453 wheat (*Triticum aestivum*) accessions against annual ryegrass (*Lolium rigidum*) by the equal-compartment-agar-method. Aust. J. Agric. Res. 51: 937-944.

Wu, H., Haig, T., Pratley, J., Lemerle, D., An, M. (2000c) Allelochemicals in Wheat (*Triticum aestivum* L.): Variation of Phenolic Acids in Root Tissues. J. Agric. Food Chem. 48: 5321-5325.

Wu, H., Haig, T., Pratley, J., Lemerle, D., An, M. (2000d) Distribution and Exudation of Allelochemicals in Wheat (*Triticum aestivum*). J. Chem. Ecol. 26: 2141-2154.

Wu, H., Pratley, J., Lemerle, D., Haig, T. (2001a) Allelopathy in Wheat (*Triticum aestivum*). Ann. Appl. Biol. 139: 1-9.

Wu, H., Haig, T., Pratley, J., Lemerle, D., An, M. (2001b) Allelochemicals in wheat (*Triticum aestivum* L.): Variation of phenolic acids in shoot tissues. J. Chem. Ecol. 27:125-135.

Wu, H., Haig, T., Pratley, J., Lemerle, D., An, M. (2001c) Allelochemicals in Wheat (*Triticum aestivum* L.): Production and exudation of 2,4-dihydroxy-7-methoxy-1,4-benzoxazin-3-one. J. Chem. Ecol. 27: 1691-1700.

Wu, H., Haig, T., Pratley, J., Lemerle, D., An, M. (2001d) Allelochemicals in Wheat (*Triticum aestivum* L.): Cultivar Difference in the Exudation of Phenolic Acids. J. Agric. Food Chem. 49: 3742-3745.

Wu, H., Haig, T., Pratley, J., Lemerle, D., An, M. (2002) Biochemical Basis for Wheat Seedling Allelopathy on the Suppression of Annual Ryegrass (*Lolium rigidum*). J. Agric. Food Chem. 50: 4567-4571.

Wu, H., Pratley, J., Haig, T. (2003) Phytotoxic Effects of Wheat Extracts on a Herbicide-Resistant Biotype of Annual Ryegrass (*Lolium rigidum*). J. Agric. Food Chem. 51: 4610-4616.

Zikmundova, M., Drandarov, K., Bigler, L., Hesse, M., Werner, C. (2002a) Biotransformation of 2-benzoxazolinone and 2-hydroxy-1,4-benzoxazin-3-one by endophytic fungi isolated from *Aphelandra tetragona*. Appl. Environ. Microb. 68: 4863-4870.

Zikmundova, M., Drandarov, K., Hesse, M., Werner, C. (2002b) Hydroxylated 2-amino-3H-phenoxazin-3-one derivatives as products of 2-hydroxy-1,4-benzoxazin-3-one (HBOA). Biotransformation by *Chaetosphaeia* spp., an endophytic fungus from *Aphelandra tetragona*. Z. Naturforsch. C 57: 660-665.

6

Ecological Relevance of the Degradation Processes of Allelochemicals

Francisco A. Macías, David Marín, Alberto Oliveros-Bastidas, Ana M. Simonet and José M.G. Molinillo

Grupo de Alelopatía, Departamento de Química Orgánica, Facultad de Ciencias, Universidad de Cádiz, C/ República Saharaui, s/n, 11510 Puerto Real (Cádiz) Spain

Some considerations are made here on the ecological relevance of the degradation processes of allelochemicals, concerning the activity of some of the degradation products produced in soil. Once the allelochemical is in contact with different organisms and environmental conditions, they can become new products. Sometimes these new compounds are less active than the initial ones, but at other times the new chemical is as active or even much more than the ones excreted by plants. Thus, in nature allelopathy occurs due to complex mixtures of compounds, some of them transformed from those exuded by plants. So, a complete understanding of the allelopathic phenomenon requires the study of degradation products of allelochemicals. An extended study of degradation of benzoxazinoids in two different wheat crop soils is described and the activity shown on seven species (*Triticum aestivum* L., *Allium cepa* L., *Lycopersicon esculentum* Will., *Lepidium sativum* L., *Lactuca sativa* L., *Echinochloa crus-galli* L., *Lolium rigidum* L., and *Avena fatua* L.) by the initial compounds (DIBOA-Glc, DIBOA and DIMBOA) as well as the main degradation products detected (BOA, MBOA, APO and AMPO). The most active compound was the degradation product APO, instead of the commonly accepted allelochemicals DIBOA and DIMBOA. It can be inferred from these data that the fate and degradation studies of allelochemicals are a necessary stage in allelopathic research if we want to have a complete understanding of allelopathy.

Keywords: *allelochemical, bioassay, benzoxazinoids, degradation, soil*

INTRODUCTION

The complexity of the allelopathic phenomenon is still far from being fully understood. The ideal situation, in which one plant exudes an allelochemical, the receptor absorbs it and the substance provokes an effect on the target plant without any other influence, cannot be found in nature. Nevertheless, this simplification of reality is widely used by researchers. When we isolate allelochemicals from plant extracts or exudates, and apply them to seedlings or isolated plants growing in a sterile medium, we are ignoring a large amount of parameters and factors. We wonder how important these environmental influences are? Can they change our point of view? Is it possible that we are considering allelochemicals that are not the real responsible of the field observations?

Allelopathy seems to be an extended defense mechanism in nature. Chemicals with allelopathic potential are present in nearly all plants (Weston *et al.*, 2003), so allelopathy should appear in the early stages of plant evolution. This can imply that many plant species have developed defense mechanisms to detoxified allelochemicals and large number of microorganisms in the soil are able to transform these substances. The amount of allelochemicals released from the donor plant and the sensitivity of the receiver plant to the allelochemical is partly dependent on the growth conditions of the donor and receiver. (Kobayashi, 2004) Moreover, the physicochemical soil properties influence the production, exudation, sorption, stability and availability for receptor species of allelochemicals.

Degradation studies are common when we are studying herbicides (Roberts, 1998). It is important to know the fate and permanence of herbicides in soil for ecological safety reasons. Nevertheless, this is not so usual in case of allelochemicals. Soil organisms do influence the allelopathic effects in all cases, especially when the release of allelochemicals is produced by decomposition of plant residues. Studies of the dynamics and fate of allelochemicals in the soil have been carried out mostly for glycosides, organic acids and phenols but rarely for lipophilic compounds. This is due to experimental simplicity for monitoring these compounds by HPLC. The phenolic allelochemicals produced by residues apple (Politycka *et al.*, 2003), Vulpia (An *et al.*, 2000), rice (Chou *et al.*, 1976a; Rimando *et al.*, 2001), corn, rye (Chou *et al.*, 1976b), or fir stump-roots (Huang *et al.*, 2000) have been studied. We can find substantial literature on the transformation of simple phenolic acid by microorganisms. (Dagley, 1971; Denis *et al.*, 1973; Turner *et al.*, 1975; Martin *et al.*, 1976;

Blum, 1998;) and some transformation routes have been elucidated. Thus, cinnamic acids (**1**) are transformed into benzoic acids (**2**), which becomes β-carboxy-*cis,cis*-musconic acid (**3**), and then β-oxoadipic acid (**4**), which in turn is broken down to acetic acid and succinic acid. Blum (2004) has analyzed these data in order to clarify whether phenolics could be or are not responsible for allelopathic interactions. This is not obvious, as the rate of degradation of these kind of compounds can provoke its persistence in the soil and its concentration would not be enough for effective interaction with the target species. This question has no simple answer, because of a number of factors that are involved in the degradation process, such as the microbial population, temperature, physical-chemical soil properties, humidity, and, even, the presence of other compounds as glucose or phenylalanine that can be preferred as carbon sources by soil microorganisms.

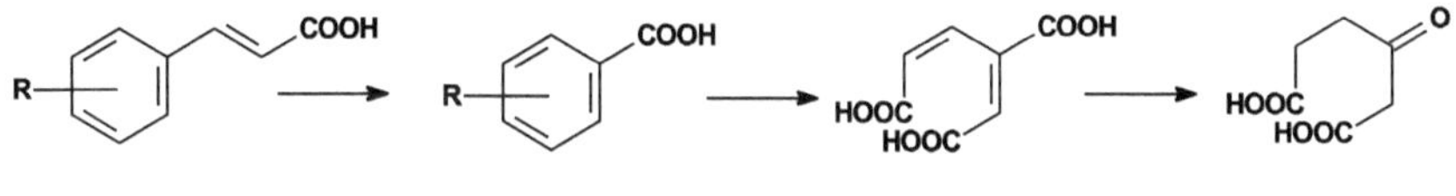

Fig. 1 Degradation pathway of cinnamic acids

Nevertheless, the fate of some lipophilic compounds such dehydromatricaria ester have been studied. This is a polyacetylene isolated from *Solidago altissima* L. and *Erigeron* species (Ito *et al.*, 1998). Both isomers (*cis* and *trans*) have been isolated and their phytotoxic activities have been demonstrated, inhibiting seedling growth of rice and hogweed in agar and soil cultures. The concentration of dehydromatricaria ester has been measured at different times in sterile and non-sterile soils. Ito and coworkers found that degradation was faster in the non-autoclaved soil than in autoclaved soil, so microbial activity should be involved (Ito *et al.*, 1998). Unfortunately, they did not consider any of the possible degradation products.

Sometimes the identification of these degradation products can show surprising results. One classical example is the allelopathic study of Florida rosemary (*Ceratiola ericoides*). This plant is allelopathic, particularly to herbaceous plants endemic to inland ridge systems (Williamson *et al.*, 1992). It releases ceratiolin (**6**) in foliar runoff of rainwater. Ceratiolin degrades to produce hydrocinnamic acid. (**7**) When the activities of both compounds were measured it was observed that the degradation product was more active than the originally exuded one and inhibited seed germination of grasses in concentrations of less than 60 ppm. Thus, we can observe both detoxification and toxification processes mediated by soil microorganisms.

Another interesting group of compounds where degradation is an important process is the case of glucosinolates, as sinigrin (**8**) (Borek *et al.*, 1994). They have been isolated from dicotyledonous species; especially those included in the order Capparales that include the family Brassicaceae (Cruciferae). Mustard is the name given for plants included in the genus *Brassica*. Glucosinolates themselves have low biological activity levels until they are hydrolyzed forming isothiocyanates, thiocyanates, nitriles, epithionitriles, oxazolidinthiones, amines and other products. (Shahidi *et al.*, 1990; Borek *et al.*, 1994; Borek *et al.*, 1995; Pecháček *et al.*, 2000) Suppression of a variety of soil pests has been attributed to the toxicity of these glucosinolate decomposition products formed in the soil environment instead of the initial metabolites exuded by the plant. (Bialy *et al.*, 1990)

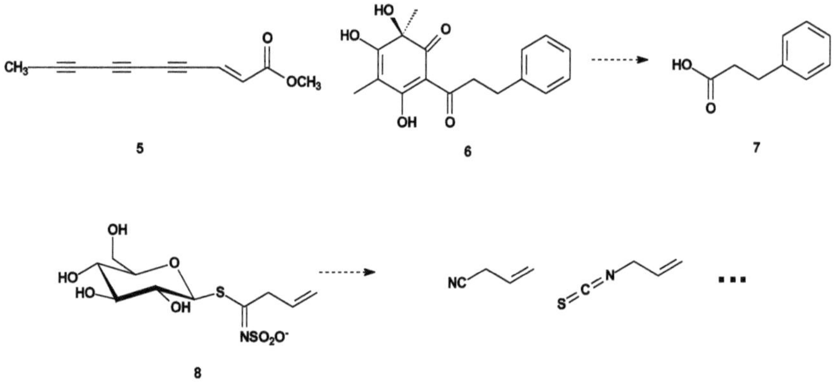

Fig. 2 Some examples of allelochemical degradations

Benzoxazinoids (cyclic hydroxamic acids) requires special attention. These compounds are involved in the defense of plants against fungi (Wahlroos *et al.*, 1959) and insects (Corcuera *et al.*, 1992), as well as in allelopathic interactions (Frey, 1997; Friebe *et al.*, 2001). The most important benzoxazinoids reported are 2,4-dihydroxy-1,4-benzoxazine-3-one (DIBOA, **9**) and 2,4-dihydroxy-7-methoxy- 1,4-benzoxazin-3-one (DIMBOA, **10**), which are present in wheat, maize and rye, and have been found in members of the families *Acanthaceae, Rannunculaceae* and *Scrophulariaceae* (Pratt *et al.*, 1995). These compounds are present as glycosides (**11**) in plants, being released as aglycones by the activity of the enzyme β-glucosidase (Cambier *et al.*, 1999; Baumeler *et al.*, 2000).

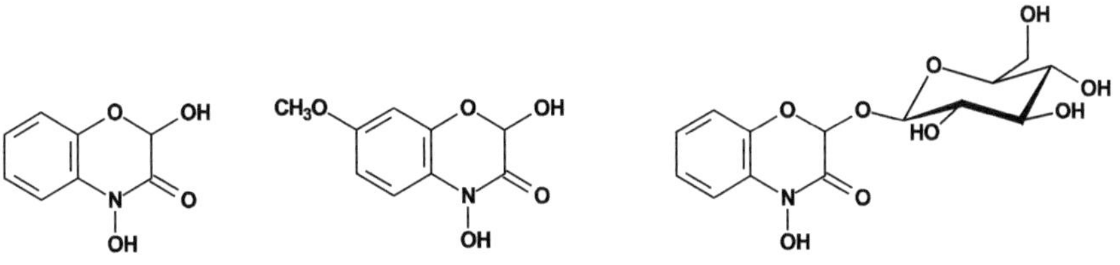

9 10 11

Fig. 3 Common benzoxazinoids

Extensive studies have been performed in order to identify the structure of degradation products of these kind of compounds. The main structural types found are benzoxazolin-2-ones (**12**), hydroxyphenylmalonamic acids (**13**) and aminophenoxazin-3-ones (**14**). With these concepts in mind, we decided to study the degradation dynamics of these compounds in two different wheat crop soils, and a evaluation of the bioactivity of each compound on a variety of plant species.

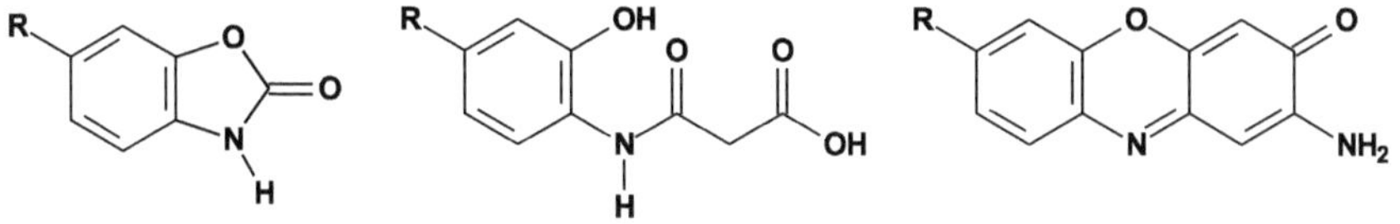

12		13		14	
R = H	BOA	R = H	HPMA	R = H	APO
R = OCH_3	MBOA	R = OCH_3	HMPMA	R = OCH_3	AMPO

Fig. 4 Main structure of degradation products from benzoxazinoids

MATERIALS AND METHODS

Origin of DIMBOA and their Derivatives

2-Benzoxazolinone (BOA) was purchased from Fluka Chemika. The benzoxazinone standards (2R)-2-O-β-D-glucopyranosyl-4-hydroxy-(2H)-1,4-benzoxazin-3(4H)-one (DIBOA-Glc) and 2,4-dihydroxy-(2H)-1,4-benzoxazin-3(4H)-one (DIBOA) were obtained from natural sources. DIBOA-Glc was isolated from 7-d-old rye (*Secale sereale* L.) seedlings. 2-Aminophenoxazin-3-one (APO) was synthesized according to previously described procedures (Zikmudová *et al.*, 2002; El-Khatib *et al.*, 2004). DIMBOA was isolated from corn (*Zea mays* L. cv. Apache) seedlings. MBOA was purchased from Lancaster Synthesis and was used as received. AMPO was synthesized from 5-methoxy-2-nitrophenol by a reductive dimerization reaction.

Soil Collection

The selected parcels for wheat cultivation contained similar physiochemical characteristics at the time of their cultivation, and did not contain any other plant species in growth. After the harvest, samples of soil from the vicinity of the plants were taken, in order to get the maximum density of the microbial population associated to the plants' underground parts. The soil was sieved at 2 mm, placed in plastic bags and preserved at -20ºC until its study. Before the degradation studies, dry vegetable material was removed from the samples as well as calcareous stones by a sieve that allows a distribution of soil particles smaller than 1 mm. Soil pH was measured in a 1:2.5 (v/v) aqueous extract and in KCl solution.

Table 1 Some physicochemical properties of soil

Exchangeable Ca2+ (meq/ 100 g)	28.96
Exchangeable Mg2+ (meq/ 100 g)	3.08
Exchangeable Na+ (meq/ 100 g)	0.29
Exchangeable K+ (meq/ 100 g)	1.58
Active limestone (% W/W)	8.81
Assimilable Phosphorus (Olsen, ppm)	8.7
Total Organic Content (ppm)	1.35
Organic Nitrogen (% W/W)	0.08
pH (KCl)	7.31
Assimilable Potassium (ppm)	635
Textural classification	Clay

Sampling and Processing of Soil

Samples (0.5 mL) of incubation solution were taken from the soil at different time periods after homogenization of the soil-water suspension. Samples were preserved at –20ºC after addition of methanol (1 mL) to avoid degradation between the end of the experiment and the moment of analysis. The sampling intervals were determined by means of preliminary studies, in which degradation of each compound was achieved (12 h for DIBOA-Glc, 12 h DIBOA, 24 h for BOA, 5 min for DIMBOA and 24 h for MBOA).

Analysis of Hydroxamic Acids in Soil

All samples, including non-inoculated soil, were analyzed on a Merck HITACHI HPLC equipped with a LaCrom L-7100 quaternary gradient pump, L-7455 LaChrom diode array detector and an L-7200 LaChrom autoinjector. Retention times (min) were obtained for each compound: DIBOA-Glc 5 min; DIBOA 9.01 min; BOA 12.56 min and APO 24.03 min,

DIMBOA 10.69, MBOA 16.56, HMBOA 7.81, APO 24.08, and AMPO 19.74. The detection was carried out at the following wavelengths: 254, 253, 271, 280, 262, 230, 260, 228, and 235 nm for DIBOA-Glc, DIBOA, BOA, APO, DIMBOA, MBOA, HMBOA, APO and AMPO respectively. For quantitative analysis, stock solutions (1 mg/mL) of each individual standard were prepared by dissolving accurate amounts of pure standard in acidified MeOH (1% AcOH). Working standard solutions were obtained by further dilution of stock solutions with MeOH: acidified H_2O (1% HOAc) (70:30). These solutions were used to generate the external standard response calibration curves for subsequent measurements of quality parameters and concentration of the hydroxamic acids derivates in the soil at different times. All the analytical procedures were validated by means intercalibration laboratory study (Eljarrat *et al.*, 2004).

The bioassay methodology has been optimized by us in the last few years, from the point of view of statistical significance and a representative selection of target species. The species used are shown in Table 2.

Table 2 Species used for bioactivity evaluation

Standard Target Species	*Weeds*
Monocots	*Monocots*
Triticum aestivum L. (wheat)	*Avena fatua* L. (wild oat)
Allium cepa L. (onion)	*Lolium rigidum* Gaud. (rigid ryegrass)
Dicots	Dicots
Lepidium sativum L. (cress)	*Echinochloa crus-galli* L. (barnyardgrass)
Lycopersicon esculentum Will. (tomato)	
Lactuca sativa L. (lettuce)	

The presence of wheat among the assayed species is especially interesting, since it helps to evaluate the possible autotoxicity phenomena due to these compounds, which are produced by wheat and related crops. Wild oat and rigid ryegrass are common weeds affecting Poaceae crops like wheat or barley. The results of these bioassays provided interesting information about the potential utilization of these natural products or analogues of them in real agronomic problems in which these weeds and crops are involved.

RESULTS

Degradation

2R-O-β-D-Glucopyranolsyl-4-hydroxy-2H-1,4-benzoxazin-3(4H)- one (DIBOA-Glc) was first analyzed in both soils, at 1 mg/g of soil (Fig. 5). The chromatographic analysis showed DIBOA as the major degradation product. The ratio DIBOA-Glc/DIBOA decreased significantly in the first 21 h. After this, it was maintained until the 31st h. Half-life of DIBOA-Glc in the soil was calculated by a first order kinetic model, showing the following result: 23 h (r^2 = 0.94; p = < 0.0001) in Astron soil and of 25 h in Ritmo soil (r^2 = 0.92, p < 0.002). (Macias *et al.*, 2005a).

A degradation experiment of 2,4-dihydroxy-7-metoxy-2H-1,4-benzoxazin-3(4H)- one (DIMBOA) in the soil was designed including two different soil samples, according to the previously reported selection process (Atkinson *et al*, 1991). DIMBOA was dosed to the soil solutions (1 and 5 mg of compound per gram of soil). Samples of soil solution were collected in a sampling period of 5 min and analyzed by HPLC. Control aqueous solution of the compounds inoculated to the soil (to discard spontaneous aqueous degradation) were used, and degradation experiments at sterile soil systems were recorded and employed to discard transformations inducted by soil material and the adsorption phenomena. (Macias *et al.*, 2004).

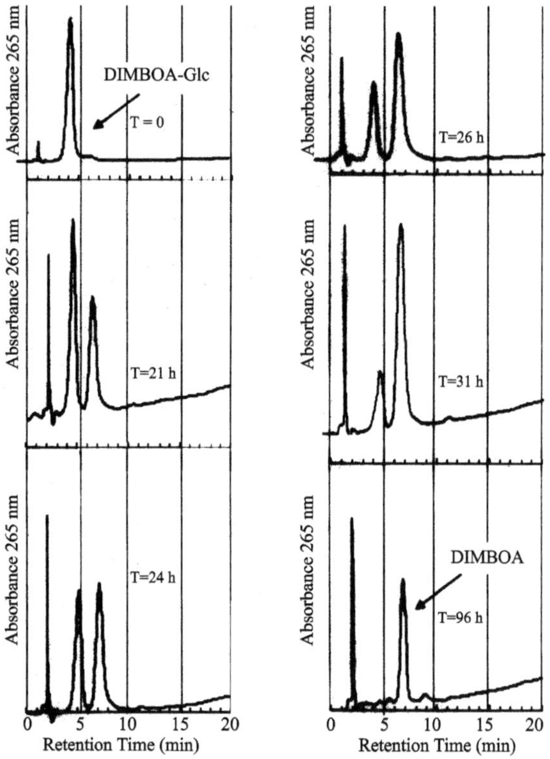

Fig. 5 HPLC analysis for crop soil solutions (*T. aestivum* cv. Astron) inoculated with DIMBOA-Glc (1 mg/g soil)

MBOA was detected as the main degradation product. A reduction of DIMBOA amount can be observed as MBOA concentration increases. Complete conversion is reached at 50 min. All analyses were repeated three times to assure accuracy of the methods employed. A second degradation product, 2-hydroxy-7-methoxy-(2H)-1,4- benzoxazin-3(4H)-one (HMBOA) is detected at trace levels in the highest dose experiment (5000 mg Kg^{-1} soil) (Fig. 6). This experiment allows us to propose the following degradation pathway for DIMBOA (Fig. 7). Nevertheless, more detailed studies to determine the final products and the possible different intermediates during degradation process are needed. A similar half life is observed in both soil types (31.5 ± 7.7 for Astron and 30.5 ± 2.3 for Ritmo). Both soils would have the same degradation capacity for DIMBOA (ANOVA, $\alpha = 0.05$, F = 5).

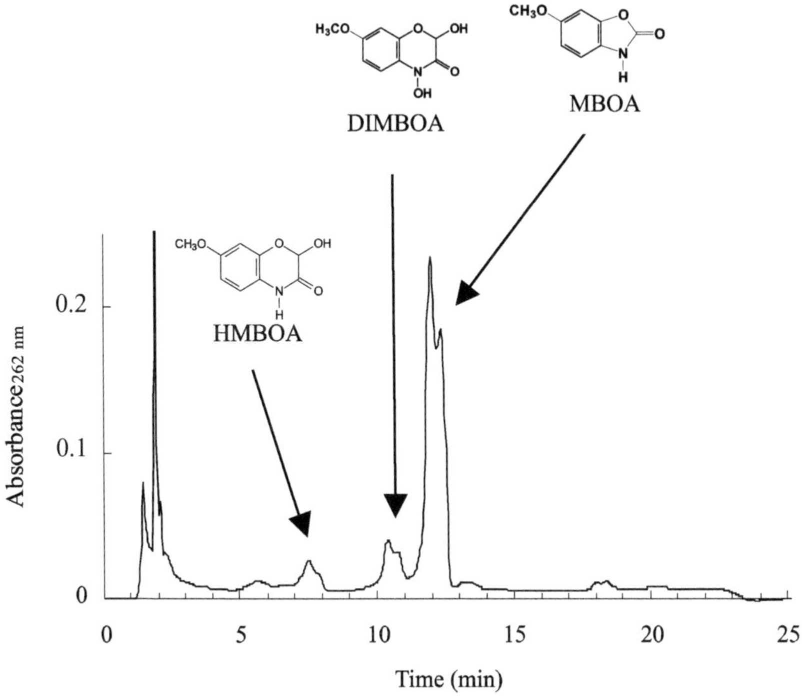

Fig. 6 HPLC analysis of DIMBOA in soil solution at an incubation time of 50 min

Similarly, a degradation experiment of 2,4-dihydroxy-(2*H*)-1,4-benzoxazin- 3(4H)-one (DIBOA) was carried out. After 24 h of incubation, the appearance in a significant amount of a degradation product could be observed. This compound was identified as 2-benzoxazolinone (BOA). But in the case of this compound, significant amounts of DIBOA could be measured even after 3.5 d. After 10 d a second degradation product was detected. This compound could be isolated and identified as 2-amine-3*H*-phenoxazin-3-one (APO).

We proceeded to study the degradation of the main degradation products of DIMBOA and DIBOA: BOA and MBOA. The first one, BOA, degraded rapidly in both soils, with a loss of 70% to 99% after 10 d. The half-life of BOA was 2.20 and 2.71 d for the biotransformation in cultivated soil with varieties Ritmo and Astron respectively. The comparison of these half-life times with those obtained for DIBOA in the two soils for a dose of 1 mg/g of soil (24 and 28 h), allowed us to conclude that BOA is much more resistant to biotransformation than its precursor DIBOA.

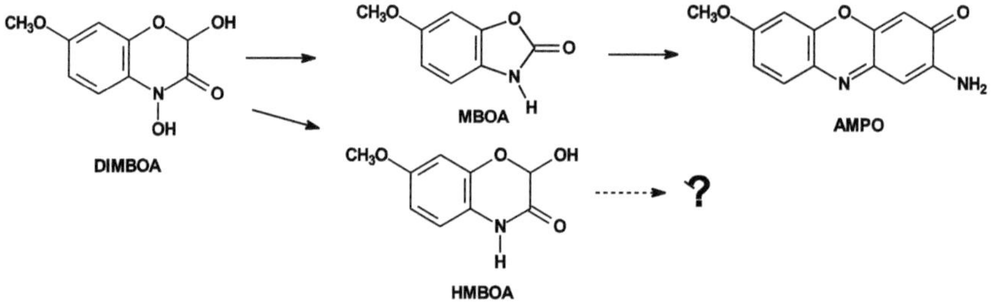

Fig. 7 Degradation pathway found for DIMBOA in wheat crop soil.

After the isolation of AMPO in soil inoculated with DIMBOA, a complete experiment was designed to characterize stability and degradation dynamics of MBOA and its conversion AMPO. Ritmo soil showed a higher conversion capacity than Astron soil (12 ± 3 y 4 ± 2% conversion respectively). For first order degradation kinetics, half-life times were 6.4 d (Ritmo) and 4.3 ds (Astron). Differences observed between both types of soil could be related to different microbial populations associated to each variety (Miethling *et al.*, 2000). The amount and structure of the chemicals liberated by different varieties would not be equal and could influence the microbes or fungi populations associated to the radicular system, producing a different degradation dynamics for other compounds released by the plant.

Lastly, we studied the final degradation products that were detected in these conditions. Several experiments were carried out in order to study APO stability in both soil types. Biotransformation was not observed after 90 d. Thus, APO would be the final product of the microbial degradation of DIBOA-Glc in these soils and conditions whereas other authors have found its transformation to the acetylated compound (Fomsgaard *et al.*,

2004). DIBOA and BOA are intermediate products of this transformation, at least for this time period and concentrations. A scale time of residence in the biotransformation shows that it is evident that the biological activities registered in bioassays of these degradable products can be due to a mixture of them with their degradation metabolites. For instance, bioassays carried out with DIBOA with incubation times of three d or more would also be measuring the activity of BOA and APO.

Bioassay Results

The standard target species, DIBOA and its glycoside showed similar inhibition profiles. *Allium cepa* was the most affected plant according to inhibition recorded at high and at low doses (70% and 91% inhibition for root length at 10^{-3} M, respectively). These compounds maintained phytotoxicity at 5.10^{-4} M for this plant (60% and 73% inhibition for root length at for DIBOA and DIBOA-Glc respectively) (Macias *et al.*, 2005c). DIBOA phytotoxic activity is higher over roots than over shoots and there is lower activity over monocotyledonous that even disappear when the target species is wheat. This fact clearly showed lack of autotoxicity for this compound. Regarding DIMBOA bioactivity, significant inhibition levels were recorded for root lengths in all assayed species except wheat and lettuce (*A. cepa*: 68%; *L. esculentum*: 86%; *L. sativum*: 74% at 5.10^{-4} M).

The bioactivity observed for natural benzoxazinones was not preserved for their degradation products. Benzoxazolinones BOA and MBOA were slightly inhibitory at the highest concentration towards lettuce (root growth, 39%, 20%), and significant inhibition values were also observed for onion (51%, 64%), tomato (70%, 68%) and cress (51%, 49%) roots at 10^{-3} M. All these effects were lost at any lower concentrations. Aminophenoxazine APO showed strong inhibition profiles for root and shoot lengths of all assayed species except wheat, in which it had a moderate effect over root length, and lettuce, in which the germination rate was the most affected parameter (76% at 10^{-3} M and 78% at 10^{-4} M), and growth was moderately inhibited (~50% at 10^{-3} M). Effects over tomato (EC_{50} = 153 μM, r^2 = 0.95), onion (EC_{50} = 72.9 μM, r^2 = 0.99) and cress (EC_{50} = 38.2 μM, r^2 = 0.99) roots are especially significant. There was also high inhibition in shoot lengths for these three species.

Phytotoxicity for AMPO is much lower than expected in comparison to its non-methoxylated analog APO.

Regarding *Echinochloa crus-galli* (barnyardgrass), all assayed benzoxazinones showed important inhibitory effects on growth parameters, some of them being equal or higher than the ones shown for the internal reference. DIBOA-Glc and its aglycon DIBOA exhibited very similar phytotoxicity, with high inhibition at the two more concentrated treatments. Although DIMBOA shows an interesting profile on shoot length, it did not inhibit root growth in a significant manner.

Regarding the phytotoxicity of benzoxazinone degradation products, benzoxazolinone BOA did not show any significant effect on barnyardgrass growth. Shoot length was slightly affected by MBOA (37 % inhibition at 10^{-3} M), this effect being lost at any lower dose.

2-Aminophenoxazin-3-one (APO) is the only aminophenoxazine to show inhibitory effects, reaching 60% at the two most concentrated treatments for both growth parameters. Nevertheless, its precursor 2-aminophenol (APH) showed potent inhibition on both growth parameters, this effect being preserved with treatment dilution especially on shoot growth (EC_{50} = 1.7.10^{-4} M, r^2 = 0.97). Although this compound provoked high root length inhibition at the two first treatments (10^{-3} and 5.10^{-4} M), the effects on this parameter were not significant at any lower concentration.

All compounds, when active, showed inhibitory profiles on *L. rigidum* and *A. fatua* roots. In general terms, the assayed compounds provoked significant effects at the three more concentrated treatments (10^{-3}, 5.10^{-4} and 10^{-4} M). The compound showing more intense effects at lower dilutions was the degradation compound APO (EC_{50} = 2.4.10^{-5} M, r^2 = 0.99, root length, *A. fatua*). The same is observed on *L. rigidum* roots, where EC_{50} values for this compound was 2.3.10^{-5} M, with good adjustment to a sigmoidal dose-response model (r^2 = 0.96). *A. fatua* was more affected by the assayed compounds than *L. rigidum*, being APO similar to the commercial herbicide Logran® in its effects (EC_{50} = 1.4.10^{-4} M, r^2 = 0.99).

Shoot length was slightly inhibited on both species, the relative behavior of the compounds being similar than in the roots, but with much lower effects. There were no significant effects on the germination rate for almost all compounds assayed.

APO was the degradation product to have higher effects of this order. Benzoxazolinones BOA and MBOA had significant activity on both species, reaching 60% inhibition at 10^{-3} M.

DISCUSSION

DIBOA half-life is too short to allow accuracy in the study without considering its degradation products, mainly BOA and APO. Both compounds are present in the soil after a relatively short time, once DIBOA is released by the plant. Thus, any allelopathic interaction involving DIBOA should also be influenced by BOA and APO. The results indicate that APO should be the most important compound in allelopathy of wheat. Allelopathic role of DIBOA has been widely researched in the few last years without taking into consideration the possible effects of its degradation metabolites. These degradation processes take place in the soil as well as in bioactivity evaluation treatment solutions, so the results obtained in those experiments have to be carefully reanalyzed. DIMBOA is degraded by soil microbial population in a relatively short time. This fact forces us to reconsider its role in chemical defense mechanisms of plants with high production of this secondary metabolite, and to take into account the possible effects of its degradation products. Depending on their potential phytotoxic effects, defense capability of the plant could be enhanced or reduced in a significant manner.

DIMBOA and DIBOA have very different half-lives. 7-Methoxy group present in DIMBOA could have an important role in its high degradability. Some authors suggest this methoxy group to influence the reactivity of DIMBOA with nucleophiles (Atkinson *et al.*, 1991). These reactions could be part of the degradation processes. Otherwise, DIBOA has a high persistence and could have a more important role in chemical defense mechanisms than DIMBOA in plants that produce large amounts of hydroxamic acids. 7-Methoxy group present in DIMBOA has a determinant role in its high degradability, chemical reactivity and biological activity. As electron density at the aromatic ring increases, the developing positive charge on nitrogen at the degradation transition state is more stabilized (Atkinson *et al.*, 1991), providing faster degradation rates. This statement, valid for aqueous and organic solutions, is in agreement with the results observed for microbial degradation. The higher persistence shown by DIBOA allows assigning a more important role in chemical defense mechanisms than DIMBOA in plants that produce large amounts of hydroxamic acids. These facts show evidence that classical allelochemicals associated with a great diversity of biological activities, can be transformed by different processes and interactions once released. Thus, more detailed studies are necessary to understand how organic materials, mineral reactivity, ionic interchanging capacity, inorganic ions and specific microbial and fungal activities could influence affordability of allelochemicals in the environment. All these phenomena closely depend on the type of soil under study. The assignment of a biological property to a compound or a family of them involves a more complex study, in which the behavior of releaser and target plants needs to be achieved (allelochemical levels and liberation capacity) together with soil dynamics of allelochemicals (diffusion through soil, chemical and biochemical interactions), and their modes of action (chemical transformations and activity measured over target plants).

Searching for the actual responsible of the allelopathic phenomenon deserved, we proceeded to assay these compounds to compare the bioactivity of each one. We assayed eight species, five standard target species (*Triticum aestivum* L., *Allium cepa* L., *Lepidium sativum* L., *Lycopersicon esculentum* Will. and *Lactuca sativa* L.) and three common weeds (*Echinochloa crus-galli* L., *Avena fatua* L. and *Lolium rigidum* Gaud.).

In STS, the most affected parameter was the root length for all the active compounds, followed by the shoot length. Effects over *Lactuca sativa* L. were diverse, but a good correspondence between concentration and effect was not observed.

All chemicals showed inhibitory profiles, except AMPO in cress and lettuce, and BOA and MBOA for wheat, although these effects were not very consistent. Natural allelochemicals were the most active compound groups. The rest of the analyzed chemicals, which are present in natural benzoxazinones degradation routes, had moderate or null activity, except APO that showed the highest inhibitory effects among the degradation products.

All assayed species were highly affected by the most active chemicals, except wheat, which was moderately inhibited. This selective behavior, especially observed for DIMBOA allows us to conclude that this natural allelochemical, together with its degradation products, is not involved in intraspecific competition phenomena for wheat. DIBOA-Glc and DIBOA, moderately affects wheat. Although these compounds and APO inhibited wheat growth, the concentrations assayed here are significantly higher than the natural ones. Regarding phytotoxicity of degradation products on *Echinochloa crus-galli*, the only one to show moderate effects was 2-aminophenoxazin-3-one (APO), which was the most active on *A. fatua*, and *L. rigidum*.

A resume of these data, the EC_{50} values of evaluated compounds of the effect provoked on root growth are shown in Table 3. It can be observed that with the exception of *T. aestivum*, in all cases the most active compound is the degradation product APO.

Table 3 EC_{50} values of compounds on root length (mM)

Species	*DIBOA*	*DIMBOA*	*BOA*	*MBOA*	*APO*	*AMPO*
L. sativum	N.C.	1.8E-1	1.60	1.25	3.82E-2	N.C.
L. sativa	4.7	N.C.	N.C.	N.C.	4.38E-1	N.C.
L. esculentum	N.C.	2.1E-1	9.18	444	1.53E-1	N.C.
A. cepa	3.6E-1	N.C.	4.06E-1	137	7.29E-2	N.C
T. aestivum	73	N.C.	3.38E-1	8.45E-2	N.C.	N.C.
L. rigidum	N.C.	8.6E-1	2.13	3.85E-1	2.35E-2	N.C.
A. fatua	3.7E-1	N.C.	2.75E-1	2.54	2.40E-2	N.C.
E. crus-galli	N.C.	N.C.	N.C.	N.C.	7.47E-1	N.C.

N.C. The effects do not fit a sigmoidal dose-response curve

CONCLUSION

We can conclude that in benzoxazinoids, degradation products could be as important or even more than those exuded by plants in the understanding of the allelopathic interaction. As in precedent examples, the fate and degradation studies of allelochemicals are a necessary stage in allelopathic research if we want to have a complete understanding of allelopathy. The results of these studies may influence the design of bioassays, and, of course, their interpretation. If active or not-active compounds are produced during the bioassay, we need to know the composition of the mixture while the bioassay is running. These data allow us to know the chemicals troly responsible for activity. If evidence from bioassays suggests that environmental and/or microbial transformation is important, it may be necessary to fractionate extracts that are exposed to environmental conditions and/or soil microorganisms

Acknowledgments

This work was financially supported by the Ministerio de Educación y Ciencia (Spain), Project n°: PPQ2002-03595. Fellowships from the Universidad de los Andes-Venezuela (A.O.B.) and from the European Commission (European Union) and Junta de Andalucía-Spain (D.M.) are also gratefully acknowledged.

References

An, M., Pratley, J.E., Haig, T. (2000) Phytotoxicity of *Vulpia* Residues: IV. Dinamics of Allelochemicals during Decomposition of *Vulpia* Residues and Their Corresponding Phytotoxicity. J. Chem. Ecol. 26: 2603-2617.

Atkinson, J., Morand, P., Arnason, J.T., Niemeyer, H.M., Bravo, H.R. (1991) Analogs of the cyclic hydroxamic acid 2,4-dihydroxy-7-methoxy-2H- 1,4-benzoxazin-3-one (DIMBOA): decomposition to benzoxazolinones and reaction with β-mercaptoethanol. J. Org. Chem. 56: 1788-1800.

Baumeler, A., Hesse, M., Werner, C. (2000) Benzoxazinoids-cyclic hydroxamic acids, lactams and their corresponding glucosides in the genus Aphelandra (Acanthaceae). Phytochemistry. 53: 213-222.

Bialy, Z., Oleszek, W., Lewis, J., Fenwick, G.R. (1990) Allelopathic potential of glucosinolates (mustard oil glycosides) and their degradation products against wheat. Plant and Soil. 129: 277-281.

Blum, U. (1998) Effects of microbial utilization of phenolic acids and their phenolic acid breakdown products on allelopathic interactions. J. Chem. Ecol. 24: 685-708

Blum, U., Shafer, S.R. (1988). Microbial populations and phenolic acids in soil. Soil Biol. Biochem. 20: 793-800.

Blum, U. (2004) Fate of phenolic allelochemicals in soils - the role of soil and rhizosphere microorganisms. *In* Allelopathy: Chemistry and Mode of Action of Allelochemicals, F.A. Macias, J.C.G. Galindo, J.M.G. Molinillo and H.G. Cutler (eds.) CRC Press LLC, Boca Raton. Fl, USA. pp. 57-76.

Borek, V., Morra, M.J., Brown, P.D., McCaffrey, J.P. (1994) Allelochemicals Produced during Sinigrin Decomposition in Soil. J. Agric. Food Chem. 42: 1030-1034.

Borek, V., Morra, M.J., Brown, P.D., McCaffrey, J.P. (1995) Transformation of the Glucosinolate-Derived Allelochemicals Allyl Isothiocyanate and Allylnitrile in Soil. J. Agric. Food Chem. 43: 1935-1940.

Cambier, V., Hance, T., de-Hoffmann, E. (1999) Non-injured maize contains several 1,4-benzoxazin-3-one related compounds but only as glucoconjugates. Phytochem. Anal. 10: 119-126.

Chou, Ch-H., Lin, H-J. (1976a) Autointoxication mechanism of *Oryza sativa*. Phytotoxic effects of decomposing rice residues in soil. J. Chem. Ecol. 2: 353-367.

Chou, Ch-H., Patrick, Z.A. (1976b) Identification and Phytotoxic activity of compounds produced during decomposition of corn and rye residues in soil. J. Chem. Ecol. 2: 369-387.

Corcuera, L.J., Argandoña, V.H., Zúñiga, G.E. (1992) Allelochemicals in wheat and barley: role in plant-insect interactions. *In* Allelopathy: Basic and Applied Aspects; S.J.H. Rizvi, V. Rizvi, (eds.). Chapman & Hall: London, UK. pp. 120-127.

Dagley, S. (1971) Catabolism of aromatic compounds by microorganisms. Advan. Microbial. Physiol. 6: 1-42.

Dennis, D.A., Chapman, P.J., Dagley, S. (1973) Degradation of protocatechuate in *Pseudomonas testosteroni* by a pathway involving oxidation of the products of metafission. J. Bact. 113: 521-523.

Eljarrat, E., Guillamon, M., Seuma, J., Mogensen, B.B., Fomsgaard, I.S., Olivero-Bastidas, A., Macias, F.A., Stochmal, A., Oleszek, W., Shakaliene, O., Barcelo, D. (2004) First European interlaboratory study of the analysis of benzoxazinone derivatives in plants by liquid chromatography. J. Chromatogr. A. 1047: 69-76.

Fomsgaard, I.S., Mortensen, A.G., Carlsen, S.C.K. (2004) Microbial transformation products of benzoxazolinone and benzoxazinone allelochemicals–a review. Chemosphere. 54: 1025-1038.

Frey, M., Chopmet, P., Glawischnig, E., Stettner, C., Grün, S., Winklmair, A., Eisenreich, W., Nacher, A., Meeley, R.B., Briggs, S., Simcox, K., Gierl, A. (1997) Analysis of a chemical plant defence mechanism in grasses. Science. 277: 696-699.

Friebe, A. (2001) Role of benzoxazinones in cereals. J. Crop Prod. 4: 379-400.

Huang, Z., Liao, L., Wang, S., Cao, G. (2000) Allelopathy of Phenolics from Decomposing Stump-Roots in Replant Chinese Fir Woodland. J. Chem. Ecol. 26: 2211-2219.

Ito, I., Kobayashi, K., Yoneyama, T. (1998) Fate of dehydromatricaria ester added to soil and its implications for the allelopathic effect of *Solidago altissima* L. Annals of Botany (London). 82: 625-630.

Kobayashi, K. (2004) Factors affecting phytotoxic activity of allelochemicals in soil. Weed Biol. Manage. 4: 1-7.

Macias, F.A., Oliveros-Bastidas, A., Marin, D., Castellano, D., Simonet, A.M., Molinillo, J.M.G. (2004) Degradation studies on benzoxazinoids. Soil degradation dynamics of 2,4-dihydroxy-7-methoxy-(2H)-1,4-benzoxazin- 3(4H)-one (DIMBOA) and its degradation products, phytotoxic allelochemicals from Gramineae. J. Agric. Food Chem. 52: 6402-6413.

Macias, F.A., Oliveros-Bastidas, A., Marin, D., Castellano, D., Simonet, A.M., Molinillo, J.M. G. (2005a) Degradation Studies on Benzoxazinoids. Soil Degradation Dynamics of (2R)-2-O-β-D-Glucopyranosyl-4-hydroxy-(2H)- 1,4-benzoxazin-3(4H)-one (DIBOA-Glc) and Its Degradation Products, Phytotoxic Allelochemicals from Gramineae. J. Agric. Food Chem. 53: 554-561.

Macias, F.A., Marin, D., Oliveros-Bastidas, A., Castellano, D., Simonet, A.M., Molinillo, J.M. G. (2005b) Structure-activity relationships (SAR) studies of benzoxazinones, their degradation products and analogues. Phytotoxicity on standard target species (STS). J. Agric. Food Chem. 53: 538-548.

Macias, F.A., Chinchilla, N., Varela, R.M., Oliveros Bastidas, A., Marin, D. Simonet, A.M., Molinillo, J.M.G. (2005c) Structure-Activity Relationships (SAR) Studiesof Benzoxazinoids and Related Compounds. Phytotoxicity on *Echinochloa crus-galli* (L.) P. Beauv. J. Agric. Food Chem. 53: 4373-4380.

Martin, J.P., Haider, K. (1976) Decomposition of specifically carbon-14-labeled ferulic acid: free and linked into model humic acid-type polymers. Soil Sci. Soc. Am. J. 40: 377-380.

Miethling, R., Wieland, G., Backaus, H., Tebbe, C.C. (2000) Variations of microbial rhyzosphere communities in response to crop species, soil origin, and inoculation with *Sinorhizobium melioti* L33. Microb. Ecol. 41: 43-56

Pecháček, R., Velíšek, J., Davídek, J. (2000) Decomposition of sinigrin by methanol/ ammonia/water treatment in model systems and mustard (*Brassica nigra* L.) seed meal. European Food Res. Tech. 210: 196-201.

Politycka, B., Adamska, D. (2003) release of Phenolic Compounds from Apple Residues Decomposing in Soil and Influence of Temperature on their Degradation. Polish J. Environ. Studies. 12: 95-98.

Pratt, K., Kumar, P., Chilton, W.S. (1995) Cyclic hydroxamic acids in dicotyledonous plants. Biochem. Syst. Ecol. 23: 781-785.

Rimando, A.M., Olofsdotter, M., Dayan, F.E., Duke, S.O. (2001) Searching for Rice Allelochemicals: An Example of Bioassay-Guided Isolation. Agronomy J. 93: 16-19.

Roberts, T. (1998) Metabolic Pathways of Agrochemicals. Part 1: Herbicides and Plant Growth Regulators. The Royal Society of Chemistry Information services. Cambridge. UK.

Shahidi, F., Gabon, J.E., Rubin, L.J., Naczk, M. (1990) Effect of methanol-ammonia-water treatment on the fate of glucosinolates. J. Agric. Food Chem. 38: 251-255.

Turner, J.A., Rice, E.L. (1975) Microbial decomposition of ferulic acid in soil. J. Chem. Ecol. 1: 41-58.

Wahlroos, O., Virtanen, A.I. (1959) The precursors of 6-methoxy- benzoxazolinone in maize and wheat plants, their isolation and some of their properties. Acta Chem. Scand. 13: 1906-1908.

Weston, L.A., Duke, S.O. (2003) Weed and Crop Allelopathy. Critical Rev. Plant Sci. 22: 367-389.

Williamson, G.B., Obee, E.M., Weidenhamer, J.D. (1992) Inhibition of *Schizachyrium scoparium* (Poaceae) by the allelochemical hydrocinnamic acid. J. Chem. Ecol. 18: 2095-2105.

7

Iron Dissolution Reaction of Mugineic Acids for Iron Acquisition of Graminaceous Plants

Syuntaro Hiradate

Biodiversity Division, National Institute for Agro-Environmental Sciences, 3-1-3 Kan-nondai, Tsukuba, Ibaraki 305-8604, Japan

Even in Iron (Fe)-deficient conditions, some plants can acquire Fe from the soil. The Fe-acquisition mechanisms are summarized by two strategies: Strategy I for dicotyledonous plants, and strategy II for graminaceous plants. In strategy II, mugineic acid (MA) and its derivatives, which are low-molecular-weight (M.W., 310 to 340) natural amino acids, are secreted from roots of Fe-deficient plants, and they can dissolve Fe in the soil by forming MA-Fe^{III} complex with high efficiency and high dissolution rate, and their Fe-dissolution ability are maximized under Fe-deficient and problematic pH soil conditions. The first step of the Fe-dissolution reaction is the adsorption reaction of MA on the surfaces of Fe-containing minerals by ligand exchange reaction. The second step is the desorption of the adsorbed MA from the minerals holding an Fe atom, resulting in the formation of soluble MA-Fe^{III} complex in solution. The MA-Fe^{III} complex is also re-adsorbed on soil minerals, although its affinity is lower than that of MA. Free MA has six vacant functional groups to bind with one Fe atom (three carboxylic groups, two imino groups, and one hydroxyl group), whereas in the MA-Fe^{III} complex, these functional groups are occupied by an Fe atom. Crystallinity of Fe minerals is the important factor in MA-promoted Fe dissolution: poorly crystalline ferrihydrite can be an Fe source of MA-Fe^{III} complex, but not the more crystalline Fe minerals (e.g., goethite, hematite, lepidocrocite and maghemite) because they cannot be dissolved by MA. This information is useful in understanding the behavior of MA and its derivatives in the soil and for predicting Fe-deficiency in graminaceous plants.

Keywords: *chelation, iron-deficiency, mugineic acids, phytosiderophores, soil iron minerals, strategy II*

INTRODUCTION

Recent research has emphasized the importance of root-secreted organic molecules on the growth of the secreting plants, as well as on the chemical interactions among plants (allelopathy). In acidic soils, phytotoxic Al^{3+} is excluded from the apoplast of the roots of Al-tolerant plants and is detoxified by forming Al-organic acid complexes (Buchanan *et al.*, 2000). In many phosphate-deficient plants, organic acids are secreted to the rhizosphere, and they enhance the phosphate acquisition of plants by releasing adsorbed phosphate on the surface of soil minerals (Gardner *et al.*, 1983; Ae *et al.*, 1990; Otani *et al.*, 1996). In calcareous and/or alkaline soils, where many plants suffer from Fe-deficiency, graminaceous plants secrete mugineic acid (MA) and its derivatives (Fig. 1) to enhance Fe uptake. The Fe acquisition mediated by MA and its derivatives is now one of the most well-defined phenomena of root-secreted organic molecules. In this chapter, Fe-acquisition mechanisms of plants and the mechanisms of the MA-promoted Fe dissolution are reviewed.

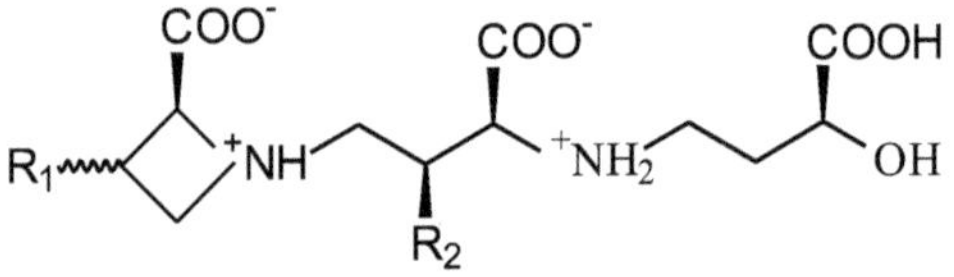

2'-deoxymugineic acid (DMA): $R_1 = R_2 = H$

mugineic acid (MA): $R_1 = H$, $R_2 = OH$

3-hydroxymugineic acid (HMA): $R_1 = R_2 = OH$

Fig. 1 Chemical structures of mugineic acids.

IRON-DEFICIENCY OF PLANTS IN THE WORLD

Iron is an essential element for plant growth. In plants, Fe controls many life-supporting reactions, such as chlorophyll synthesis and oxide-reductive reactions, etc. Mean content of Fe in the soil and the earth's crust ranges from 2.6 to 4.1% (Sparks, 2003), and the amount of Fe required for sufficient plant growth is at the 0.01% level (Taiz and Zeiger, 1998), but Fe-deficiency in plants frequently occurs, especially in calcareous soils. Calcareous soils occur where limestone is the parent material of the soil and where the secondary carbonate and/or bicarbonate minerals are accumulated on the surface of the soil in arid and semi-arid zones. In these conditions, the pH of soil solution is kept in the neutral to alkaline range by the presence of the carbonate and/or bicarbonate minerals. Solubility of Fe^{3+} decreases 1/1000-fold per unit increase of the solution pH (Lindsay, 1988). In neutral to alkaline conditions, therefore, the concentration of Fe^{3+} in soil solution becomes lower than the critical concentration required for the Fe absorption of plants, and plants face difficulty in Fe acquisition (Lindsay and Schwab, 1982). This Fe-deficient and potentially Fe-deficient soil (problematic high pH soil) amounts to 25 to 30% of world land area (Wallace and Lunt, 1960).

IRON-ACQISITION MECHANISMS OF PLANTS UNDER IRON-DEFICIENT CONDITIONS

Even from calcareous and/or alkaline soil, several plant species can acquire Fe. These Fe acquisition mechanisms are summarized into strategies I and II (Buchanan *et al.*, 2000). Strategy I is a Fe-acquisition mechanism of dicotyledonous plants, and is explained by the reduction of Fe^{3+} to Fe^{2+} by the reductase on the surface of roots and by the decrease of rhizosphere pH by H^+ release. Iron in the reduced form (Fe^{2+}) has much higher solubility than that in the oxidized form (Fe^{3+}) at the same pH. Decrease of rhizosphere pH is also effective in dissolving Fe. Therefore, strategy I is characterized by the release of e^- and H^+ from roots. In strategy II, which is the Fe acquisition mechanism of monocotyledonous plants, amino acids having chelating ability with Fe^{3+} are secreted to the rhizosphere. The released amino acids dissolve soil Fe by forming complexes with Fe^{3+}, and these complexes are absorbed by the plant roots (Fig. 2). A similar Fe acquisition mechanism has been reported in several microorganisms, and the Fe-dissolving agent has been termed siderophore, which means Fe transporter. In the case of strategy II, since the Fe^{3+}-chelating organic molecules have a similar function with that of siderophores and are produced by plants, they are named phytosiderophores. Until now, several phytosiderophores have been reported, and they are all MA and its derivatives (Fig. 1). The MA-family of phytosiderophores were first found and named by Dr. S. Takagi (Takagi, 1976), emeritus professor of Iwate University, where "mugi" and "ne" means wheat and root in Japanese, respectively. These phytosiderophores are biosynthesized from three methionines, and this transformation is initiated by an Fe-deficiency signal. The ability of phytosiderophore secretion by graminaceous plants explains well their Fe-acquisition ability (Kawai *et al.*, 1988). Other physiological aspects of MA and its derivative secreted from graminaceous plants are summarized in Takagi (1993).

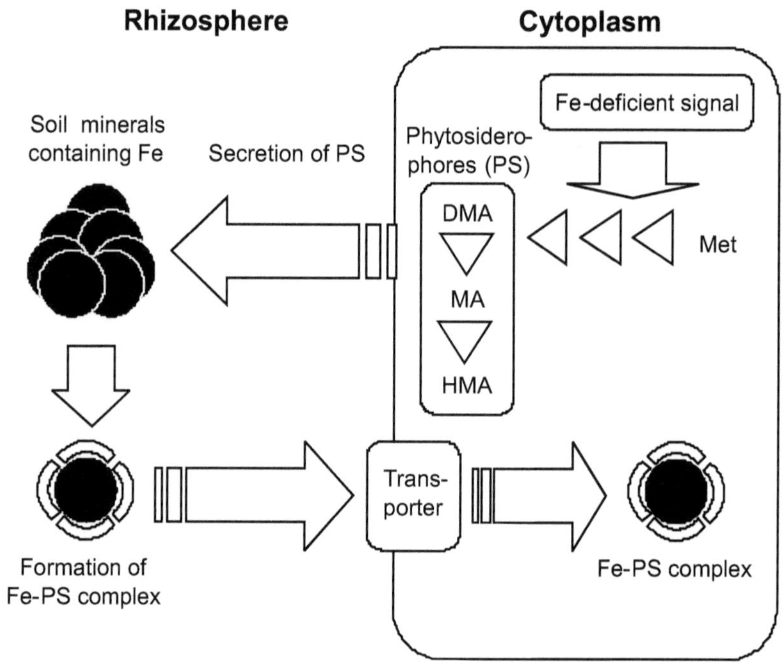

Fig. 2 Schematic representation of strategy II

CHEMICAL FORM OF IRON IN SOIL

In soil, Fe is involved in primary minerals, secondary minerals and humus-Fe complexes. Among them, secondary Fe minerals play an important role in controlling the concentration of Fe in soil solution. In primary minerals, Fe is present in chemically stable forms, resulting in extremely low dissolution rate and consequently low bio-availability. In humus-Fe complexes, Fe has high reactivity, but the solubilities of the complexes are strongly influenced by soil factors, such as solubility of coexisting secondary Fe minerals and soil pH. Furthermore, the amount of humus-Fe complexes is quite smaller than that of primary and secondary minerals, and they are localized at surface horizons.

Figure 3 shows major secondary Fe minerals in the soil and their solubility products. Goethite and hematite are common in calcareous and/or alkaline soils, and they have relatively high crystallinities and low Fe-solubilities. Lepidocrocite is frequently found in paddy soil, and it has moderate crystallinity and moderate Fe-solubility. Ferrihydrite, a poorly crystalline Fe mineral, is composed of porous and fine particles, and it has the highest Fe-solubility and the largest specific surface area among these

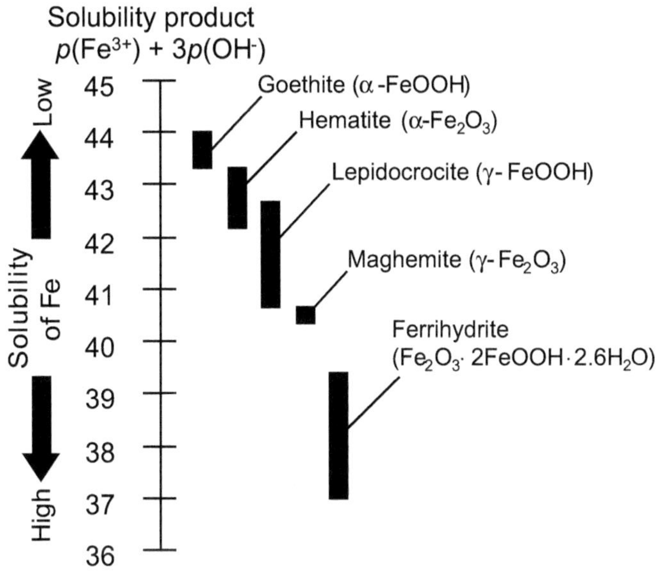

Fig. 3 Major secondary Fe minerals in soil and their solubility products. Data are cited from Schwertmann and Taylor (1989).

Fe minerals. Ferrihydrite has a large amount of active surface hydroxyls to adsorb organic acids, phosphate, fluorine, etc., by ligand exchange reaction, irreversibly. Ferrihydrite is ubiquitous in less-weathered soil (e.g., volcanic ash soil), but it is also found in calcareous and/or alkaline soil although its amount is very small.

In general, several kinds of secondary Fe minerals are present in a soil. This means that these Fe minerals do not reach thermodynamic equilibrium, and thereby the concentration of Fe in soil solution is controlled by the Fe-containing minerals having the highest solubility in the soil.

REACTIONS OF MUGINEIC ACID WITH SOIL MINERALS AND SOILS

To clarify how MA and its derivatives work in the Fe acquisition of plants, laboratory experiments were conducted to react pure MA and pure soil minerals. In this section, the reaction characteristics of MA with these soil minerals and the soil are summarized.

Kinetics of the Interaction between Mugineic Acid and Ferrihydrite

When MA was added to ferrihydrite suspension, concentrations of MA (Fig. 4, ○) and Fe (●) in the reacting solution were gradually decreased and increased with the reaction time, respectively. In this experimental condition (equilibrium pH 8 to 9), free Fe^{3+} ion cannot exist in a detectable concentration, therefore, the increased concentration of Fe detected in Fig. 4 is attributable to the formation of MA-Fe complex. Formation of MA-Fe^{III} (1:1) complex and its formation constant (log K = 18) are reported by Sugiura *et al.* (1981). Therefore, Fig. 4 indicates that MA dissolves Fe from ferrihydrite by forming MA-Fe^{III} complex. Since the ratio of MA to Fe in the MA-Fe^{III} complex is 1:1, Fe in Fig. 4 (●) cannot exist beyond the MA concentration (○) in the solution. Fig. 4 also shows that MA can dissolve ferrihydrite quickly (reach maximum dissolution after 4 h) with high efficiency (30 to 40% of added MA dissolves Fe by forming MA-Fe^{III} complex). Such a quick dissolution reaction of Fe by MA might be advantageous for Fe-absorbing plants, because MA and its derivatives being secreted to the rhizosphere are easily decomposed by microorganisms (Watanabe and Wada, 1989). On the other hand, the decrease of MA concentration in Fig. 4 (○) is attributable to the adsorption of MA on ferrihydrite. Thus, in the interaction between MA and ferrihydrite, dissolution of Fe and adsorption of MA occur at the same time and at the same concentration level.

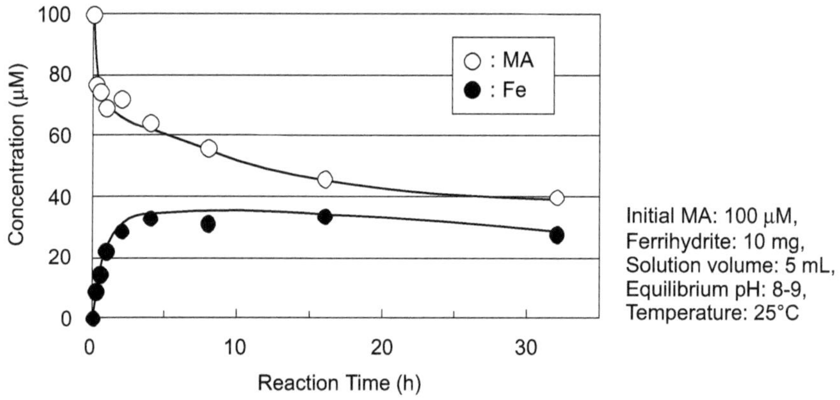

Fig. 4 Kinetics of interaction between mugineic acid (MA) and ferrihydrite. Concentrations of MA (○) and Fe (●) in the reacting solution are shown as a function of reaction time (Inoue *et al.*, 1993).

Interactions of Mugineic Acid with Ferrihydrite as Affected by pH

As the stability of a ligand-metal complex depends on pH, the Fe-dissolution ability of MA also depends on pH. When MA reacted with ferrihydrite for 4 h, almost no Fe-dissolution was observed at equilibrium pH higher than 10, and maximum Fe-dissolution was found at equilibrium pH of 7 to 8 (Fig. 5A). Adsorption of MA on ferrihydrite is more prominent at lower equilibrium pH. These interaction features between MA and ferrihydrite are also true when the initial concentration of MA and the amount of added ferrihydrite are changed (Inoue *et al.*, 1993). It is notable that MA, which is secreted under Fe-deficient conditions, exerts its maximum Fe-dissolution ability under problematic pH conditions (7 to 8). Mugineic acid and its derivatives are the functional natural products that enhance Fe uptake under Fe-deficient conditions for graminaceous plants.

Interactions of Mugineic Acid with Various Fe Minerals

Mugineic acid can dissolve Fe from poorly crystalline ferrihydrite (Fig. 5A), but the Fe dissolution is reduced exremely in crystalline Fe minerals,

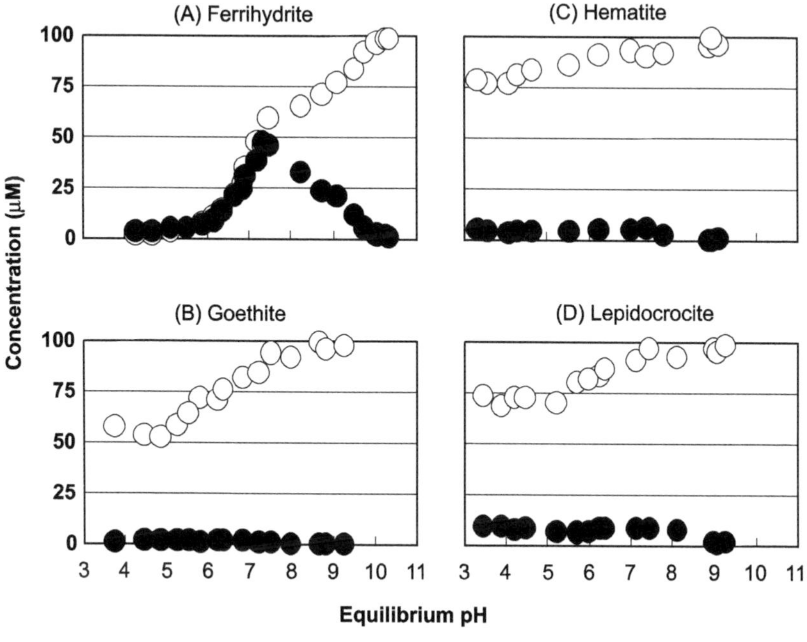

Fig. 5 Interactions of mugineic acid (MA) with four representative Fe minerals as a function of equilibrium pH. Concentrations of MA (○) and Fe (●) in solution are determined after 4 h of reaction time (Inoue *et al.*, 1993).

such as goethite (Fig. 5B), hematite (Fig. 5C), and lepidocrocite (Fig. 5D). The amount of Fe dissolved by MA is also small in other crystalline Fe minerals, magnetite and maghemite. Iron dissolution by MA depends on the crystallinity of Fe minerals: poorly crystalline ferrihydrite can be an Fe source of MA-Fe^{III} complex but more crystalline Fe minerals cannot be dissolved by MA (Hiradate and Inoue, 1998a). On the other hand, adsorption of MA does not depend on the crystallinity but depends on the specific surface area of the Fe minerals. At the same equilibrium pH, the amount of adsorbed MA rises with increasing specific surface area of Fe minerals (Inoue *et al.*, 1993). Thus, Fe-dissolution by MA and MA-adsorption by Fe minerals are regulated by different parameters of Fe minerals, which are crystallinity and specific surface area, respectively.

Mechanisms of Interaction between Mugineic Acid and Fe Minerals

When MA reacted with ferrihydrite, the maximum Fe dissolution occurred at equilibrium pH 7 to 8 (Fig. 5A). This feature resulted from the adsorption reaction of MA on ferrihydrite and the formation reaction of MA-Fe^{III} complex. The adsorption reaction of MA on Fe minerals is a ligand exchange reaction: the surface of the Fe minerals is covered with the active surface hydroxyls, and the hydroxyls are exchanged by MA, resulting in strong covalent bonds between MA molecule and Fe minerals (Fig. 6). As the Fe minerals have high affinity to hydroxyl ion (OH^-) in solution, adsorption of MA decreases with increasing solution pH by the competitive adsorption between MA and OH^- (see Fig. 5). Mugineic acid-Fe^{III} complex is also adsorbed on Fe minerals by the similar mechanism (ligand exchange reaction), although the affinity of MA-Fe^{III} complex to the Fe minerals is lower than that of free MA. Thus, reduced Fe dissolution at equilibrium pH lower than 7 to 8 is caused by the adsorption of MA and MA-Fe^{III} complex on Fe minerals.

Stability of MA-Fe^{III} complex is also affected by solution pH. At equilibrium pH 7 to 8 or higher, MA-Fe^{III} complex is not stable and begins to dissociate becoming free MA and Fe, and the dissociated Fe is transformed into Fe hydroxide (precipitate) in such a high pH condition (Fig. 7). At equilibrium pH 10 or higher, only a little MA-Fe^{III} complex can keep its chemical form in solution. This feature can also be estimated from the formation constant of MA-Fe^{III} complex ($\log K = 18$, Sugiura *et al.*, 1981). Therefore, as shown in Fig. 5A, the Fe dissolution from ferrihydrite by MA is decreased at equilibrium pH higher than 7 to 8 because of the limitation in MA-Fe^{III} complex formation and the dissociation of MA-Fe^{III} complex.

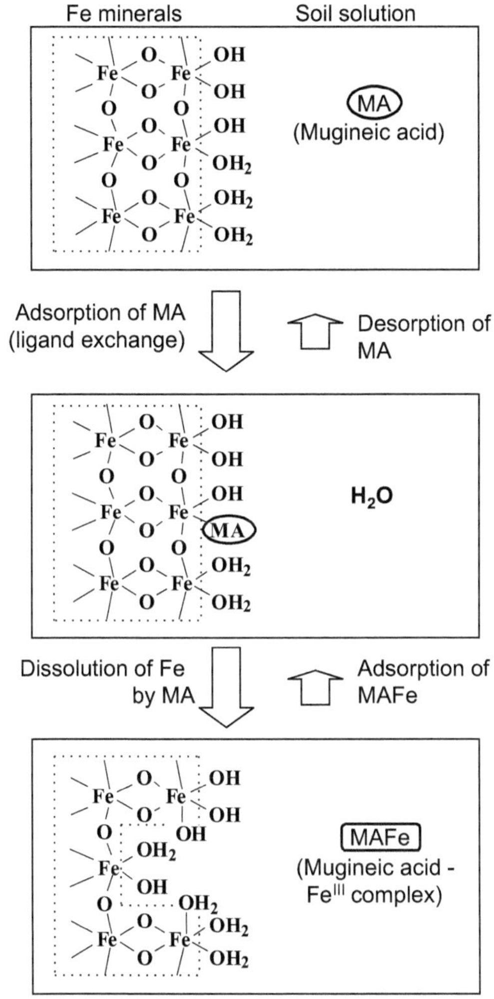

Fig. 6 Schematic representation of interactions of mugineic acid (MA) with Fe minerals.

The mechanisms of the interaction between MA and Fe minerals can be summarized as shown in Fig. 6. The first step of the Fe dissolution from Fe minerals by MA is the adsorption reaction of MA on Fe minerals. If the MA adsorption is inhibited, the subsequent Fe dissolution reaction is also suppressed (Hiradate and Inoue, 1998b). The formation of MA-Fe^{III} complex (Fe dissolution) occurs when the adsorbed MA is desorbed from the Fe minerals holding an Fe atom.

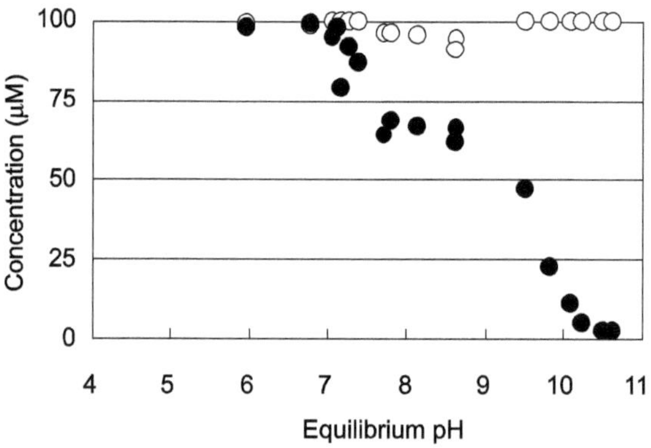

Fig. 7 Changes of chemical form of mugineic acid (MA)-Fe^{III} complex as a function of equilibrium pH. ○; total MA in solution, ●; Fe (MA-Fe^{III} complex) in solution. A solution containing 100 μM MA-Fe^{III} complex is incubated at arbitrary pH for 4 h (Inoue *et al.*, 1993).

The desorbed MA-Fe^{III} complex is also re-adsorbed on Fe minerals, although the affinity of MA-Fe^{III} complex to the Fe minerals is lower than that of MA (Inoue *et al.*, 1993). This is because free MA has six vacant functional groups to bind with Fe atom (three carboxylic groups, two imino groups, and one hydroxyl group), whereas in MA-Fe^{III} complex, these functional groups are already occupied by an Fe atom (Fig. 8). This also explains why the adsorption of MA on ferrihydrite is faster than that of the MA-Fe^{III} complex (Fig. 9).

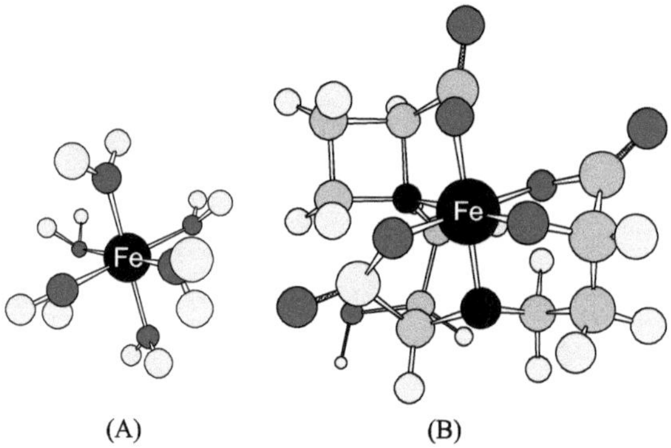

Fig. 8 Molecular image of Fe^{3+} ion (A) and mugineic acid (MA)-Fe^{III} complex (B). ○; H, ◎; C, ◉; O, and ●; N.

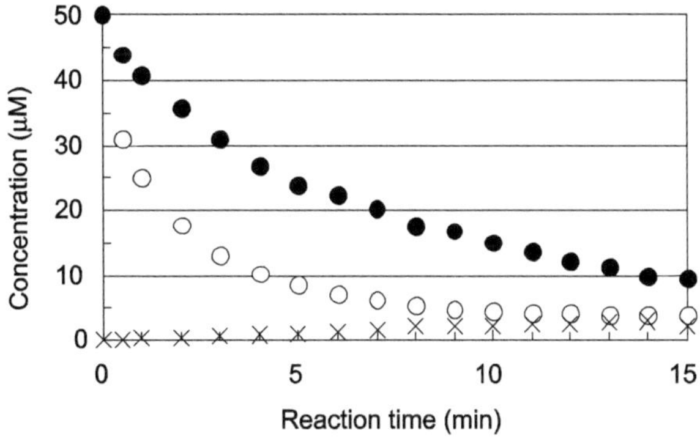

Fig. 9 Comparison of the adsorption rate of mugineic acid (MA, ○) on ferrihydrite with that of MA-Fe^{III} complex (●). Initial concentrations of MA and MA-Fe^{III} complex are both 50 μM. Equilibrium pH: 5.1 ± 0.1. ×: Fe dissolved by the addition of MA.

Influences of Coexisting Ions on the Interactions between Mugineic Acid and Fe Minerals

Phosphate, which is a macro nutrient in plants and a common component of fertilizer, is also adsorbed on Fe minerals by the ligand exchange reaction. The affinity of phosphate on the Fe minerals is greater than that of MA. Therefore, phosphate inhibits the adsorption reaction of MA on Fe minerals and the subsequent Fe dissolution reaction by MA. Sulfate is also adsorbed on Fe minerals by the ligand exchange reaction, but its affinity to the Fe minerals is less than that of MA and greater than that of MA-Fe^{III} complex. Sulfate, therefore, does not inhibit the adsorption of MA, but it does suppress the re-adsorption of MA-Fe^{III} complex on Fe minerals, resulting in the increase of MA-Fe^{III} complex concentration in solution. Nitrate and chloride ions are also common in the soil, but they do not participate in the ligand exchange reaction. Therefore, influences of these anions on the interactions of MA and MA-Fe^{III} complex with soil minerals are minimal.

Interactions between Mugineic Acid and other Soil Components

Mugineic acid can interact with any soil components having active surface hydroxyls. For example, allophane and imogolite, which are common aluminosilicates in volcanic ash soils, have active surface hydroxyls coordinated to aluminum (Al) on their surfaces, as well as gibbsite, a common aluminum hydroxide. Therefore, these soil minerals adsorb MA on their surfaces. The affinity of MA to silicon (Si) atom, however, is very low, so hydroxyls coordinated on Si atom are inactive and cannot be replaced by MA by the ligand exchange reaction. Similarly, layer aluminosilicates, whose surfaces are covered with silicon tetrahedral sheet (e.g., smectite), are relatively inactive in adsorbing MA. In the case of aluminosilicates minerals, therefore, not only the specific surface area, but also the other factors, such as Si/Al molar ratio and stereo-structure of the minerals, affect the amount of MA adsorption (Hiradate and Inoue, 1999).

Poorly crystalline aluminosilicates, such as allophane and imogolite, can be dissolved by MA, and MA-Al complex is formed following the MA adsorption reaction. The amount of MA adsorption on these aluminosilicates is smaller than that on ferrihydrite, and the formation of MA-Al complex from these aluminosilicates is also smaller than that of MA-Fe^{III} complex from ferrihydrite. This means that MA has higher affinity to Fe compared with Al. Mugineic acid cannot dissolve Al from crystalline Al minerals (e.g., gibbsite), as it does with Fe in crystalline Fe minerals (Hiradate and Inoue, 1999).

It is well known that humic substances can adsorb hydrophobic organic compounds by hydrophobic interactions. However, MA and its derivatives are extremely hydrophilic. They are not retained on C18 reversed-phase HPLC column without ion-pair reagents (Hiradate and Inoue, 1996). Therefore, little MA is adsorbed on H^+-type of humic acids prepared from volcanic ash soil (A-type), calcareous soil (Rp-type), and peat soil (P-type). When the surface cation of the humic acids (H^+) is exchanged to Fe^{3+} or Al^{3+}, MA can be adsorbed through these metal bonds, as well as Fe and Al minerals.

Interactions of Mugineic Acid with Crude Soil

It is possible to estimate the behavior of MA in the soil if the kind and amount of soil components and soil pH value are known. The amount of poorly crystalline components in the soil can be estimated by acid-oxalate extraction (Schwertmann, 1964). In an acidic soil, the amount of adsorbed MA and formed MA-Fe^{III} complex increases and decreases, respectively, because poorly crystalline soil components prevail. In a calcareous and/or alkaline soil, where plant Fe deficiency is common, the amount of poorly crystalline minerals is extremely small. Since poorly crystalline Fe minerals (mostly ferrihydrite) are the dominant Fe source for the formation of MA-Fe^{III} complex by the MA-promoted Fe-dissolution reaction, the amount of poorly crystalline Fe minerals in the soil (can be estimated by the acid-oxalate extractable Fe) correlates with the amount of MA-Fe^{III} complex formed in calcareous and/or alkaline soils (Hiradate and Inoue, 2000).

To predict Fe-deficiency in plants, the DTPA Soil Test is frequently applied for calcareous and/or alkaline soil. In this soil test, Fe is extracted with a solution containing 5 mM diethylenetriaminepentaacetic acid (DTPA) in the presence of 10 mM $CaCl_2$ buffered at pH 7.3, and Fe is dissolved by forming DTPA-Fe^{III} complex in solution. Although the formation constant of DTPA-Fe^{III} complex (log K = 27.5, Lindsay and Norvell, 1969) is extremely larger than that of MA-Fe^{III} complex (log K = 18), the fundamental mechanism of the Fe-dissolution by DTPA Soil Test could be similar to that of strategy II. The amount of DTPA Soil Test-extractable Fe is closely correlated with that of MA-extractable Fe (Hiradate and Inoue, 2000). It should be noted that the DTPA Soil Test is effective in predicting Fe-deficiency in graminaceous plants (strategy II plants) but not so effective in dicotyledonous plants (strategy I plants) (Loeppert and Inskeep, 1996). It is likely that the DTPA Soil Test would be a good Fe-deficiency predictor in graminaceous plants simulating strategy II.

CONCLUSION

Iron-deficient graminaceous plants secrete MA and its derivatives to the rhizosphere to acquire Fe from the soil by strategy II. Although these compounds are low-molecular-weight (M.W., 310 to 340) amino acids, they can dissolve Fe from the soil with high efficiency and high dissolution rate, and their Fe-dissolution ability are maximized under Fe-deficient and problematic pH soil conditions. These reaction characteristics can be understood from the adsorption and dissolution reaction concept between MA and its derivatives and soil minerals. This information is useful in understanding the behavior of MA and its derivatives in the soil, and for predicting Fe-deficiency in graminaceous plants.

References

Ae, N., Arihara, J., Okada, K., Yoshihara, T., Johansen, C. (1990) Phosphorus uptake by pigeon pea and its role in cropping systems of the Indian subcontinent. Science. 248: 477-480.

Buchanan, B.B., Gruissem, W., Jones, R.L. (2000) Biochemistry and Molecular Biology of Plants. American Society of Plant Physiologists, Rockville, MD, USA.

Gardner, W.K., Barber, D.A., Parbery, D.G. (1983) The acquisition of phosphorous by *Lupinus albus* L. Plant Soil. 70: 107-124.

Hiradate, S., Inoue, K. (1996) Determination of mugineic acid, 2′-deoxymugineic acid, 3-hydroxymugineic acid, and their iron complexes by ion-pair HPLC and colorimetric procedures. Soil Sci. Plant Nutr. 42: 659-665.

Hiradate, S., Inoue, K. (1998a) Dissolution of iron from iron (hydr)oxides by mugineic acid. Soil Sci. Plant Nutr. 44: 305-313.

Hiradate, S., Inoue, K. (1998b) Interaction of mugineic acid with iron (hydr)oxides: sulfate and phosphate influences. Soil Sci. Soc. Am. J. 62: 159-165.

Hiradate, S., Inoue, K. (1999) Interactions of mugineic acid with allophane, imogolite, montmorillonite, and gibbsite. *In* Effect of Mineral-Organic-Microorganism Interactions on Soil and Freshwater Environments. J. Berthelin, *et al* (eds) Kluwer Academic/Plenum Publishers, NY, USA. pp. 89-96.

Hiradate, S., Inoue, K. (2000) Dissolution of iron by mugineic acid from soils and comparison with DTPA soil test. Soil Sci. Plant Nutr. 46: 673-681.

Inoue, K., Hiradate, S., Takagi, S. (1993) Interaction of mugineic acid with synthetically produced iron oxides. Soil Sci. Soc. Am. J. 57. 1254-1260.

Kawai, S., Takagi, S., Sato, Y. (1988) Mugineic acid-family phytosiderophores in root-secretions of barley, corn and sorghum varieties. J. Plant Nutr. 11: 633-642.

Lindsay, W.L. (1988) Solubility and redox equilibria of iron compounds in soils. *In* Iron in Soil and Clay Minerals, J.W. Stucki *et al* (eds). D. Reidel Publishing Company, Dordrecht, Holland. pp. 37-62.

Lindsay, W.L., Norvell, W.A. (1969) Equilibrium relationships of Zn^{2+}, Fe^{3+}, Ca^{2+}, and H^{+} with EDTA and DTPA in soils. Soil Sci. Soc. Am. J. 33: 62-68.

Lindsay, W.L., Schwab, A.P. (1982) The chemistry of iron in soils and its availability to plants. J. Plant Nutr. 5: 821-840.

Loeppert, R.H. and Inskeep, W.P. (1996) Iron. *In* Methods of Soil Analysis. Part 3. Chemical Methods. D.L. Sparks (ed). Soil Sci. Soc. Am, Madison, WI, USA. pp. 639-664.

Otani, T., Ae, N., Tanaka, H. (1996) Phosphorous (P) uptake mechanisms of crops grown in soils with low P status: II significance of organic acids in root exudates of pigeonpea. Soil Sci. Plant Nutr. 42: 553-560.

Sparks, D.L. (2003) Environmental Soil Chemistry. 2nd ed Academic Press, Amsterdam, The Netherlands.

Schwertmann, U. (1964) Differentzierung der Eisenoxide des Bodens durch Extraktion mit Ammoniumoxalat-Loesung. Z. Pflanzenernaehr. Dueng. Bodenkd. 105: 194-202.

Schwertmann, U., Taylor, R.M. (1989) Iron oxides. *In* Minerals in Soil Environments. 2nd ed. J. B. Dixon and S.B. Weed. (eds) Soil Sci. Soc. Am., Madison, WI, USA. pp. 379-438.

Sugiura, Y., Tanaka, H., Mino, Y., Ishida, T., Ota, N., Inoue, M. (1981) Structure, properties and transport mechanism of iron (III) complex of mugineic acid, a possible phytosiderophore. J. Am. Chem. Soc. 103: 6979-6982.

Taiz, L., Zeiger, E. (1998) Plant Physiology. 2nd ed, Sinauer Association, MA, USA.

Takagi, S. (1976) Naturally occurring iron-chelating compounds in oat- and rice-root washings: 1. activity measurement and preliminary characterization. Soil Sci. Plant Nutr. 22: 423-433.

Takagi, S. (1993) Production of phytosiderophores. *In* Iron Chelation in Plants and Soil Microorganisms. L.L. Barton, B.C. Hemming (eds). Academic Press, CA, USA. pp. 111-131.

Wallace, A., Lunt, O.R. (1960) Iron chlorosis in horticultural plants: a review. Proc. Am. Soc. Hort. Sci. 75: 819-841.

Watanabe, S., Wada, H. (1989) Mugineic acid-decomposing bacteria isolated from rhizoplane of iron-deficient barley. Jpn. J. Soil Sci. Plant Nutr. 60: 413-417 (in Japanese with English summary).

8

Chemical and Biological Analysis of Novel Allelopathic Substances, Lepidimoide and Lepidimoic Acid

Kosumi Yamada[1], Kensuke Miyamoto[2], Nobuharu Goto[3], Hisashi Kato-Noguchi[4], Seiji Kosemura[5], Shosuke Yamamura[5] and Koji Hasegawa[1]

[1]*Graduate School of Life and Environmental Sciences, University of Tsukuba, Tsukuba, Ibaraki 305-8572, Japan*
[2]*Department of Biological Science, Graduate School of Science, Osaka Prefecture University, Sakai, Osaka 599-8531, Japan*
[3]*Department of Biology, Miyagi University of Education, Sendai, Miyagi 980-0845, Japan*
[4]*Department of Biochemistry and Food Science, Faculty of Agriculture, Kagawa University, Miki, Kagawa 760-0795, Japan*
[5]*Department of Chemistry, Keio University, Yokohama, Kanagawa 223-8522, Japan*

It was recently reported that a novel allelopathic substance, which has potent stimulating activity for the growth of other plant species was isolated from the exudates of germinating cress seeds. The substance was identified and its absolute configuration was determined to be 2-*O*-[(1*S*,2*R*,3*S*)-4-deoxy-enopyranosyluronic acid sodium salt]-L-rhamnopyranose (lepidimoide, LM), and the structure-activity relationship of LM was also studied. LM has shown multiple physiological functions for the growth and development of plants. It promotes leaf development, frequency of flowering and seed production, as well as shoot growth in Arabidopsis plants. It inhibits the loss of total chlorophyll (Chl) in excised oat leaf segments during leaf senescence, and delays abscission in bean petiole explants. LM also promotes Chl accumulation by affecting the level of 5-aminolevulinic acid (ALA) in sunflower seedlings. In addition, the free carboxylic acid form of LM (lepidimoic acid, LMA) affects glycolytic metabolism in Amaranthus seedlings by affecting the activity of pyrophosphate-dependent phosphofructokinase (PFP) due to increasing levels of fructose 2,6-bisphosphate. These observations suggest that LM and LMA may play the role of a natural hormone-like substance that controls various aspects of physiological growth and development in plants.

Keywords: *allelopathy, lepidimoic acid, lepidimoide, oligosaccharide, phytohormone, plant growth promotion*

INTRODUCTION

The term "allelopathy" is defined as any direct or indirect detrimental or beneficial biochemical interaction among all classes of plants, including microorganisms. When seeds or seedlings of various plant species were cultured together with cress (*Lepidium sativum* L.) seeds in a Petri dish, the shoot growth of the plants was significantly promoted. This phenomenon suggests that the germinating cress seeds secreted growth-promoting substance(s) as an allelopathic factor to their environment. As a candidate for this allelopathic factor, a potent novel growth-promoting substance was isolated from the mucilage of germinating cress seeds, and its structure was identified (Hasegawa *et al.*, 1992). This substance was named "lepidimoide (LM)", the botanical name for cress. LM could be synthesized from D-glucose and α-L-rhamnose (Kosemura *et al.*, 1993), and its absolute configuration was unambiguously determined to be 2-*O*-[(1*S*,2*R*,3*S*)-4-deoxy-enopyranosyluronic acid sodium salt]-L-rhamnopyranose (Fig. 1). LM-like activity was found in the exudates from the seeds of various plant species (Hasegawa *et al.*, 1993). The occurrence of LM was also determined using a HPLC in a variety of weeds and crops, although its amounts did not differ greatly among genera or families (Yamada *et al.*, 1995). In addition, the free carboxylic acid form of LM (lepidimoic acid, LMA) was also isolated from the exudate of germinating Arabidopsis (*Arabidopsis thaliana* (L.) Heynh.) seeds (Fig. 1; Yokotani-Tomita *et al.*, 1998). On the other hand, the mode of exudation of LM from the germinating seeds was also studied. LM, already present in dry seeds, as well as those newly produced in the kernels following imbibition, was released into the environment in relation to seed germination (Yamada *et al.*, 1997). On some preliminary aspects of its biological activities, LM has been reported to promote shoot growth significantly, but inhibits root growth weakly in the seedlings of various plant species, especially of cockscomb (*Cerosia argentea* var. *cristata* (L.) Kuntze). These findings, together with the wide distribution of LM(s) in the plant kingdom, suggest that it is possible to use LM(s) to control various physiological phenomena of plants as a new type of endogenous plant growth regulator, like phytohormones.

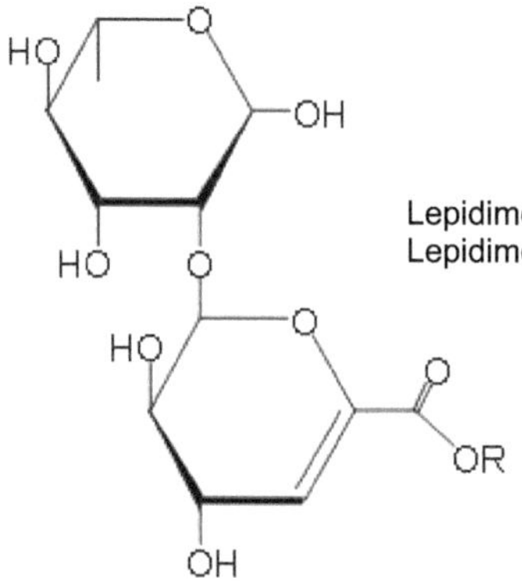

Lepidimoide (**1**: R = Na)
Lepidimoic acid (**2**: R = H)

Fig. 1 Structures of lepidimoide (LM) and lepidimoic acid (LMA)

In this chapter, we concentrate on some recent advances in our understanding of the structure-activity relationship and physiological functions of LM and LMA in several developmental stages of plants.

STRUCTURE-ACTIVITY RELATIONSHIP

To study the structural requirement of LM(s), the structure-activity relationship of LM, its analogues and some sugars were studied using the cockscomb hypocotyl elongation test (Figs. 1-3; Yamada *et al.*, 1996). The most active compound was compound 4, which was composed of rhamnose and glucose. It exhibited about 200% or greater promotion at 3×10^{-3} M. However, when the R_1 and R_2 were changed to acetyl group and benzyl group, respectively, the activity was completely lost. LMA (2), which is the free carboxylic acid form of LM (1), showed as high growth-promoting activity as 1. The methyl ester (14) of LM exhibited no activity. However, the acetylation of compound 13 revealed activity. On the other hand, the hydroxylated compound (8) showed lower activity by about 10% compared with 1. The carboxylic acid free form (7) of compound 8 showed as low activity as compound 8 and its methylated compound (6) decreased the activity, but the decreased activity was not recovered by acetylation. The deoxy forms (9-12) of compounds 5 to 8 showed higher activity than the actual compounds, except the methylated compound which showed no activity, although the deoxy forms showed lower activities than lepidimoide, respectively. The activities of the monosaccharides were also studied. 1-Methoxy-4-deoxyuronic acid (17) and its sodium form (18) showed about 30% lower activity than that of 1.

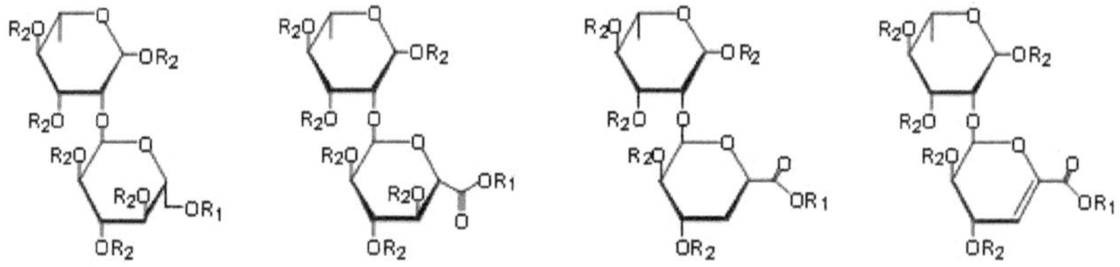

3: R_1 = Ac
R_2 = Bn
4: R_1 = R_2 = H

5: R_1 = Me
R_2 = Ac
6: R_1 = Me
R_2 = H
7: R_1 = R_2 = H
8: R_1 = Na
R_2 = H

9: R_1 = Me
R_2 = Ac
10: R_1 = Me
R_2 = H
11: R_1 = R_2 = H
12: R_1 = Na
R_2 = H

13: R_1 = Me
R_2 = Ac
14: R_1 = Me
R_2 = H

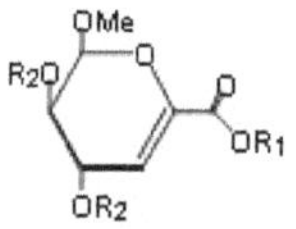

15: R_1 = Me
R_2 = Ac
16: R_1 = Me
R_2 = H
17: R_1 = R_2 = H
18: R_1 = Na
R_2 = H

19: α-D-Galacturonic acid
20: L-(+)-Rhamnose
21: D-(+)-Glucose
22: Saccharose
23: 19 + 20

Bn = Benzyl group
Ac = Acetyl group
Me = Methyl group

Fig. 2 Structures of LMs analogues applied to cockscomb

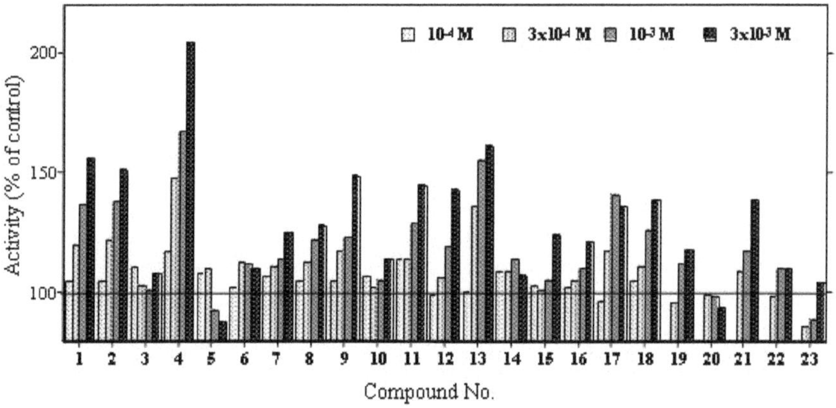

Fig. 3 Promoting activities of LMs, their analogues and some sugars on the growth of cockscomb hypocotyls

The methylated form (16) and the acetylated form (15) of compound 17 had minimal activity. In addition, the activities of *α*-D-galacturonic acid and L-(+)-rhamnose, which are component sugars of 1, were also studied.

α-D-Galacturonic acid (19), L-(+)-rhamnose (20) and their mixture (23) exhibited weak or no activity, respectively. These data suggest that the structural requirements of LM for plant growth-promoting activities, obtained by the cockscomb hypocotyl test, are the uronic acid derivative bearing an α, β-unsaturated carboxylate bonded to rhamnose via an α-glucoside linkage, and a double bond in the C-4,5 position in the uronic acid. Furthermore, the sodium form or free carboxylic acid form does not seem to be a structural requirement for exhibiting the promoting activity.

ARABIDOPSIS BIOASSAY

Arabidopsis has a short life cycle among species of higher plants, and has had several major advantages as a model plant in recent molecular and physiological studies. The effect of LM on some developmental stages has been shown in several wild types of Arabidopsis (Goto *et al.*, 1995). Among the ecotypes of wild type studied, Columbia (Col) and Wassilewskija (WS) were generally sensitive to lepidimoide. LM increased hypocotyl lengths, the number of leaves, height, mass (fresh and dry weight at maturation), and seed production in several wild types of Arabidopsis. Here, we describe some of the effects of LM on various events throughout the developmental process in Col and WS. The effect on form at the rosette stage in Col was investigated. The development of leaves was accelerated during the 30 d of the culture period (Fig. 4). The leaf was markedly enlarged, in which both the length and width of the leaf blade, as well as the length of the petiole, increased (Goto *et al.*, 1995). In addition, the plant form at the rosette developmental stage, with and without LM, was

Col (Control)

Col (+LM)

Fig. 4 LM activity in the Col ecotype of Arabidopsis. Upper: Control, Lower: LM (100 m/L) was applied every 2 wk. Picture was taken on day 30 after transfer to the growth chamber.

investigated. LM exhibited a promoting effect on the frequency of flowering (Fig. 5). It also hastened flowering without decreasing the number of rosette leaves, suggesting that the substance accelerates the development of those leaves within a shorter period. LM also increased the number of flowers resulting in an increment of pod and seed production (Fig. 6). The increment was about 1,200 seeds/plant with 100 mg/L LM treatment, whereas only about 700 seeds/plant in the control. In addition, LM increased plant height in a dose-dependent manner. Both the fresh and dry weight also increased with lepidimoide treatment, but was larger at 10 mg/L rather than at the higher concentration of 100 mg/L.

WS (Control) WS (+ LM)

Fig. 5 LM activity in the WS ecotype of Arabidopsis. Left: Control, Right: LM (100 mg/L) was applied every 2 wk. Picture was taken on day 30 after transfer to the growth chamber.

From the present study, we can conclude that LM promotes the growth of the shoot and leaf area, and increases the fresh and dry weights of plants, although the effect is greatly different between ecotypes. At the same time, it hastens the life cycle of Arabidopsis through the acceleration of leaf and flower development. Activation by LM on reproductive growth, such as seed production, was also strongly indicated.

EFFECT ON SENESCENCE

Senescence may be associated with the complex deterioration processes that naturally terminate the functional life of plants. In this process, an important contributing factor is the decrease in total Chl content and/or the formation of abscission in the leaves. The effects of LM and LMA on Chl preservation in oat (*Avena sativa* L.) leaf segments in the dark was studied (Miyamoto *et al.*, 1997a). In this bioassay system, the Chl content of control leaf segments significantly decreased soon after detachment. LMA greatly inhibited the loss of total Chl at concentrations higher than 10 μM,

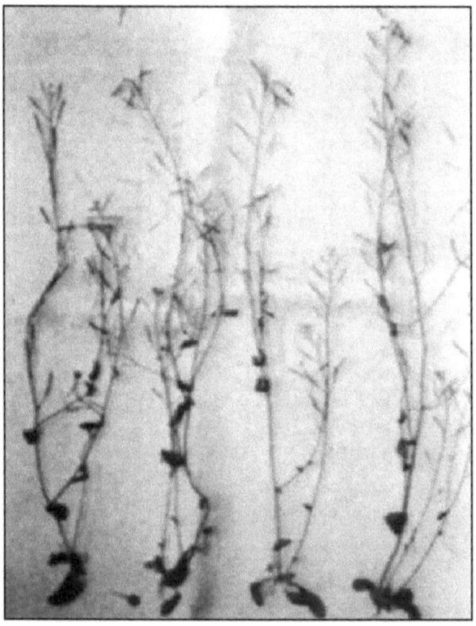

Fig. 6 LM activity in the WS ecotype of Arabidopsis. LM was applied every 2 wk. Picture was taken on day 50 after transfer to the growth chamber.

tested 24 h after the incubation (Fig. 7). Most of the Chl (about 90% of the initial level) treated with LMA was preserved, whereas in the control leaf segment about 35% of the initial Chl was degraded. Forty eight hours after the incubation, the inhibitory effect of LMA on Chl degradation was reduced. In general, it has been well known that cytokinins are the most active compounds for Chl preservation in most plants (Beevers, 1976). The loss of total Chl was also retarded in the presence of cytokinin (kinetin) in this assay system. LMA synergistically preserved the Chl content with kinetin (Miyamoto *et al.*, 1997a). Furthermore, LM preserved the Chl content in the presence of kinetin to a similar extent as the LMA. These results suggest that both LM and LMA have inhibitory effects on leaf senescence, which is characterized by the loss of total Chl in the leaves in cooperation with cytokinin.

EFFECT ON ABSCISSION

Abscission is considered to be the last phenomenon during the senescence period in plants. When the petiole explant is excised from an intact plant, cells at the base of the leaf (pulvinus) immediately begin to form an abscission layer (Osborne, 1973). The effects of LM and LMA on the formation of abscission was studied in bean (*Phaseolus vulgaris* L.) petiole

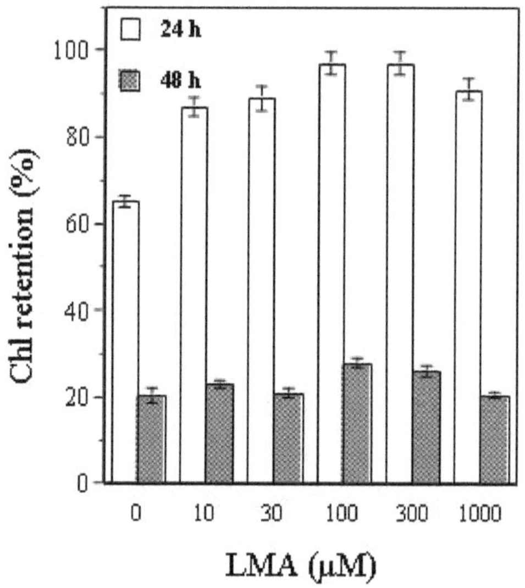

Fig. 7 Effect of LMA on chlorophyll (Chl) retention in oat leaf segments in the dark. Oat leaf segments were incubated with or without LMA for 24 h (open columns) and 48 h (solid columns).

explants, with the junction between the petiole and the pulvinus in the primary leaves exposed to light (Miyamoto *et al.*, 1997b). Bean petiole explants were incubated with various concentrations of LMA for 48 and 60 h in light conditions (Fig. 8). LMA at concentrations higher than 1 μM significantly affected the inhibition of abscission. Sixty hours after the incubation, about 80% of the explants incubated with water formed abscission, whereas only 30% of the explants treated with 1 mM of lepidimoic acid abscised. Moreover, LM also exhibited a similar inhibitory effect in the same bioassay system (Miyamoto *et al.*, 1997b). These results indicate that LM and LMA have an inhibitory effect on leaf senescence in terms of the formation of abscission.

EFFECT ON CHL BIOSYNTHESIS

In our previous study, it was shown that lepidimoide markedly promoted greening in the cotyledon of sunflower (*Helianthus annuus* L.) seedlings exposed to light. The effect of LM on light-induced Chl accumulation was studied (Yamada *et al.*, 1998). LM at concentrations higher than 100 μM significantly promoted light-induced Chl accumulation (Fig. 9A). It is well known that 5-aminolevulinic acid (ALA) is a rate-limiting step in the formation of Chl (Beale and Weinstein, 1990). LM, at concentrations higher

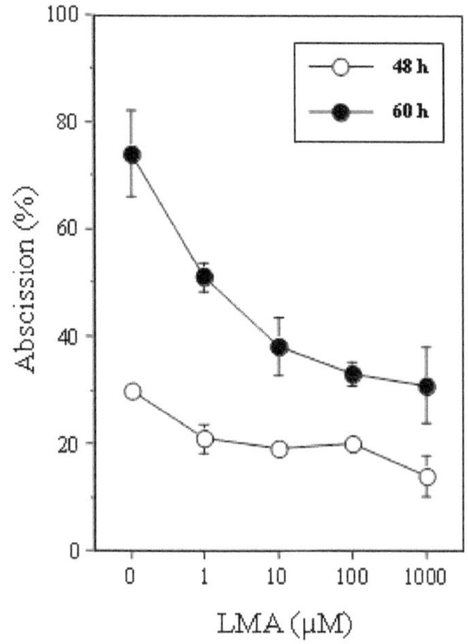

Fig. 8 Effect of LMA on the abscission of bean petiole explants exposed to light. Bean petiole explants were incubated with or without LMA for 48 h (open circles) and 60 h (solid circles).

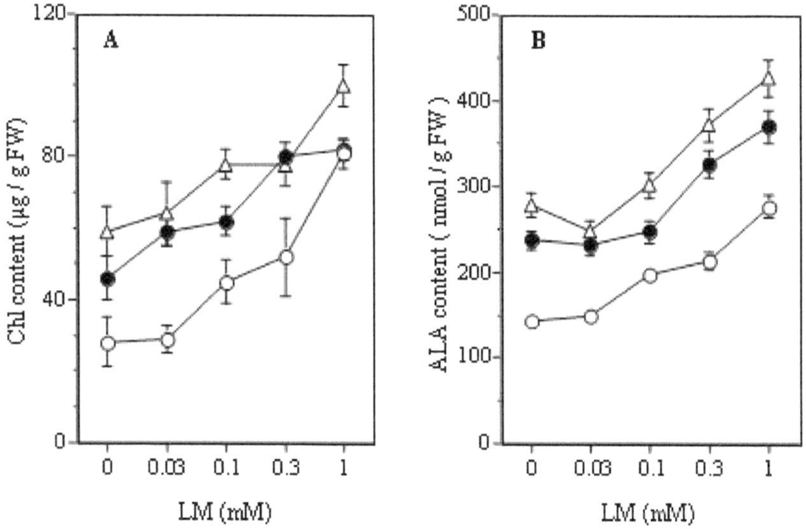

Fig. 9 Effect of LM on Chl (A) and ALA (B) accumulation in the cotyledons of sunflower seedling. Four-day-old etiolated sunflower seedlings were pre-treated with LM and/or benzyladenine (open circles: control; solid circles: 30 μM, open triangles: 100 μM) for 6 h in the dark and then incubated for 6 h in the light.

than 100 μM, significantly promoted light-induced ALA accumulation in the cotyledons of sunflower seedlings (Fig. 9B). These results suggest that the LM-induced enhancement of Chl accumulation is related to the enhancement of ALA synthesis. On the other hand, cytokinin is known to stimulate Chl accumulation in various plants (Davies, 1987). Benzyladenine at concentrations of 30 μM and 100 μM enhanced the levels of both Chl and ALA in the cotyledons of sunflower seedlings (Fig. 9). LM acted somewhat additively with benzyladenine, but not synergistically, on the increase of Chl and ALA, as with LM alone. Moreover, LMA also exhibited a promoting activity on both Chl and ALA levels in relation to benzyladenine in an additive manner, as expected (Yamada *et al.*, 1998). These results suggest that LM, in cooperation with cytokinins, causes light-induced Chl accumulation by affecting the level of ALA.

EFFECT ON GLYCOLYTIC METABOLISM

As described above, LM(s) promoted shoot growth among several plant species. This growth promotion may be due to the acceleration of respiratory metabolism to produce metabolic energy and biosynthetic precursors. Upon the regulation of sugar metabolism between glycolysis and gluconeogenesis in plants, the interaction of pyrophosphate-dependent phosphofructokinase (PFP) and fructose 2,6-bisphosphate (Fru-2,6-P_2), which is an important cytosolic signal metabolite, is considered to be an important mechanism. The effect of LMA on the level of fructose Fru-2,6-P_2, which can activate PFP, and on glycolytic metabolism, was studied in Amaranthus (*Amaranthus caudatus* L.) seedlings (Kato-Noguchi *et al.*, 2001). LMA at concentrations higher than 0.3 mM increased the Fru-2,6-P_2 level in Amaranthus seedlings (Fig. 10A). Moreover, it affected the levels of glycolytic intermediates in the seedlings. In general, PFP activity was measured in the glycolytic direction from Fru-6-P to Fru-1,6-P_2 with various amounts of exogenously supplied Fru-2,6-P_2. The effect of Fru-2,6-P_2 on PFP activity in a crude enzyme solution of Amaranthus seedlings was also studied (Fig. 10B). Fru-2,6-P_2 markedly activated PFP activity in a dose-response manner as in previous reports (Van Schaftingen *et al.*, 1982; Kesy and Kowalczyk, 1987). The increased activity of PFP may have been due to the increased level of Fru-2,6-P_2, suggesting a promotive effect of LMA on glycolytic metabolism in Amaranthus seedlings.

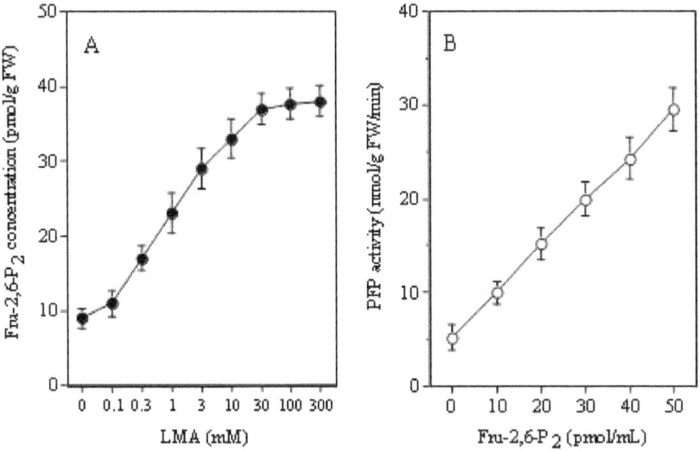

Fig. 10 Effect of LM on glycolytic metabolism in Amaranthus seedlings. A) Effect of LMA on Fru-2,6-P_2 levels. B) Effect of Fru-2,6-P_2 on PFP activity in the extracts of Amaranthus seedlings.

CONCLUSIONS AND PROSPECTS

During the past decade, many significant advances concerning the biological activities of LMs have been made, including the results described here. It seems likely that LMs might be a new type of plant growth regulator with multiple physiological functions in the regulation of growth and development of plants. Despite this progress, however, the biosynthetic mechanism of LMs and the biological meaning for its exudation into the environment remain to be determined. To clarify whether newly produced LM(s) from the germinating seed is synthesized from sugars, such as rhamnose and glucose, or from compounds produced by the hydrolysis of polysaccharides, such as rhamnogalacturonan I, further studies are needed. It is of vital interest to determine the origin and the enzymes concerning the production of LM(s). Moreover, new approaches may be needed to elucidate the mode of action of LM(s), including the discovery of new biological effects alone and in collaboration with other phytohormones, and more thorough physiological descriptions of its known effects. Studies with Arabidopsis mutants may lead to breakthroughs in the studies of transport, binding, action and turnover of LM(s) in plants, since this plant is now the most important tool in genetic and biochemical studies. We look forward to the application of LM(s) in agriculture in the near future.

Acknowledgments

We thank Dr. Kaori Yokotani-Tomita (University of Tsukuba, Japan) for many helpful discussions, and Dr. Katsutoshi Hirose (Kobe Natural Products and Chemicals Co. Ltd., Japan) for many useful discussions and kindly providing the LMs.

References

Beale, S.I., Weinstein, J.D. (1990) Tetrapyrrole metabolism in photosynthetic organisms. *In* Biosynthesis of Heme and Chlorophylls. H.A. Dailey. (ed), McGraw-Hill, NY, USA. pp. 287-391.

Beevers, L. (1976) Senescence. *In* Plant Biochemistry 3rd ed. J. Bonner and J.E. Varner, (eds), Academic Press, NY, USA. pp. 775-790.

Davies, P.J. (1987) The plant hormones: their nature, occurrence and functions. *In* Plant Hormones and Their Role in Plant Growth and Development. P.J. Davies. (ed), Martinus Niihoff, Dordrecht, The Netherlands. pp. 1-11.

Goto, N., Sando, S., Sato, Y., Hasegawa, K. (1995) Effects of lepidimoide on growth and development of *Arabidopsis thaliana.* Weed Res., Japan. 40: 87-94.

Hasegawa, K., Mizutani, J., Kosemura, S., Yamamura, S. (1992) Isolation and identification of lepidimoide, a new allelopathic substance from mucilage of germinated cress seeds. Plant Physiol. 100: 1059-1061.

Hasegawa, K., Amano, M., Asakawa, C., Kakuta, H., Mizutani, J. (1993) Elongation-promoting activity of *Amaranthus* hypocotyls in mucilage of germinated seeds. Weed Res., Japan (in Japanese). 34: 109-111.

Kato-Noguchi, H., Yamada, K., Hasegawa, K. (2001) Lepidimoic acid increases fructose 2,6-bisphosphate in *Amaranthus* seedlings. Phytochemistry. 56: 499-503.

Kesy, J., Kowalczyk, S. (1987) Comparison of the activity of the enzymes participating in fructose-6-phosphate/fructose-1,6-bisphosphate interconversion during early stages of mung bean (*Phaseolus coccineus* L.) development. Measurement of pyrophosphate and fructose-2,6-bisphosphate content. J. Plant Physiol. 127: 411-421.

Kosemura, S., Yamamura, S., Kakuta, H., Mizutani, J., Hasegawa, K. (1993) Synthesis and absolute configuration of lepidimoide, a high potent allelopathic substance from mucilage of germinated cress seeds. Tetrahedron Lett. 34: 2653-2656.

Miyamoto, K., Ueda, J., Yamada, K., Kosemura, S., Yamamura, S., Hasegawa, K. (1997a) Inhibitory effect of lepidimoide on senescence in *Avena* leaf segments. J. Plant Physiol. 150: 133-136.

Miyamoto, K., Ueda, J., Yamada, K., Kosemura, S., Yamamura, S., Hasegawa, K. (1997b) Inhibition of abscission of bean petiole explants by lepidimoide. J. Plant Growth Regul. 16: 7-9.

Osborne, D.J. (1973) Internal factors regulating abscission. *In* Shedding of Plant Parts. T.T. Kozlowski (ed). Academic Press, NY, USA. pp. 125-147.

Van Schaftingen, E., Lederer, B., Bartrons, R., Hers, H.G. (1982) A kinetic study of pyrophosphate:fructose-6-phosphate phosphotransferase from potato tubers. Application to a microassay of fructose 2,6-bisphosphate. Eur. J. Biochem. 129: 191-195.

Yamada, K., Anai, T., Hasegawa, K. (1995) Lepidimoide, an allelopathic substance in the exudates from germinated seeds. Phytochemistry. 39: 1031-1032.

Yamada, K., Anai, T., Kosemura, S., Yamamura, S., Hasegawa, K. (1996) Structure-activity relationship of lepidimoide and its analogues. Phytochemistry. 41: 671-673.

Yamada, K., Kosemura, S., Yamamura, S., Hasegawa, K. (1997) Exudation of an allelopathic substance lepidimoide from seeds during germination. Plant Growth Regul. 22: 189-192.

Yamada, K., Matsumoto, H., Ishizuka, K., Miyamoto, K., Kosemura, S., Yamamura, S., Hasegawa, K. (1998) Lepidimoide promotes light-induced chlorophyll accumulation in cotyledons of sunflower seedlings. J. Plant Growth Regul. 17: 215-219.

Yokotani-Tomita, K., Goto, N., Kosemura, S., Yamamura, S., Hasegawa, K. (1998) Growth-promoting allelopathic substance exuded from germinating *Arabidopsis thaliana* seeds. Phytochemistry. 47: 1-2.

SECTION 3

Allelopathy in Potential Invasive Weeds

9

Allelopathic Activity of White Rocket [*Diplotaxis erucoides* (L.) DC.]

Jamal R. Qasem
Plant Protection Department, Faculty of Agriculture, University of Jordan, Amman, Jordan

White rocket [*Diplotaxis eruciodes* (L.) DC.] is an annual widely spread weed species in field crops, fruit trees and in rangelands in Jordan. This study was carried out to investigate any possible allelopathic activity of aqueous extracts, volatiles, foliage leachates, root exudates and shoot residues of this weed on cabbage, cucumber, tomato and wheat under laboratory and glasshouse conditions. Aqueous shoot extract of *D. erucoides* was highly phytotoxic and significantly reduced seed germination and inhibited seedling growth of the tested crops. The degree of inhibition was extract concentration-dependent. Roots were more sensitive to low extract concentration than shoots, with cucumber being most sensitive and severely affected by volatiles and foliage leachates of *D. erucoides*. Root exudates of *D. erucoides* reduced seed germination of cabbage and tomato and growth of cabbage seedlings. In pot experiments, surface-placed shoot residues of *D. erucoides* reduced seed germination of the three vegetable crops and wheat, with cucumber being the most affected. However, its soil incorporated residues reduced growth of all crops especially when used at high rates. Decayed residues at 32 g kg^{-1} were also toxic to cucumber, tomato and wheat, while its foliage leachates or root exudates added or released into the soil mixture reduced seedling growth of the cucumber plants that followed. Results showed that *D. erucoides* is of a high allelopathic effect against crops tested; with cucumber and wheat being the most affected.

Keywords: *allelopathy, cabbage, cucumber, Diplotaxis erucoides, tomato, white rocket, wheat*

INTRODUCTION

Many annual and perennial weeds have allelopathic activities and can affect crop survival and productivity (Putnam and Duke, 1978; Rice, 1979; Qasem, 1994). Different studies have shown that water extracts of weeds can inhibit or reduce seed germination (Barens and Putnam, 1986; Qasem, 1993; Sahid and Sugau, 1993) and seedling growth and development of different crops (Drost and Doll, 1980; Mason-Sedun et al., 1986; Sahid and Sugau, 1993).

Allelochemicals produced by plants may be released into the surrounding environment in sufficient amounts with enough persistence to affect neighboring and succession species (Akram *et al.*, 1990). Different studies showed that some allelopathic agents are volatiles, emanated from different plant parts (Oleszek, 1987; Bradow and Connick, 1988), others indicated that they exuded from roots to the root zone and interfere in root growth and functions (Qasem and Hill, 1989b; Rovira, 1969), or inhibit seed germination (Rovira, 1969). Plant residues and their decomposition products are also implicated in virtually all biochemical processes (Patrick et al., 1963; Bhowmik and Doll, 1984; Qasem, 2002). Some allelochemicals are water-soluble leached from foliage parts by rain, mist, dew, or fog drip (Lovett and Lynch, 1979; Qasem, 1994) leading to colonies that several perennial weeds form in nature (Rice, 1984). However the inhibitory materials may be autotoxic or heterotoxic (Kumari and Kohli, 1987), some can be highly selective (Stachon and Zimdahl, 1980; Sahid and Sugau, 1993) and their effects are concentration dependent (Qasem, 1993). *Diplotaxis erucoides* is an annual weed which belongs to Cruciferae (Plate 1). It is widely spread in cultivated and fallow lands in Jordan. It invades field crops as well as orchards (Plate 2).

Plate 1 White rocket (*Diplotaxis erucoides* L.)

Plate 2 White rocket (*Diplotaxis erucoides* L.) invading olive fields choking out other weed species

The weed has the potential to produce a huge number of seeds that enable this species to occupy a wide area and to become difficult to control. It persists and tolerates semi-arid conditions, and significantly spreads in the area mainly allocated for wheat and barley production (Plate 3).

D. erucoides tends to dominate other species in these habitats and if left uncontrolled, it soon colonizes a large area, pushing out other plants that are present. It has strong competition and tolerance to low soil moisture and poor soil conditions. There were no reports on its competition or allelopathic effects.

Plate 3 General view of white rocket (*Diplotaxis erucoides* L.) invasion of cultivated land by wheat and barley crops in Jordan

The objective of the present work was to investigate any possible allelopathic role of *D. erucoides* through shoot extracts, volatile materials, root exudates, foliage leachates and shoot residues, on germination and growth of the selected crop species.

MATERIALS AND METHODS

Laboratory Experiments

Three hundred g of fresh shoots of *D. erucoides* were washed with tap water, then with distilled water, and allowed to dry for 2 h. Shoots were then chopped, added to a liter of distilled water, and homogenized in a Waring blender for five min. at room temperature. The mixture was allowed to stand for 30 min. before the supernatant was filtrated through Whatman No. 1 filter paper. The filtrate (considered full strength concentration) was stored at 4°C until used. Seeds of all crops were bought from local markets. The following experiments were carried out:

Experiment 1. Effect of full-strength aqueous extracts

Twenty seeds of cabbage (*Brassica oleracea* L. var. *Capitata* cv. Pronzwik), cucumber (*Cucumis sativus* L. cv. Beithalpha), tomato (*Lycopersicon esculentum* Mill cv. Special Back) and wheat (*Triticum durum* L cv. Stork.) were placed separately on filter paper in each of the four Petri dishes (11 cm in diameter) per treatment. Ten ml of full strength fresh shoot extract of *D. erucoides* were added to each Petri dish. Dishes to which 10 ml of distilled water were added served as controls. Petridishes were incubated in the dark at 24°C for 2 wk. Data on germination at 48 h and at harvest, and on seedlings growth were recorded.

Experiment 2. Effect of extract concentrations

Twenty seeds of each crop were sown separately on a filter paper in each of the four Petri dishes. Ten ml of solution containing 0.5, 1, 2, 3, 4 and 5 ml of full strength extract diluted with distilled water were added to the dishes. For the control treatment, 10 ml of distilled water was added per Petri dish. Dishes were incubated as described above and data on germination and on seedling growth were recorded.

Experiment 3. Effect of root exudates

Ten cm diameter plastic pots were filled with 500 g of soil mixture (clay: sand: peat 3:1:1 of a pH 7.7) and sown with *D. erucoides* seeds. After emergence, seedlings were thinned to 10 per pot, irrigated with tap water when needed and left to grow for two mon before being harvested. At harvest, the soil was loosened, cleaned up from weed roots then mixed with an equal volume of distilled water and was thoroughly shaken for 2 h on a shaker. The mixture was passed through filter paper and immediately assayed for phytotoxicity. For the control treatment 500 g of weed-free soil was mixed with the same volume of distilled water and then similarly treated before use.

The effect of soil filtrate was studied by placing 20 seeds of cabbage (*Brassica oleracea* L. var. *Capitata* cv. Pronzwik), cucumber (*Cucumis sativus* L. cv. Beithalpha), or tomato (*Lycopersicon esculentum* Mill cv. Special Back) on a filter paper in each of the four Petri dishes at which 10 ml of the filtrate of weed species was added per Petri dish. In another treatment 10 ml of soil-weed-free filtrate was added per Petri dish and treated as a control.

Dishes were incubated for germination in the dark at 24°C for 2 wk, before the experiment was terminated. Data on germination at 48 h incubation and at harvest and on growth of different crops were recorded.

Experiment 4. Effect of foliage leachates

a. Effect of full strength leachate

   Twenty seeds of wheat were sown on a filter paper in each of the four Petri dishes. Ten ml of foliage leachates were added per Petri dish. In the control, 10 ml of distilled water were added per Petri dish and considered as a control. Dishes were incubated as described above and data on germination at 48 h incubation and at harvest and on seedling growth were recorded.

b. Effect of leachate concentration

   Ten ml of solution containing 1, 2, 3, 4, or 5 ml of full strength foliage leachates were added per Petri dish sown with wheat as before and the volume was completed to 10 ml by adding distilled water. For the control treatment, 10 ml of distilled water was added per Petri dish.

Dishes were incubated as described above and similar data on germination and seedlings growth were obtained.

Experiment 5. Effect of volatile materials

Ten seeds of each crop species tested were placed separately in each of the four sterilized (9-cm diameter) Petri dish lined with moistened filter paper. Thirty g of fresh, healthy and clean shoots of *D. erucoides* were placed at the bottom of a wide mouth 500-ml cup (11-cm diameter), using four cups per weed species. Uncovered Petri dishes (containing seeds) were placed over weed shoots inside the cups, which were then tightly closed to eliminate any air diffusion. For the control treatment, dishes sown with crop seeds were placed in similar cups, without any weed materials added. All cups were incubated, as in the previous experiments, for 2 wk. Data on germination and growth of different crops were recorded.

Glasshouse Experiments

Experiment 6. Effect of soil incorporated residues

Dried shoots (oven dried at 80°C for 48 h) of *D. erucoides* were ground to a fine powder, until they passed through 1.72 mm mesh, then were thoroughly mixed with a constant weight of 500 g of the soil mixture at a rate of 0, 1, 2, 4, 8 and 16 g kg^{-1}. The soil residue mixture was placed in plastic pots of 10 cm diameter. Pots were sown with 10 seeds of a single vegetable crop or 20 seeds of wheat, and then were irrigated with tap water when needed. In another treatment, a similar number of seeds of each crop was sown in each of the four pots filled with 500 g of the soil mixture containing no weed residue and treated as a control. Plants were grown in the glasshouse for 5 wk, before being harvested and data on germination, shoot length and root and shoot dry weights were recorded.

Experiment 7. Effect of surface placed residues

Plastic pots of 10-cm diameter filled with 500 g of the soil mixture were sown with 10 seeds of each vegetable crop, or 20 seeds of wheat. The ground dried shoot materials of *D. erucoides* were added at a rate of 16 g kg^{-1} soil on the soil surface. For the control treatment, seeds of each crop species were sown in the potted soil but without any weed residue added. Plants were grown in the glasshouse for 5 wk after emergence, and then similar data as in the above experiments were taken at harvest.

Experiment 8. Effect of root exudates

Pots of 10-cm diameter were filled with 500 g of soil mixture and planted with seeds of *D. erucoides*. After emergence, weed seedlings were thinned to 10 per pot, and then irrigated with tap water when necessary. The weed was grown for 2 mon, before being harvested from the above soil surface. The soil in each pot was loosened, weed roots and their parts were removed then all the pots were re-sown with seeds of the tested vegetable crops, using 10 seeds per pot. Fresh soil mixture (not used before) was sown by the same number of seeds of vegetable crops tested and regarded as controls. Irrigation with 100 ml of full strength Hoagland nutrient solution (Hewitt, 1966) was carried out twice a week and with tap water when needed. The experiment was terminated at 5 wk after emergence. Germination percentage was recorded at different intervals, and then at harvest, data on stem length and root and shoot dry weights were recorded.

Experiment 9. Effect of foliage leachates added to the soil

One kg of fresh shoots of *D. erucoides* was sprinkled with tap water, then with distilled water, to remove dust and soil particles. Shoots were immersed in 1 liter of distilled water for 5 min, and then leached water was collected and filtered through a Whatman No.1 filter paper. The filtrate was considered as full strength leachates and immediately tested.

Vegetable crops were sown in plastic pots (10-cm diameter) containing 500 g of the soil mixture. After emergence, seedlings were thinned to one per pot, then, 150 ml leachates were applied to each of the four pots grown by a single vegetable species and at the cotyledon stage. Pots irrigated only with tap water were included and considered as a control. After 4 wk, the experiment was harvested, when stem length and shoot and root dry weights of plants were determined.

Experiment 10. Effect of decayed residues

Ground dried shoots of *D. erucoides* were added and mixed thoroughly with the potted soil mixture at a rate of 32 g kg^{-1}. The soil shoot residue mixture was placed in pots of 10-cm diameter, and frequently irrigated with tap water for a month to allow natural decay of weed residues. The soil was then loosened, and each pot was sown with 10 seeds of a single vegetable crop or 20 seeds of wheat. There were four pots per crop. Pots filled with the same soil mixture, without any residue added and sown by similar number of seeds of each crop were included and considered as a control. All pots were irrigated with tap water when needed. The experiment continued for 5 wk, then plants were harvested from the above soil surface and their stem length, fresh and dry weights, and root dry weights were determined.

Statistics and data analysis

Treatments in all experiments were laid out in a randomized complete block design with four replicates. Data were taken on germination percentage, stem and total root lengths, and shoot and root dry weights after being oven dried at 80° C for 48 h. All data were statistically analyzed by ANOVA, and treatments means were compared using the least significant differences (LSD) at $p = 0.05$.

RESULTS

Laboratory Experiments

Experiment 1. Effect of full-strength aqueous extracts

Results on the effect of full strength extract on different crops are shown in Fig. 1. Germination and growth of all species were inhibited when 10 ml aqueous shoot extracts were added per Petri dish compared with the control. However, differences in the sensitivity of different species were

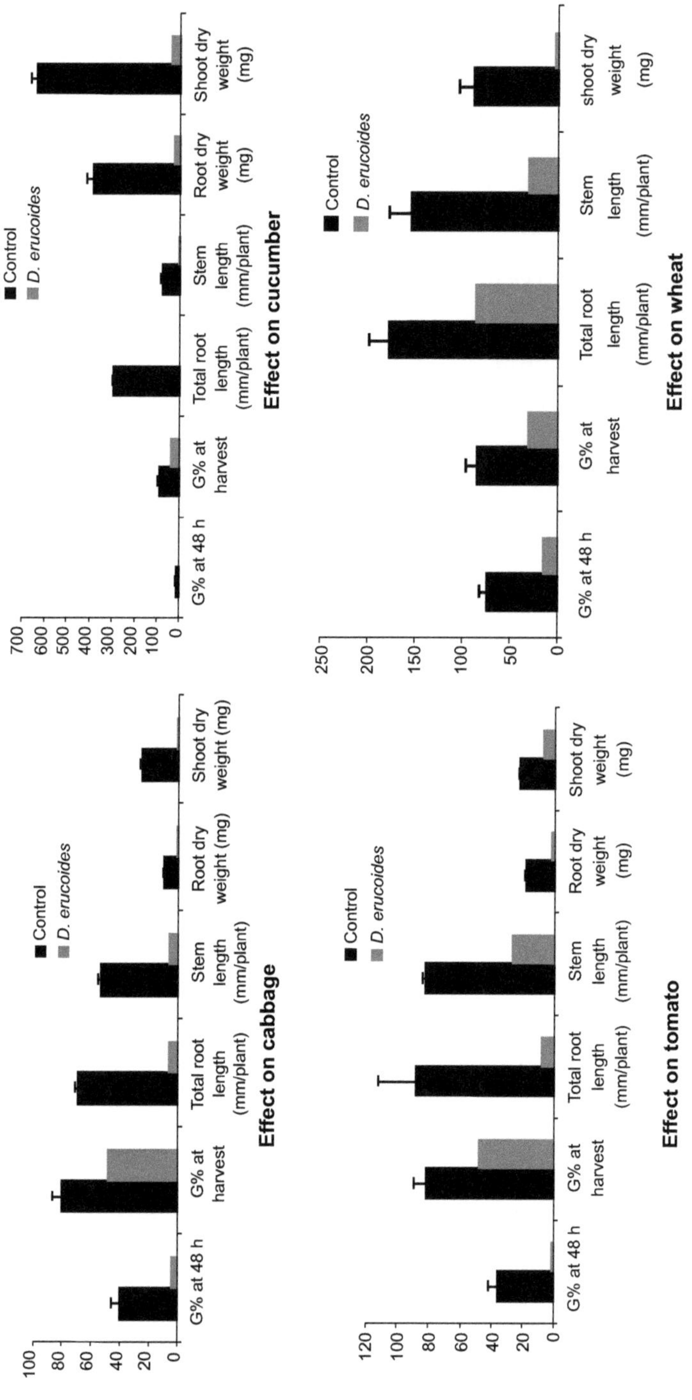

Fig. 1 Effect of full extract concentration of *D. erucoides* on germination and growth of different vegetable crops and wheat grown in Petri dishes.

evident. Cucumber and wheat were severely affected while cabbage and tomato were less so. Roots appeared more affected than shoots while extract effect was more pronounced on germination early after incubation.

Experiment 2. Effect of extract concentrations

The effect of extract on germination of all species was pronounced and increased with concentration (Table 1). Roots appeared greatly affected in terms of total length and dry weight with extract concentration and were clearly demonstrated on all species. Variations between species in their responses to increase in extract concentration are well demonstrated. Delay in seed germination of cabbage and cucumber was more pronounced compared with tomato and wheat, while germination and growth of both weed species were reduced with extract concentration.

Experiment 3. Effect of root exudates

The effect of root exudates added to Petri dish was less pronounced than any other treatment and on three vegetable crops. This treatment delayed germination after 48 h of incubation compared with the control and for all crops tested (Fig. 2).

Experiment 4. Effect of foliage leachates on wheat

The effect of full strength leachates on germination and growth of wheat is shown in Fig. 3. Foliage leachates of *D. erucoides* reduced germination, root length, stem length, and shoot and root dry weights of wheat compared with the control. Root length and dry weight were severely reduced with weed leachate and more than shoots. Leachate effect showed gradual decrease in germination and seedling growth with concentration and up to 5 ml per Petri dish.

However, shoots appeared less affected than roots in terms of length and dry weight (Table 2).

Experiment 5. Effect of volatile materials

Cucumber was the most affected vegetable crop with volatile materials of *D. erucoides* (Fig. 4). Germination, shoot length and root dry weight of cucumber and total root length of tomato were significantly reduced compared with the control. In contrast, cabbage was not affected by this treatment.

Glasshouse Experiments

Experiment 6. Effect of soil incorporated residues

Dried shoot residues of *D. erucoides* greatly reduced germination and growth of cucumber, tomato and wheat but not cabbage germination (Table 3).

Table 1 Effect of aqueous fresh shoot extract concentration of *D. erucoides* on germination, total root length, stem length, root dry weight and shoot dry weight of cabbage, cucumber, tomato and wheat grown in Petri dishes at 24°C for two wk.

Extract volume	*Cabbage*				*Cucumber*			
(ml/Petri dish)	*G%*	*Total root length (mm/plant)*	*Root dry weight (mg)*	*Shoot dry weight (mg)*	*G%*	*Total root length (mm/plant)*	*Root dry weight (mg)*	*Shoot dry weight (mg)*
Control (dH_2O)	85	46	9	25	94	280	68	280
0.5	85	48	11	28	90	270	65	252
1	72	44	10	27	90	275	61	255
2	75	40	9	20	88	260	58	252
3	72	36	9	18	88	255	68	252
4	71	30	8	17	70	235	66	235
5	65	28	6	16	80	230	74	280
LSD (P = 0.05)	7	5	1	2	9	20	8	50

Extract volume	*Tomato*				*Wheat*			
(ml/Petri dish)	*G%*	*Total root length (mm/plant)*	*Root dry weight (mg)*	*Shoot dry weight (mg)*	*G%*	*Total root length (mm/plant)*	*Root dry weight (mg)*	*Shoot dry weight (mg)*
Control (dH_2O)	90	95	12	22	80	80.1	142	200
0.5	82	88	19	20	82	79.6	140	220
1	88	86	12	19	80	76.4	133	215
2	84	65	8	21	70	61.1	106	213
3	75	55	10	20	75	52.5	91	193
4	86	45	12	19	65	12.0	85	194
5	80	40	7	17	65	5.7	73	190
LSD (P = 0.05)	12	11	4	1	9	11.3	14	22

G% = Germination percentage

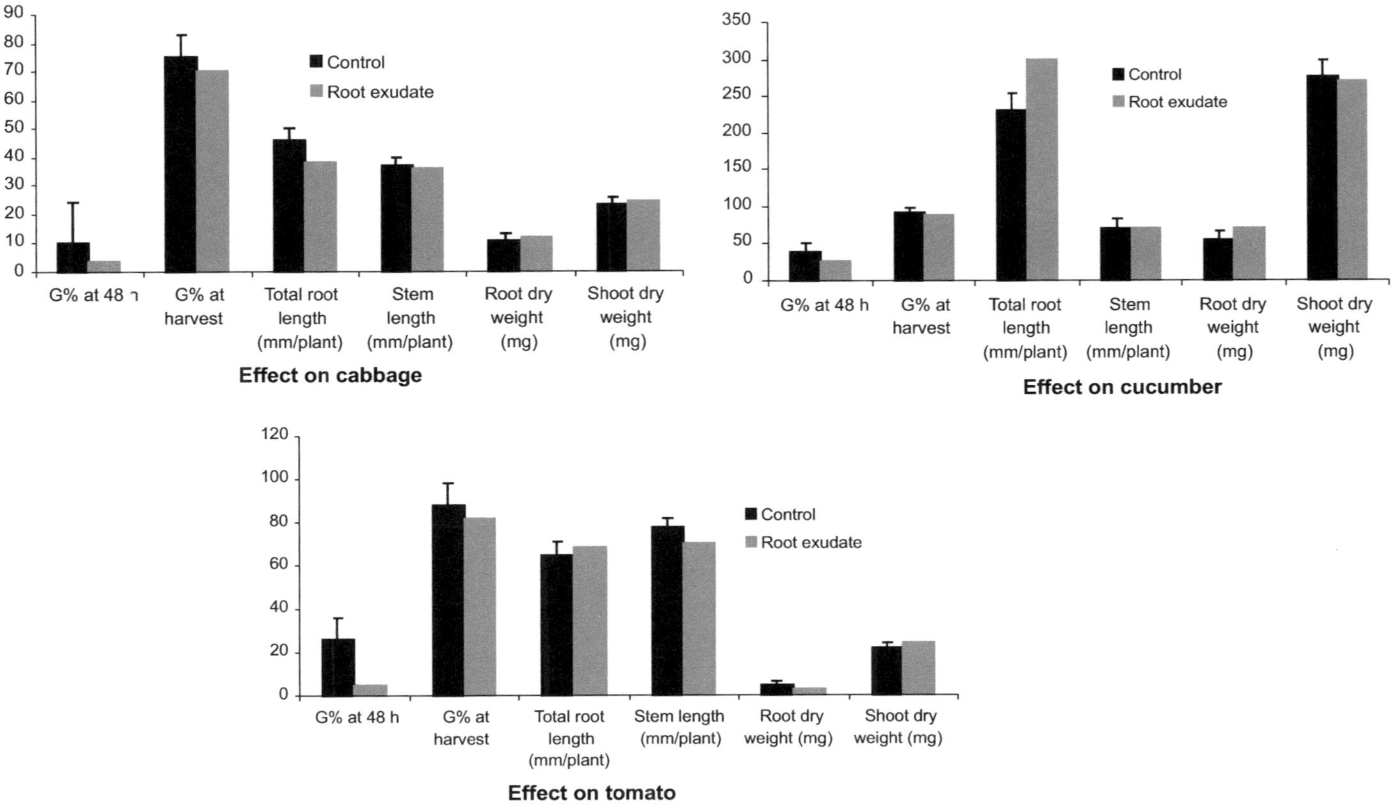

Fig. 2 Effect of root exudates of *D. erucoides* added to Petri dishes on germination and growth of different vegetable crops grown under laboratory conditions

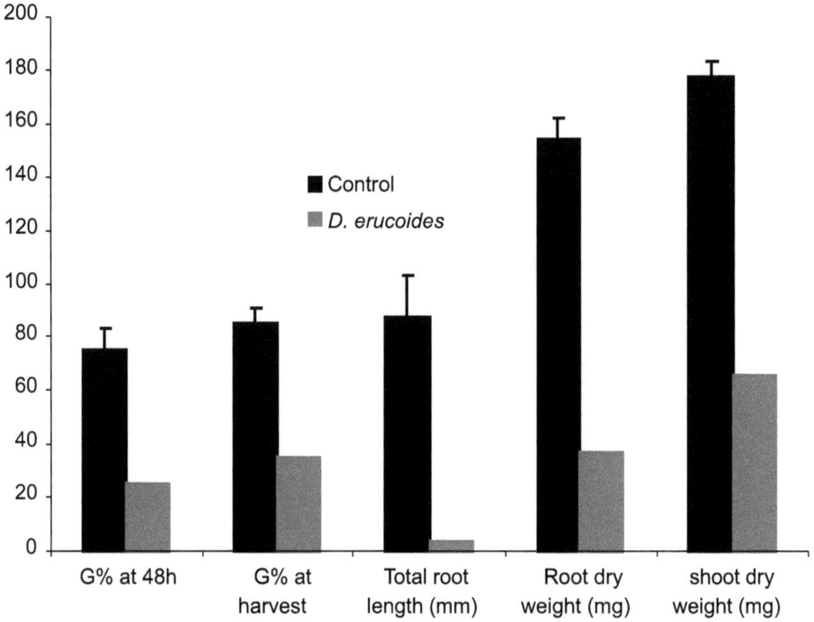

Fig. 3 Effect of *D. erucoides* full strength foliage leachates on germination and growth of wheat seedlings grown in Petri dishes for 10 d after incubation at 24°C

Table 2 Effect of foliage leachates concentration of *D. erucoides* on germination and growth of wheat at 10 d of incubation at 24°C

Leahates volume (ml/Petri dish)	*G% at 48 h*	*G% at harvest*	*Total root length (mm/plant)*	*Stem length (mm/plant)*	*Root dry weight (mg)*	*Shoot dry weight (mg)*
Control (dH_2O)	75	85	87.5	154.0	154	178
0.5	65	80	39.1	134.8	94	155
1	55	70	36.9	124.8	82	145
2	55	65	28.6	122.8	80	165
3	35	55	13.9	124.9	63	170
4	25	55	12.3	126.4	69	176
5	10	45	9.7	128.8	60	154
LSD (P = 0.05)	8	12	11.5	15.4	20	14

G% = Germination percentage

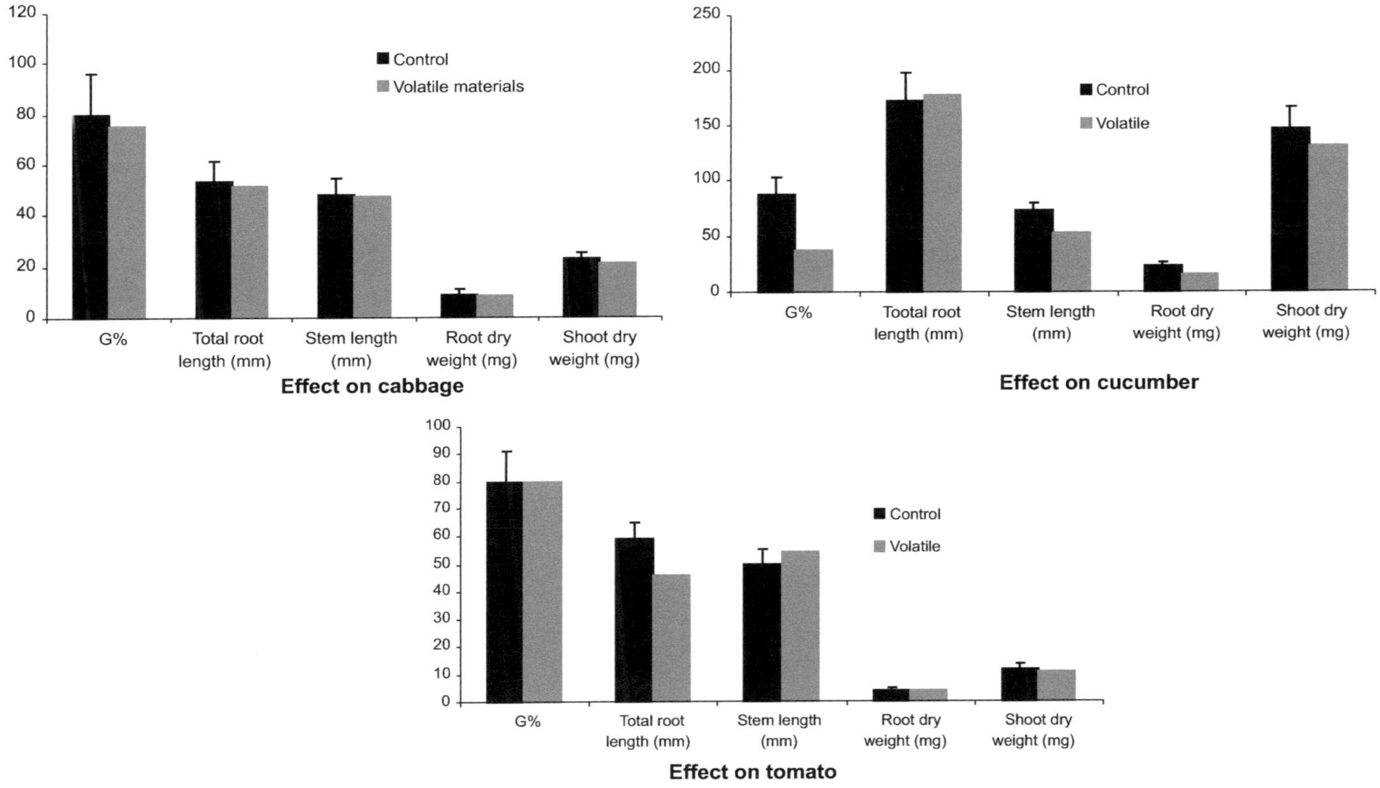

Fig. 4 Effect of volatile materials of *D. erucoides* on germination and growth of different vegetable crops grown under laboratory conditions

Table 3 Effect of soil incorporated shoot residues of *D. erucoides* on germination, shoot dry weight and root dry weight of cabbage, cucumber, tomato and wheat grown in pots

Residue rate (g/kg^{-1} soil)	*Cabbage*			*Cucumber*		
	G% at harvest	*Shoot dry weight (mg)*	*Root dry weight (mg)*	*G % at harvest*	*Shoot dry weight (mg)*	*Root dry weight (mg)*
0 (no residue)	75	600	300	90	540	280
1	73	450	280	90	480	340
2	63	400	240	83	450	330
4	80	420	230	85	480	175
8	73	400	270	85	480	170
16	74	350	200	78	400	120
LSD (P = 0.05)	3	70	6	10	9	63

Residue rate (g/kg^{-1} soil)	*Tomato*			*Wheat*	
Residue rate (g/kg^{-1} soil)	*G% at harvest*	*Shoot dry weight (mg)*	*Root dry weight (mg)*	*G % at harvest*	*Shoot dry weight (mg)*
0 (no residue)	88	320	268	95	1010
1	86	452	247	90	1110
2	76	452	250	93	1130
4	77	210	190	95	1220
8	63	230	168	60	840
16	68	200	118	45	380
LSD (P = 0.05)	11	103	75	8	125

G% = Germination percentage

Cucumber and wheat were the most affected and the effect increased with residue rate added (Plate 4).

Plate 4 Allelopathic effect of soil incorporated dried residues (rates increased from left to right) of white rocket (*Diplotaxis erucoides* L.) on germination and growth of wheat grown under glasshouse conditions

Experiment 7. Effect of surface placed residues

Surface placed residues of *D. erucoides* significantly reduced germination of all crops tested (Fig. 5). Wheat and cucumber were affected more than cabbage and tomato. While shoot length of cabbage was significantly reduced, and root dry weight of cucumber and shoot dry weight of wheat, no reduction was detected on any growth parameter of tomato plants compared with the control.

Experiment 8. Effect of root exudates

Root exudates of *D. erucoides* severely affected cabbage and to a lesser extent the tomato crop (Fig. 6). The effect on germination of cabbage was more pronounced at 48 h of incubation.

Stem length, root dry weight and shoot dry weight of cabbage were all reduced compared with the control. In contrast, the same treatment stimulated germination and growth of cucumber compared with the control.

Experiment 9. Effect of foliage leachates

The effect of foliage leachates of *D. erucoides* added to the soil on growth of the three vegetable crops is shown in Fig. 7. While all growth parameters of cucumber were reduced compared with the control, only shoot dry weight of cabbage was significantly affected. In contrast, tomato growth was better with leachates than that of untreated plants.

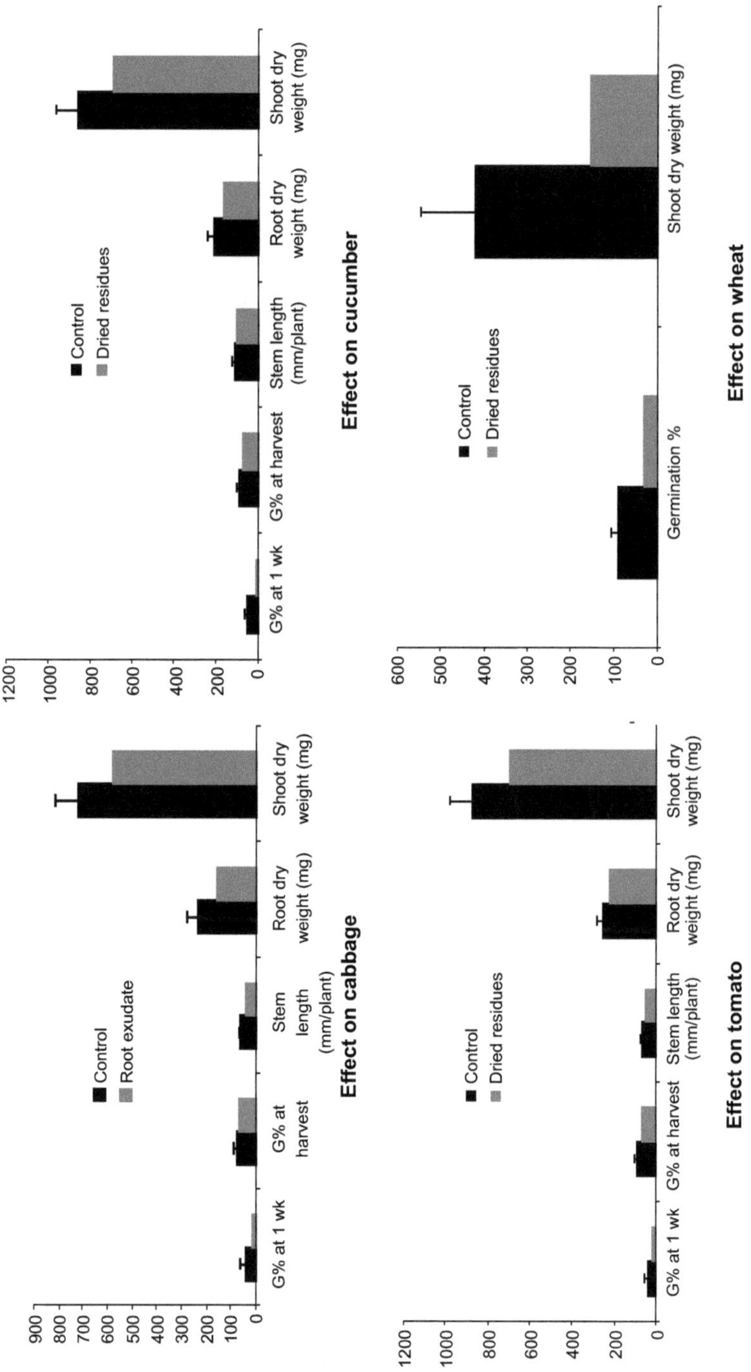

Fig. 5 Effect of surface placed residues of *D. erucoides* on germination and growth of different crops grown in pots under glasshouse conditions

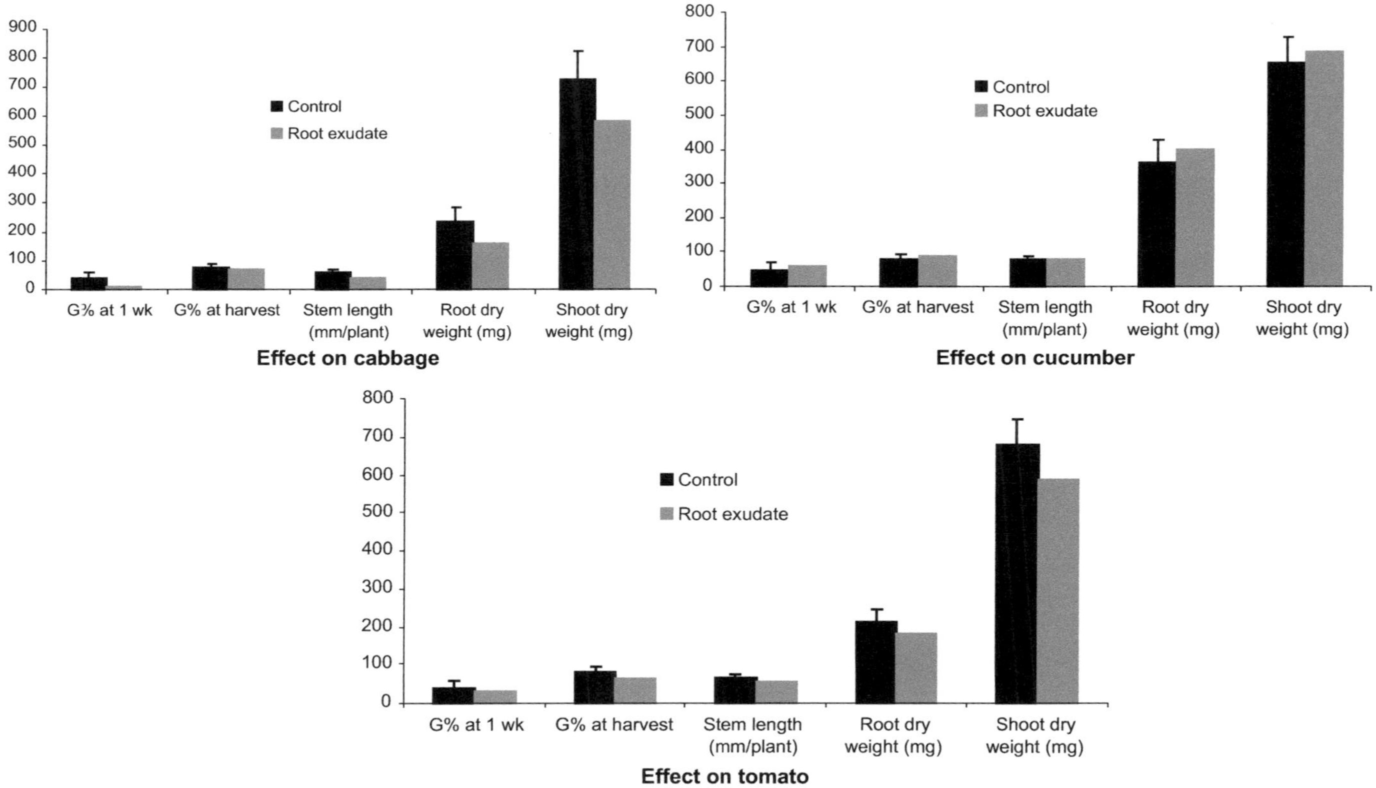

Fig. 6 Effect of root exudates of *D.erucoides* released into the soil on germination and growth of different vegetable crops grown under glasshouse conditions

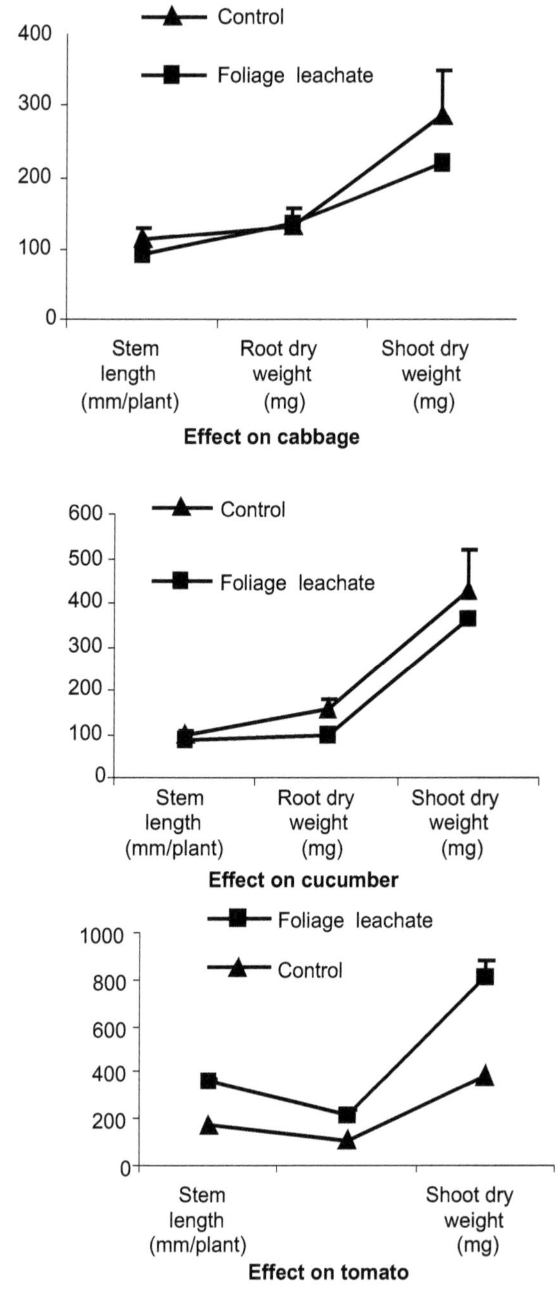

Fig. 7 Effect of foliage leachates of *D. erucoides* added to the soil on germination and growth of different vegetable crops grown under glasshouse conditions

Experiment 10. Effect of decayed residues

Residues of *D. erucoides* left to decay in the soil were phytotoxic to all crops tested except cabbage (Fig. 8). Germination, shoot length and shoot and root dry weights of cucumber and tomato, and germination and shoot dry weight of wheat were significantly reduced compared with the control. Reduction was more pronounced on cucumber and tomato than on other crops (Plates 5 and 6).

DISCUSSION

The present work was conducted to investigate any possible allelopathic activity of *D. erucoides* and its role in the spreading of this weed in different parts of Jordan. Results obtained using different experimental techniques clearly showed a great potential effect of weed on targeted plant species through this mechanism of plant interference and agreed with results reported by different workers on allelopathic activity of different annual and perennial weeds on different plant species (Qasem, 2001; Qasem and Foy, 2001; Rice, 1984). Testing the effect of aqueous shoot extract of *D. erucoides* revealed strong inhibitory action against all crops tested (Fig. 1, Table 1). The effect of extract was concentration- dependent and results showed great variations from sensitive to tolerant species in their responses to this treatment (Table1). However, full strength extract was phytotoxic to all tested species and completely inhibited germination and /or growth of certain tested crops. Different workers reported high phytotoxicity of weed extracts to different plant species (Qasem and Foy, 2001; Obaid and Qasem, 2002). However, this technique cast a lot of doubt on the results obtained and the opposite action may be found in soil culture or under field conditions (Qasem and Hill, 1989a). Plant extracts may contain chemicals not readily released in nature and may not be possible to dissolve in water.

Results on the effect of root exudates showed that *D. erocoides* might diffuse allelochemicals with root exudates released into the soil that may harm other plant species (Figs. 2, 6). This however, was confirmed in the present work through two different techniques used in which exudates were isolated from the soil and tested in Petridishes sown by target crops and by planting crops in the same soil used to grow *D. erucoides* after plants of this weed were harvested. Results obtained were similar to those reported for other weed species using similar techniques to investigate possible allelopathic effects of their root exudates (Qasem, 1995). Although, evidence is accumulating and has shown that root exudates of certain plant species are toxic to the roots of neighboring plants, and

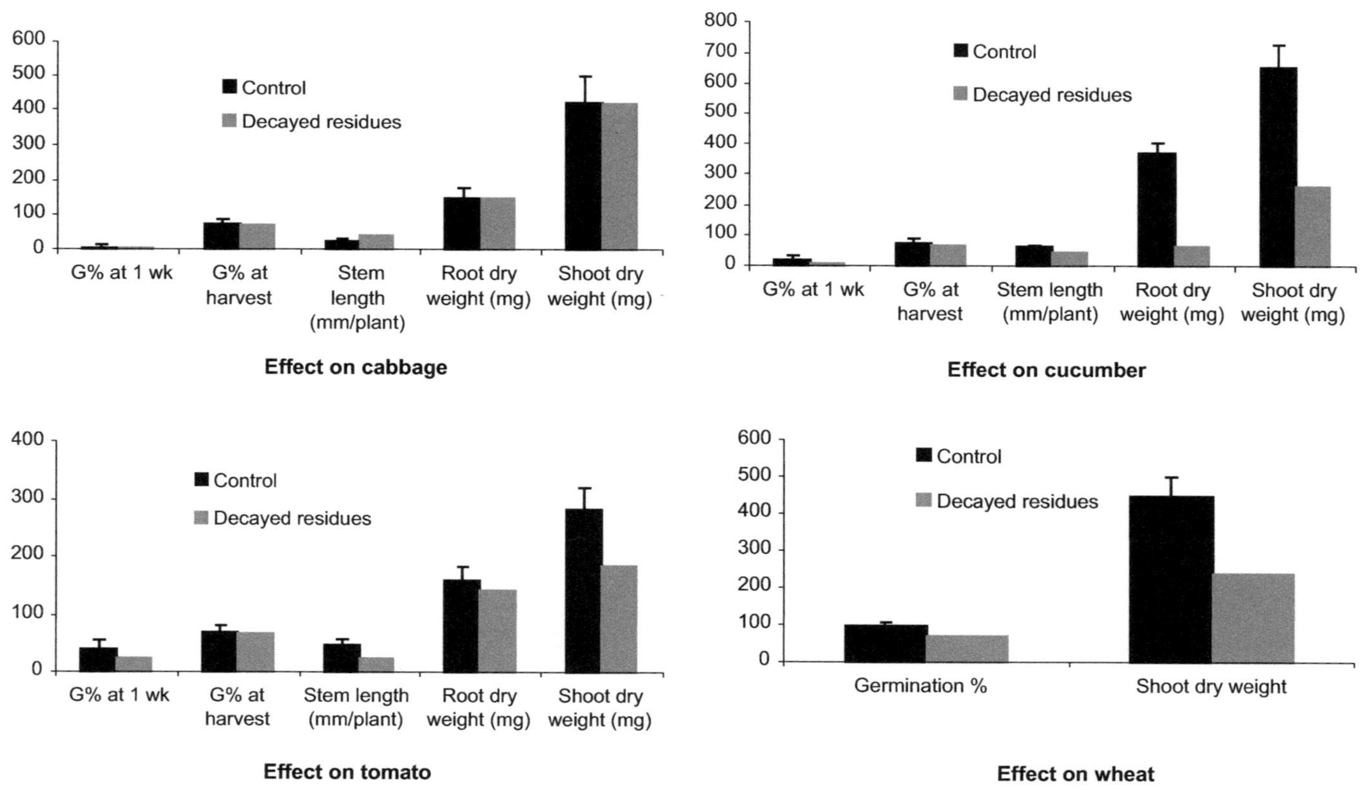

Fig. 8 Effect of soil incorporated decayed residues of *D. erucoides* on germination and growth of different crops grown under glasshouse conditions

Plate 5 Allelopathic effect of decayed soil incorporated dried residues of white rocket (*Diplotaxis erucoides* L.) on germination and growth of cucumber grown under glasshouse conditions

Plate 6 Allelopathic effect of decayed soil incorporated dried residues of white rocket (*Diplotaxis erucoides* L.) on germination and growth of tomato grown under glasshouse conditions

inhibits germination, growth or nutrient accumulation of other species (Rovira, 1969; Tukey, 1969; Brown *et al.*, 1983; Qasem and Hill, 1989b; Qasem, 1995). However, root exudates and hence allelochemicals released through this venue are mostly released in minute amounts and may not be enough to express strong allelopathic influence. Appreciable amounts of allelochemicals diffused may be also adsorbed to soil colloids that mask their effects or could have been exposed to microbial degradation and thus showed low activity on receiver species. Results showed high variations between vegetable crops exposed to root exudates of *D. erucoides* (Figs. 2, 6). Plants are different in their sensitivities to allelochemicals and therefore allelopathy is considered as a selective mechanism of plant interference in nature.

D. erucoides is a widely spread weed in areas cultivated by wheat and barley crops in Jordan. It is felt that the weed has taken over these crops and spread at their expense under field conditions. This phenomenon , is well demonstrated in semi-arid regions in Jordan. Results on the effect of foliage leachates of *D. erucoides* on wheat germination and seedlings growth confirmed our speculation on allelopathic activity of this weed against wheat. It has been experimentally proved that allelochemicals collected with water leached from its foliage parts are also phytotoxic to wheat (Fig. 3). Leachates of different weeds were reported to be phytotoxic to different plant species, and were mentioned to be responsible on the inhibitory zones they form in nature. Qasem (1994, 2001) reported that water leached from foliage parts of *Lepedium draba* was highly toxic to different wheat and barley lines tested under glasshouse conditions, while allelochemicals were reported to leach out from shoots of different plants by rain water or dew and can cause severe injuries to receiver plants (Qasem, 1994, 2001; Qasem and Foy, 2001). The concentration of leached material and the effect of microorganisms may be important factors in determining the stimulatory or the inhibitory action of the leached compounds (Patterson, 1981). The effect of foliage leachates of *D. erucoides* was confirmed under laboratory conditions in Petridishes sown by wheat seeds (Fig. 3, Table 2) and in pot experiments sown by vegetable crops and irrigated with foliage leachates of this weed under glasshouse conditions (Fig. 7).

Further tests carried out on the possible effect of volatiles emanated from shoots of *D. erucoides* showed possible allelopathic effect of the weed through this mechanism. However, it has been shown that vegetable crops tested are greatly different in their sensitivities to this treatment, with cucumber being the most affected (Fig. 4). Different weed species showed strong allelopathic effects through volatile materials have been already reported (Muller *et al.*, 1964; Lovett and Lynch, 1979; Qasem, 1999, 2001), while differences between various plant species including cultivated crops to allelochemicals have been also documented (Qasem, 1995, 2001). These differences may reflect mechanical, physiological or biochemical variations between receiver species and their ability to utilize allelochemicals. Oleszek (1987) concluded that the toxicity of compounds from crucifers was species-dependent, and the acceptor species react in different ways to volatiles of different donor species.

Residues of *D. eruoides* left on the soil surface, incorporated or left to decay for certain periods in the soil before being sown by crop plants, were all found to be highly toxic to the tested vegetable and wheat crops (Figs. 5, 8, Table 3). Toxicity effects increased with residue rate added. These results were in agreement with results reported on different weed or crop species showing inhibitory effects through their residues left on or in the soil (Bhowmik and Doll, 1984; Qasem, 1994, 2001; Teasdale and Mohler, 2000; Araki and Hatano, 2002). This method of residue application on the soil was exploited in weed control programs against certain noxious weed species (Wu *et. al.*, 1999; Jakkola, 2002; Teasdale, 2002). In addition, crucifer species were reported to have strong allelopathic effects against different weed species and could be considered as an effective tool in weed management (Oleszek, 1987; Jakkola, 2002). However, selectivity of the straw mulch used is important for less residual action of these allelochemicals against sensitive useful plant species. Our results clearly showed that *D. erucoides* residue left on the surface or incorporated in the soil may yield phytotoxic chemicals and harm the crops tested. However, surface placed residues appeared more toxic than soil incorporated materials, while decayed residues proved to be highly toxic to all crops except cabbage. Grinding plant tissues may not represent the exact situation occurring in nature and may enhance the release of allelochemicals. This procedure however, is followed in the present experiment to allow better coverage of pot surface with the relatively small amount of weed residue used. In nature, the amount of plant residues is expected to be higher and the decaying process of these residues may take place faster than that under glasshouse conditions due to different environmental factors, which prevail better under field conditions, leading to the fast release of allelochemicals. Dried residues of *D. erucoides* delayed seedling emergence of different crops and significantly affected growth of cucumber and tomato. This may indicate that certain allelochemicals may be released during the decaying process in the soil, or their concentration was substantially increased, or certain non-phytotoxic chemicals have been transformed to toxic forms by soil factors including microorganisms. The phytotoxic effect of *D. erucoides* residues may be important under arid or semi-arid conditions under which cereal crops are grown in Jordan. Drought stress may increase allelochemical concentration in weed tissues and/or in the soil and may lead to more suffering for the crop plants under such conditions.

CONCLUSION

In conclusion, *D. erucoides* is an annual weed widely spread in cultivated and fallow lands in Jordan. Results obtained using different experimental techniques showed that this species has an allelopathic activity against different vegetable crops and wheat. Its aqueous shoot extract, foliage leachates, volatiles, root exudates, soil incorporated and decayed shoot residues in the soil were all phytotoxic to different crops and negatively affected their germination and seedling growth. However, the degree of inhibition varied with the source of inhibitors in different weed parts, receiver species and concentration/rate of inhibitors added to the growing medium. In general, cucumber and wheat were severely harmed, while decayed shoot residues of *D. erucoides* in the soil were extremely toxic to cucumber, tomato and wheat.

References

Akram, M., Ahmad N., Hussain F., Aslam K.M. (1990) Allelopathic potential of four species of *Ficus*. Pakistan J. Sci. Ind. Res. 33: 52-54.

Araki, H., Hatano, Y. (2002) Weed control effect of cover crop mulch in the soil with different weed density. *In* the Proceedings of the Third World Congress on Allelopathy. August 26-30, Tsukuba-Japan. p. 116.

Barens, J.P., Putnam, A.R. (1986) Evidence for allelopathy by residue and aqueous extracts of rye (*Secale cereale*). Weed Sci. 34: 384-390.

Bhowmik, P.C., Doll, J.D. (1984) Allelopathic effects of annual weed residues on growth and nutrient uptake of corn and soybean. Agron. J. 76: 383-388.

Bradow, J.M., Connick, W.J. (1988) Seed germination inhibition by volatile alcohols and other compounds associated with *Amaranthus palmeri* residues. J. Chem. Ecol. 14: 1633-1648.

Brown, R.L., Tang, C.S. Kishimoto, R.K. (1983) Growth inhibition from guava root exudates. HortScience, 18: 316-318.

Drost, D.C., Doll, J.D. (1980) The alleopathic effect of yellow nutsedge (*Cyperus esculentus*) on corn (*Zea mays*) and soybean (*Glycine max*). Weed Sci. 28: 229-233.

Hewitt, E.J. (1966) Sand and Water Culture Methods Used in the Study of Plant Nutrition. 2nd ed. Commonwealth Agricultural Bureaux. Bucks, UK.

Jakkola, S. (2002) Weed control with cruciferous plant material. *In* the Proceedings of the Third World Congress on Allelopathy. August 26-30, Tsukuba-Japan. p. 81.

Kumari, A., Kohli, R.K. (1987) Autotoxicity of ragweed parthenium (*Parthenium hysterophorus*). Weed Sci. 35: 629-632.

Lovett, J.V., Lynch, J.A. (1979) Studies on *Salvia reflexa* Hornem. I. Possible competitive mechanism. Weed Res. 19: 351-357.

Mason-Sedun, W., Jessop, R.S., Lovett, J.V. (1986) Differential phytotoxicity among species and cultivars of the genus *Brassica* to wheat. Plant and Soil. 93: 3-16.

Muller, C.H., Muller, W.H., Hains, B.L. (1964) Volatile growth inhibitors produced by shrubs. Science. 143: 471-473.

Obaid, K.A., Qasem, J.R. (2002) Inhibitory effects of *Cardaria draba* and *Salvia syriaca* extracts to certain vegetable crops. Dirasat. 29: 247-259.

Oleszek, W. (1987) Allelopathic effects of volatiles from cruciferae species on lettuce, barnyard grass and wheat. Plant and Soil. 102: 271-273.

Patrick, Z.A., Toussoun, T.A., Snyder, W.C. (1963) Phytotoxic substances in arable soils associated with decomposition of plant residues. Phytopathol. 53: 152-161.

Patterson, D.T. (1981) Effects of allelopathic chemicals on growth and physiological responses of soybean (*Glycine max*). Weed Sci. 29: 53-59.

Putnam, A.R., Duke, W.B. (1978) Allelopathy in agroecosystems. Ann. Rev. of Phytopathol. 16: 431-451.

Qasem, J.R. (1993) Allelopathic effects of some common weed species in cereal crops. Dirasat. 20B: 7-28.

Qasem, J.R. (1994) Allelopathic effects of white top (*Lepidium draba*) on wheat and barley. Allelopathy J. 1: 29-40.

Qasem, J.R. (1995) Allelopathic effect of *Amaranthus retroflexus* and *Chenopodium murale* on vegetable crops. Allelopathy J. 2: 49-66.

Qasem, J.R. (1999) Biological Activity of Corn Buttercup (*Ranunculus arvensis* L.). *In* Recent Advances in Allelopathy: A Science for the Future. Vol 1. F.A. Macias, J.C.G. Galindo, J.M.G. Molinillo, H.G. Cutler (eds.) Servicio De Publicaciones Universidad De Cadiz, Cadiz, Spain. pp. 287-300.

Qasem, J.R. (2001) Allelopathic potential of White top and Syrian sage on vegetable crops. Agron. J. 93: 64-71.

Qasem, J.R. (2002) Allelopathic effects of selected medicinal plants on *Amaranthus retroflexus* and *Chenopodium murale*. Allelopathy J. 10: 105-122.

Qasem, J.R., Foy, C.L. (2001) Weed Allelopathy: Its Ecological Impact and Future Prospects. *In* Allelopathy in Agro-ecosystem. R.K. Kohli, H.P. Singh, D.R. Batish (eds). Haworth Press, USA. pp. 43-119.

Qasem, J.R., Hill, T.A. (1989a) On difficulties with allelopathy methodology. Weed Res. 29: 345-347.

Qasem, J.R., Hill, T.A. (1989b) Possible role of allelopathy in the competition between tomato, *Senecio vulgaris* L. and *Chenopodium album* L. Weed Res. 29: 349-356.

Rice, E.L. (1979) Allelopathy–an update. The Bot. Rev. 45: 17-109.

Rice, E.L. (1984) Allelopathy. 2nd ed. Orlando, FL, USA.

Rovira, A.B. (1969) Plant root exudates. Bot. Rev. 35: 35-59.

Sahid, I.B., Sugau, J.B. (1993) Allelopathic effect of lantana (*Lantana camara*) and siam weed (*Chromolaena odorata*) on selected crops. Weed Sci. 41: 303-308.

Stachon, W.J., Zimdahl, R.L. (1980) Allelopathic activity of Canada thistle (*Cirsium arvense*) in Colorado. Weed Sci. 28: 83-86.

Teasdale, J.R., Mohler, C.L. (2000) The quantitative relationship between weed emergence and physical properties of mulches. Weed Sci. 48: 385-392.

Teasdale, J.R. (2002) Weed management with cover crop mulches. *In* the Proceedings of the Third World Congress on Allelopathy. August 26-30, Tsukuba- Japan. p. 115.

Tukey, H.B. (1969) Implications of allelopathy in agricultural plant science. Bot. Rev. 35: 1-15.

Wu, H., Naber, H., Oomen, G. (1999) The allelopathic effect of dead and living mulches from perennial ryegrass (*Lolium perenne* L.) on *Calystegia sepium* (L.). R. Br. *In* Recent Advances in Allelopathy: A Science for the Future. Vol. 1 F.A. Macias, J.C.G. Galindo, J.M.G. Molinillo, H.G. Cutler (eds.). Servicio De Publicaciones Universidad De Cadiz, Cadiz, Spain. pp. 351-362.

10

Weed-crop Interferences in Hungary

Gabriella Kazinczi, Imre Béres and Joseph Horvath

Office for Academy Research Groups Attached to Universities and Other Institutions, Virological Group, Pannon University, Georgikon Faculty of Agricultural Sciences, Plant Protection Institute
H-8361 Keszthely, P.O. Box 71, Hungary

Laboratory (bioassay), glasshouse and field experiments were set up in order to study allelopathic potential of two annual (*Ambrosia artemisiifolia, Abutilon theophrasti*) and two perennial (*Cirsium arvense, Asclepias syriaca*) weed species serious and widespread in Hungary. In bioassays, inhibitory effect on germination and radicle length of the test plants greatly depended on donor and recipient species, plant parts, extracts (water, ethanol and alcoholic ones) and the concentration of the extracts. In pot experiments under glasshouse conditions, water extract of *A. theophrasti* and *C. arvense* shoots, and roots residues of *A. syriaca* had a promoting effect on the development of crop plants. It could be due to utilization of nutrients from the plant extracts and residues and/or the microbiological degradation of phytotoxins in the soil. The opposite effect was observed in the case of shoot and root residues of *A. theophrasti* incorporated into the soil. It has been concluded that serious weeds, widespread in Hungary have allelopathic potential. Besides their resistance to some herbicides and considerable competitive ability, allelopathy must also be considered.

Keywords: *allelopathy, annuals, fresh weight, germination, radicle length, perennials*

INTRODUCTION

Abutilon theophrasti is a principal weed in Canada and the United States and causes major yield losses in soybean and cotton (Oliver, 1979; Dekker and Meggitt, 1983). It is a noxious weed that affects the growth of corn and sunflower in Hungary. Due to its hard seed mechanism it germinates two or three times a year, therefore, herbicides without long term persistence in soil are not effective (Czimber, 1970). Its allelopathy has been intensively studied (Gressel and Holm, 1964; Colton and Einhellig, 1980; Elmore, 1980a,b; Bhowmik and Doll, 1982; Sterling *et al.*, 1987, Mikulas *et al.*, 1990; Kazinczi *et al.*, 2001a). Among *Ambrosia* species the most common and widespread is *A. artemisiifolia.* It is native of North America and is also found in Europe, Asia and South America (Holm *et al.*, 1977) in crops and waste lands. The plant or its pollen may cause dermatitis and allergy in some people (Frankton and Mulligan, 1970). Till now, little information was available on allelopathy of *A. artemisiifolia* (Béres *et al.*, 1998, 2002). Among dicot perennials serious and widespread all over the world two species: *Cirsium arvense* and *Asclepias syriaca* are also common in Hungary (Toth *et al.*, 1989; Toth and Spilak, 1998). Among them only allelopathy of *C. arvense* has been intensively studied (Bendall, 1975; Moore, 1975; Stachon and Zimdahl, 1980; Wilson, 1981; Kovács *et al.*, 1988; Béres and Csorba, 1992; Erkelsen and Crabtree, 1995; Solymosi and Nagy, 1999; Kazinczi *et al.*, 2001b). The aim of our study was to examine allelopathy of two annual (*A. theophrasti, A. artemisiifolia*) and two perennial weeds in laboratory and pot experiments.

MATERIALS AND METHODS

Laboratory Experiments (Bioassays)

Water, acetone and and alcoholic extracts were made from fresh shoots and roots of *C. arvense, A. syriaca, A. theophrasti* and *A. artemisiifolia* using 25 g fresh plant parts/100 ml distilled water as a stock solution. The plant parts were cut into small pieces in a grinder. After keeping the cut pieces for 1 d the mixtures were filtered, followed by 4 × 100 seeds of corn and sunflower, respectively germinated in Petri dishes. Ten ml extracts were added to each Petri dish. Stock solutions in 2-, 5- and 10-fold dilutions were also used in order to determine the effect on the radicle length of cucumber.

Pot Experiments

Dried root, stem and leaf (0.8 kg) of *A. theophrasti* and *A. syriaca* were mixed in 10 kg of soil mixture (sand:peat in a ratio of 1:1). After two month of decomposition, the plastic pots were filled with the soil mixtures containing the plant residues and sown with four seeds each of three test plants (wheat, corn and sunflower). In other experiments, water extracts from fresh leaves of *A. theophrasti* and *C. arvense* were made (25 g fresh leaves/100 ml distilled water). The pots were irrigated with the water extracts as needed. The fresh weight of the test plants was recorded 42 d after sowing.

RESULTS AND DISCUSSION

Laboratory Experiments (Bioassays)

It is concluded, that inhibitory effect on germination greatly depended on donor and recipient species, plant parts, and the type of the extracts (water, acetone or alcoholic ones). Considerable (98, 95 and 92%) reduction in germination of sunflower was observed, due to the water extracts of *A. syriaca* shoots, alcoholic root extracts of *C. arvense* and shoot water extract of *A. theophrasti*, respectively (Figs. 1 and 2).

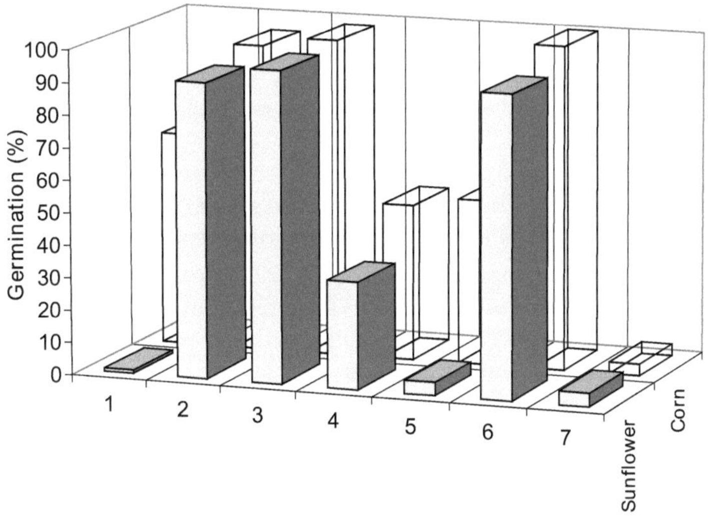

Fig. 1 The effect of *A. syriaca* and *C. arvense* on the germination of test crops (1. ASCSY shoot water extract; 2. ASCSY root water extract; 3. CIRAR shoot water extract; 4. CIRAR root water extract; 5. CIRAR root alcoholic extract; 6. Control; 7. $LSD_{5\%}$)

All extracts significantly reduced the radicle length of cucumber at higher concentrations (Table 1).

Pot Experiments

In pot experiments water extracts of *A. theophrasti* and *C. arvense* shoots, and root residues of *A. syriaca* had a promoting effect on the development of the test plants. It could be due to the utilization of nutrients from the

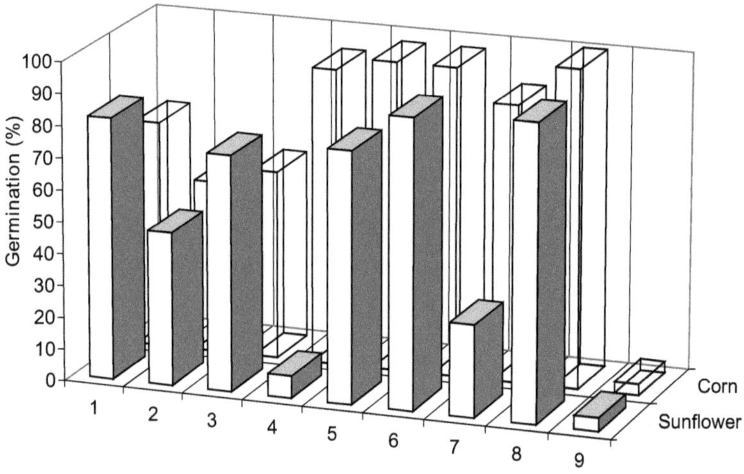

Fig. 2 The effect of *A. artemisiifolia* and *A. theophrasti* extracts on the germination of the test crops (1. AMBAR shoot water extract; 2. AMBAR shoot alcoholic extract; 3. AMBAR shoot acetone extract; 4. ABUTH shoot water extract; 5. ABUTH shoot alcoholic extract; 6. ABUTH root water extract; 7. ABUTH root alcoholic extract; 8. Control; 9. $LSD_{5\%}$).

plant parts/extracts and/or the rapid microbiological degradation of phytotoxins in the soil. The opposite effect was observed in some cases of plant residues of *A. theophrasti* incorporated into the soil (Table 2).

It can be concluded that noxious weeds widespread in Hungary have allelopathic potential, therefore besides their resistance to some herbicides and considerable competitive ability, allelopathy must also be considered. Nevertheless, future investigations are necessary in order to evaluate the role of their allelopathy in weed-crop interactions.

Table 1 The effect of plant extracts on the radicle length of cucumber

C. arvense shoot water extract						*C. arvense* root water extract					
S	2x	5x	10x	C	$LSD_{5\%}$	S	2x	5x	10x	C	$LSD_{5\%}$
Radicle length (mm)											
30	39	62	59	64	12	45	67	60	78	64	17
A. syriaca leaf water extract						*A. syriaca* stem water extract					
S	2x	5x	10x	C	$LSD_{5\%}$	S	2x	5x	10x	C	$LSD_{5\%}$
Radicle length (mm)											
27	55	64	67	62	8	11	53	61	61	62	12

S, stock solution; 2x, two-fold dilution; 5x, five-fold dilution; 10x, 10-fold dilution; C, control

Table 2 The effect of weeds on the fresh weight of the crops in pot experiments

	A. syriaca			*A. theophrasti*					*A. theophrasti*			*C. arvense*		
	rr	c	$LSD_{5\%}$	root	stem	leaf	c	$LSD_{5\%}$	fl	c	$LSD_{5\%}$	fl	c	$LSD_{5\%}$
	Fresh weight (g/plant)													
Corn	49.8	48.5	4.7	32.1	33.9	39.8	36.7	4.7	3.6	1.5	1.6	3.2	2.8	0.56
Sunflower	44.8	43.8	3.6	26.1	28.1	24.9	27.1	3.3	6.9	2.2	3.6	3.2	2.6	0.47
Wheat	93.3	91.8	3.9	-	-	-	-	-	-	-	-	0.7	0.48	0.17

rr root residues; c, control; fl, fresh leaves; -, not investigated

Acknowledgements

The authors wish to express their thanks to the Hungarian Scientific Research Fund (OTKA project No. T 037931, T 049093) for their financial support.

References

Bendall, G.M. (1975) The allelopathic activity of Canada thistle (*Cirsium arvense* (L.) Scop) in Tasmania. Weed Res. 15: 77-81.

Béres, I., Csorba, R. (1992) Data on the biology of *Cirsium arvense* (L.) Scop. Növényvédelem 28: 322-327.

Béres, I., Sárdi, K., Kámán, S. (1998) Allelopathic effets of *Ambrosia artemisiifolia* L. on seed germination and seedling growth of field crops. *In*: Comptes-rendus 6éme Symposium Mediterranéen EWRS, Montpellier, France. pp. 89-90.

Béres, I., Kazinczi, G., Narwal, S.S. (2002) Allellopathic plants. 4. Common ragweed (*Ambrosia elatior* L. syn. *A. artemisiifolia*). Allelopathy J. 9: 27-34.

Bhowmik, P.C., Doll, J.D. (1982) Corn and soybean response to allelopathic effects of weed and crop residues. Agron. J. 74: 601-606.

Colton, C.E., Einhellig, F.A. (1980) Allelopathic mechanisms of velvetleaf (*Abutilon theophrasti* Medic.) on soybean. Amer. J. Bot. 67: 1407-1413.

Czimber, Gy. (1970) Resistance of hard-coates seeds of *Convolvulus arvensis* L. to various herbicides. Acta Agron. Acad. Sci. Hung. 19: 321-329.

Dekker, J., Meggitt, W.F. (1983) Interference between velvetleaf (*Abutilon theophrasti* Medic.) and soybean (*Glycine max* (L.) Merr.) I. Growth. Weed Res. 23: 91-101.

Elmore, C.D. (1980a) Free amino acids of *Abutilon theophrasti* seed. Weed Res. 20: 63-64.

Elmore, C.D. (1980b) Inhibition of turnip (*Brassica rapa*) seed germination by velvetleaf (*Abutilon theophrasti*) seed. Weed Sci. 28: 658-660.

Erkelsen, S.R., Crabtree, G.D. (1995) The role of allelopathy in buckwheat (*Fagopyrum sagittatum*) inhibition of Canada thistle (*Cirsium arvense*). Weed Sci. 43: 70-74.

Frankton, C., Mulligan, G.A. (1970) Weeds of Canada. Canada Department of Agriculture, Publ. No. 948.

Gressel, J.B., Holm, L.G. (1964) Chemical inhibition of crop germination by weed seeds and the nature of inhibition by *Abutilon theophrasti*. Weed Res. 4: 44-53.

Holm, L.G., Plucknett, D.L., Pancho, J.V., Herberger, J.P. (1977) The World's Worst Weeds. University Press, Honolulu, Hawaii, USA.

Kazinczi, G., Béres, I., Narwal, S.S. (2001a) Allelopathic plants. 3.Velvetleaf (*Abutilon theophrasti* Medic.). Allelopathy J. 8: 179-188.

Kazinczi, G., Béres, I., Narwal, S.S. (2001b) Allelopathic plants. 1. Canada thistle [*Cirsium arvense* (L.) Scop]. Allelopathy J. 8: 29-40.

Kovács, L., Mikulás, J., Pölös, E. (1988) Allelopathic activity of *Cirsium arvense* (L.) Scop in Hungary. Acta Agron. Hung. 37: 65-69.

Mikulás, J., Váradi, Gy., Pölös, E. Kazinczi, G., Béres, I. (1990) Studies on allelopathy of some weeds. Z. PflKrankh. PflSchutz, Sonderh. 12: 265-277.

Moore, R.K. (1975) The biology of Canadian weeds. 13. *Cirsium arvense* (L.) Scop. Canadian J. Plant Sci. 55: 1033-1048.

Oliver, L.R. (1979) Influence of soybean (*Glycine max*) planting date on velvetleaf (*Abutilon theophrasti*) competition. Weed Sci. 24: 43-46.

Solymosi, P., Nagy, P. (1999) Phytotoxicity of rhizome and leaf extracts of *Cirsium arvense* taxa. Növényvédelem 35: 245-251.

Stachon, W.J., Zimdahl, R.L. (1980) Allelopathic activity of Canada thistle (*Cirsium arvense*) in Colorado. Weed Sci. 28: 83-86.

Sterling, T.M., Houtz, R.L., Putnam, A.R. (1987) Phytotoxic exudates from velvetleaf (*Abutilon theophrasti* Medic.) glandular trichomes. American J. Bot. 74: 543-550

Toth, A., Spilak, K. (1998) The fourth national weed survey in Hungary. *In* the Proceedings of the 8th Forum on Plant Protection, Keszthely, Hungary. p. 49.

Toth, A., Molnar, J., Torok, T., Fekete, A. (1989) Preliminary results about the third national weed survey in Hungary. Növényvédelem. 25: 374-377.

Wilson, J.R. (1981) Effect of Canada thistle (*Cirsium arvense*) residue on growth of some crops. Weed Sci. 29: 159-164.

SECTION 4

Allelopathic Cover Crops to Suppress Weeds

11

Allelopathic Activity of Buckwheat: A Ground Cover Crop for Weed Control

Zahida Iqbal, Habib Nasir and Yoshiharu Fujii

Department of Biological Safety Science
National Institute for Agro-Environmental Sciences
3-1-3 Kan-nondai, Tsukuba, Ibaraki 305-8604, Japan

Allelopathy plays an important role in agro-ecosystems through a wide variety of interactions between plants. Use of ground cover crops and smother crops are one of the oldest and traditional practices. Several studies have shown that cover crops reduce soil erosion, conserve moisture, improve soil nutrients and suppress weeds. Cover crops with allelopathic effect can inhibit the growth of noxious weeds and thus maintain the sustainability of the agro-ecosystem. This chapter reviews the allelopathic potential of buckwheat, isolation and identification of the bioactive phenolics and piperidine alkaloids and role of these allelochemicals as to how these bioactive compounds from buckwheat suppress weeds.

Keywords: *buckwheat, allelopathy, plant growth inhibitors, phenolics, piperidine alkaloids*

INTRODUCTION

Allelopathy is a phenomenon of chemical interaction and is probably of wide spread significance in the functioning of natural communities (Rice, 1974; 1984). In plant-plant interactions, allelopathy is generally used to denote the process by which plants release phytotoxic compounds (allelochemicals) in the rhizosphere, resulting in a harmful effect on neighboring plants. Recently, utilization of the allelopathic potential of plants for weed control, instead of herbicide application, has been given great emphasis, because it can help reduce the risk of environmental toxicity (Shilling *et al.*, 1985; Weston, 1996; Chou, 1999). Both crops and weeds have been found to contain compounds that can be considered allelopathic and crop species with allelopathic potential have been given greater attention during the last two decades. However, only a few crops have been studied thoroughly in terms of their weed suppressing activity (Rice, 1979; Einhellig, 1996). These include crops such as barley, oat, wheat, rye, cucumber, canola, mustard species, sunflower, sweet clover, hairy vetch, creeping red fescue, tall fescue, perennial and rye grass (Overland, 1962; Fay and Duke, 1977; DeFrank and Putnam,1978; Lockerman and Putnam, 1979, 1981; Putman and DeFrank 1979, 1982, 1983; Barnes and Putnam, 1981, 1983, Iron and Burnside, 1982; Leather, 1983; Barnes *et al.*, 1986; White *et al.*, 1989; Fujii, 1999a, 1999b). Cover crops are mostly used in no tillage production system because they provide surface residue which reduces soil erosion, conserve water retention and soil nutrients (Guenzi and McCalla, 1962; McCalla and Norstadt, 1974; Teasdale *et al.*, 1991). Identification of cover crops that provide weed suppression may increase acceptance and utilization of these soil conserving tillage techniques (Leather, 1982; Weston and Duke, 2003).

Buckwheat is a grain that has been eaten for hundreds of years in the Far East. China, Japan, Korea and other Asian countries have long enjoyed noodles made from buckwheat flour. In fact, Japan is the largest consumer of buckwheat and annually imports 100,000 metric tons of buckwheat. Buckwheat is also used for a variety of baked products, including pancakes, breads, muffins, crackers, bagels, cookies and tortillas among others. Buckwheat is an agronomic species of the polygonaceae family that belongs to the genus *Fagopyrum*. There are about 15 species in the genus *Fagopyrum*, which occur in temperate areas of Euro-Asia. Plants in this family produce a wide array of biologically active constituents. As a fast-growing cover crop, buckwheat is most useful for weed suppression (Tsuzuki *et al.*, 1987a; Tsuzuki and Yamamoto, 1987b). Japanese farmers have long used *Fagopyrum esculentum* (common buckwheat) as a cover crop for limiting soil erosion. There are some reports about its allelopathic activity in the field and in greenhouses (Tominaga and Uezu, 1995). Experiments conducted under field conditions have shown that buckwheat shades and smothers weeds, or out competes them for soil moisture and nutrients. Both living buckwheat and buckwheat residue have an allelopathic effect on weed germination. The best way to take advantage of allelopathy is to cut or mow buckwheat at the flowering stage and to use it as a mulch instead of incorporating it in the soil.

Research shows that allelopathic effects can last for about 30-60 d. A recent report showed that the aqueous extract of aerial parts of buckwheat inhibited germination, or seedling growth of lettuce and other weeds (Isojima *et al.*, 2000). Buckwheat shoots produce and release a variety of bioactive chemicals, which are toxic to weeds (Iqbal *et al.*, 2002, 2003).

The present chapter discusses the role of allelochemicals from the viewpoint of total activity and describes how these bioactive substances contribute towards the allelopathic activity of buckwheat and make it a potentially successful candidate as an allelopathic ground cover crop.

ALLELOPATHIC SUBSTANCES ISOLATED FROM AERIAL PARTS OF BUCKWHEAT

Field Experiments

Crop plants are known to release a vast variety of allelochemicals into the environment, and nearly every class of secondary metabolites is implicated in allelopathic interactions (Harborne, 1980; Putnam 1988; Macias *et al.*, 1993; Qasem and Foy, 2001). Experiments conducted in the field showed that the total number of weeds, which appeared in plots, were not different between buckwheat-weed plots and weed plots (data not shown), however, the total weed dry weight in buckwheat-weed plots was only 25% of that in the weed plots alone (Table 1). The reduction rate of each weed was different among species (data not shown) suggesting differential sensitivity to buckwheat. The current experiments have shown that buckwheat is a successful competitor in the field against weeds and can effectively suppress weed growth. Although, buckwheat did not influence the number of the weeds that emerged, it reduced the weed biomass to a considerable extent.

To suppress weeds, the cover crop of buckwheat should be planted 30-50 d before crop planting. Two consecutive plantings of buckwheat will have a greater effect on weeds. One week after flowering, buckwheat can be mowed or cut and crops can directly be planted into the buckwheat mulch.

Laboratory Experiments

Aqueous extract of aerial parts of buckwheat showed high plant growth inhibitory activity. Bioactivity guided extraction, various chromatographic purification techniques (silica gel, Sephadex gel, HPLC and thin layer chromatography) and advanced instrumentation (mass and NMR spectroscopy) led to the isolation and identification of five phytotoxins implicated as allelochemicals (Fig. 1). Growth and development of test plants was affected significantly and selectively when

Table 1 Effect of buckwheat on the growth of weeds in field

Treatments	*Weed Fresh wt (g)*	*Weed Dry wt (g)*	*Buckwheat Fresh wt (g)*	*Buckwheat Dry wt (g)*	*Weed Suppression (%)*
Control	253.9 7 ± 59.7	78.9 ± 22.4	–	–	–
Buckwheat	39.5 ±1.8	12.8 ± 1.4	1,369.2 ±10.9	431.3 ± 44.7	75.1 ± 2.7

The values within columns represent the average of 3 data points
± Represents standard error (SE) $N = 3$

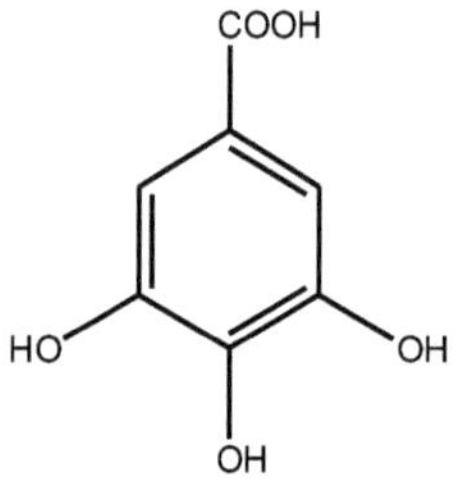

1

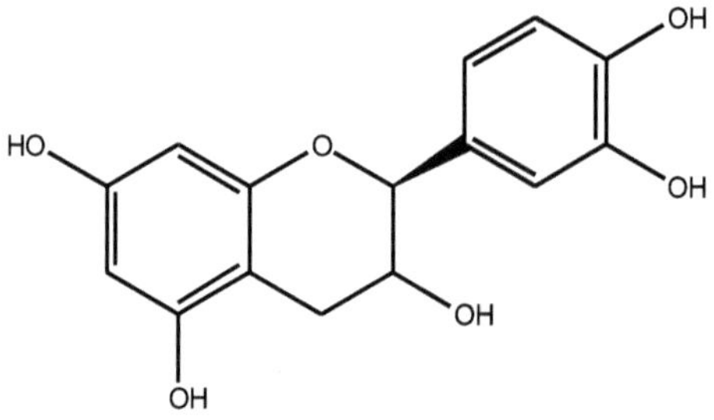

2

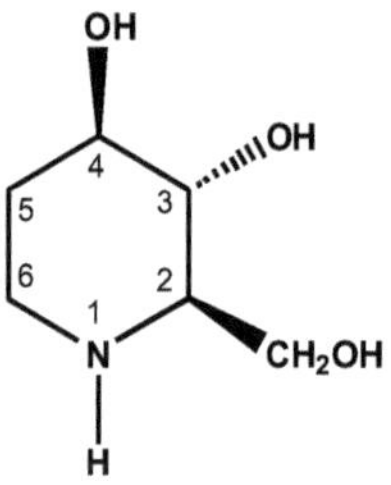

3

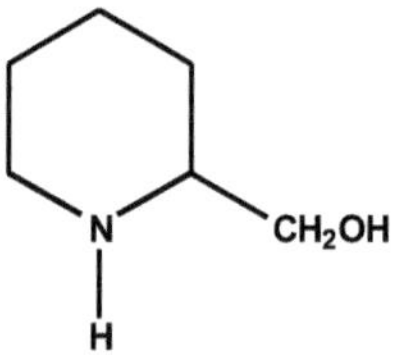

4

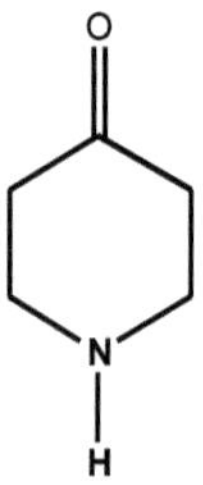

5

Fig. 1 Structures of allelochemicals isolated from buckwheat (*F. esculentum)*

exposed to these allelochemicals. The readily visible morphological abnormalities included reduced radicle or shoot extension, severe browning, and swelling or necrosis of root tips, and discoloration, lack of root hairs and curling of root axis indicating that different classes of compounds have been isolated which are exerting their influence on test plants through different mechanisms.

The growth inhibitory activity of gallic acid (**1**) on the roots of the test plant (*Lactuca sativa* L.) was as much as 10-fold higher than on the hypocotyls indicating that root growth was initially more sensitive to the presence of this natural product (Fig. 2a). The concentration of gallic acid required for a 50% inhibition of elongation of radicals and hypocotyls of lettuce were estimated to be 5 and 50 μgml^{-1}, respectively. This compound inhibited the growth of broad leaf species more than the grasses (Iqbal *et al.*, 2003).

On the contrary, (+)-catechin (**2**) did not show strong inhibitory effects on the root and shoot growth of lettuce (Fig. 2b) but significantly inhibited

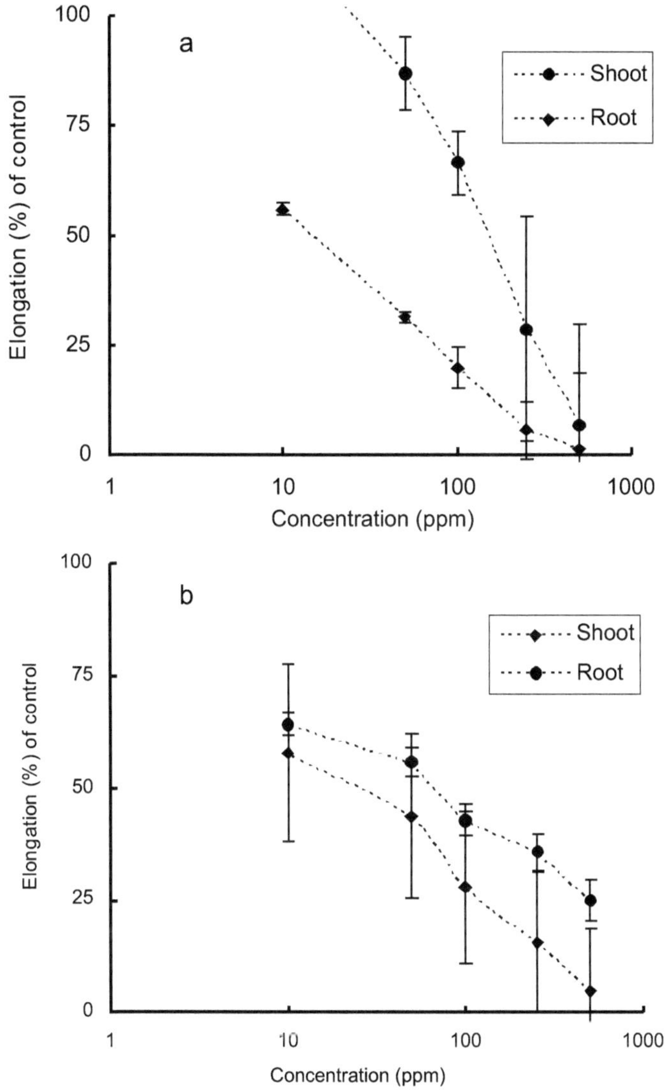

Fig. 2 Effect of gallic acid (a) and (+)-catechin (b) on the root (◆) and shoot (●) elongation of lettuce (*Lactuca sativa* L.) seedlings. Plotted values are means of three experiments. Error bars represent ± standard deviation.

growth of various other weeds and crop species (Fig. 3). Red rice and white rice were found relatively resistant to (+)-catechin, while onion, timothy grass, white clover and yellow mustard showed significant inhibition at the highest applied concentrations.

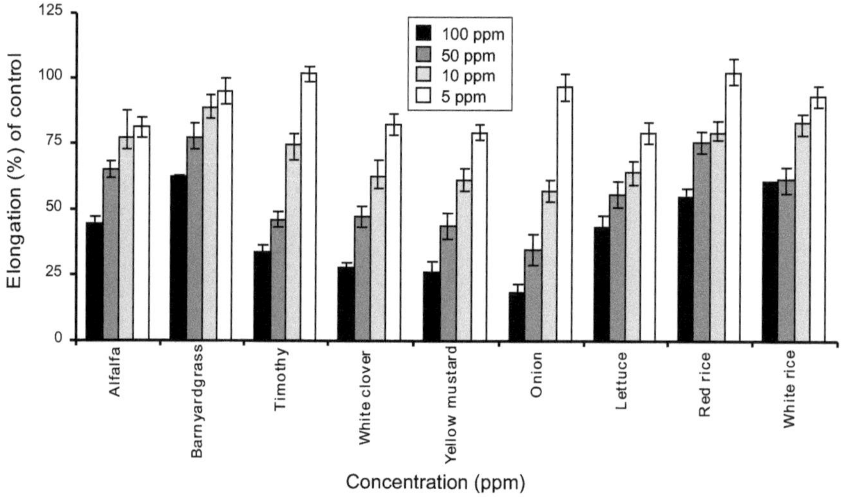

Fig. 3 Effect of (+)-catechin on the root growth various weeds at 100, 50, 10, and 5 ppm. Plotted values are means of three experiments. Error bars represent ± standard deviation.

The other three isolated compounds identified, as fagomine (**3**), 2-piperidine methanol (**4**) and 4-piperidone (**5**) are piperidine alkaloids. The bioassays of these compounds (**3-5**) revealed that the root and shoot growth was completely inhibited at 500 ppm. It was observed that at a concentration of 100 ppm, fagomine inhibited the root elongation of lettuce seedlings by 80% (Figure 4a). The inhibitory potency of compounds **3** and **4** was almost the same. 4-Piperidone (**5**), however, was less potent (Fig. 4b).

The concentrations of compounds **1-5** in aerial parts of buckwheat range from 0.01%-0.04% on fresh weight basis. Their individual threshold concentration to induce plant growth inhibition of target plant are shown in Table 2. These compounds were particularly effective at inhibiting common amaranth. It is closely related to redroot pigweed, which is one of the few weeds in Northeastern buckwheat fields in the USA. The compounds were also effective on lettuce, alfalfa and timothy grass. White clover and yellow mustard were moderately sensitive and crabgrass was relatively insensitive.

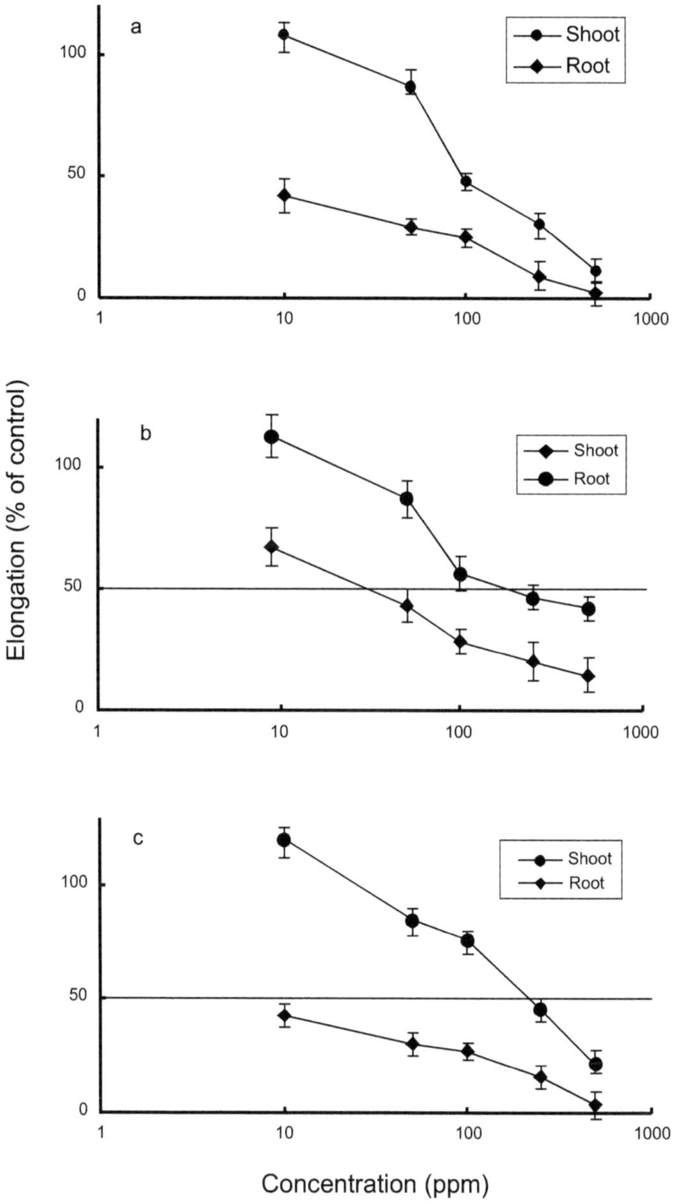

Fig. 4 Effect of fagomine (a), 4-piperidone (b) and 2-piperidine methanol (c) at 500, 250,100, and 10 ppm on the root (◆) and shoot (●) growth of lettuce seedlings. Each symbol represents mean of six data points and error bars represent ± standard deviation (N = 3).

Allelochemicals are present in all parts of plants including leaves, stems and roots and under appropriate conditions, these are released into the environment to affect the growth of neighboring plants (Weston and Duke, 2003). The aerial portions of buckwheat possessed selective plant growth inhibitors such as alkaloids (Iqbal *et al.*, 2002), and simple phenolics (Iqbal *et al.*, 2003) which may be released from plant tissues during the process of degradation. Phenolic compounds have been widely reported to be highly potent inhibitory substances to seed germination and seedling root and shoot elongation (Rice, 1984; Yamada *et al.*, 1995; Kushima *et al.*, 1998). Alkaloids are considered to have a major role in the plant defense strategy (Swain 1977, 1993; Lovett *et al.*, 1981; Wink and Twardowski, 1992; Kato-Noguchi *et al.*, 1994).

It has recently been reported that the purified compounds of low specific inhibitory activity may play a greater role in the expressed inhibitory potential than those of high specific inhibitory activity by virtue of being present in larger quantity (Hiradate *et al.*, 2004). Table 2 shows the inhibitory activity of isolated allelochemicals (**1-5**) in terms of specific activity and total activity. It appears that among these compounds, gallic acid plays a pivotal role in the total allelopathic activity of buckwheat in combination with all the other compounds, which are not present in significant amounts. Indeed, the interaction of a number of allelochemicals may be required for the potent activity observed in the crude extract. These studies suggest the allelopathic potential of buckwheat in biological weed management as a ground cover crop. The compounds isolated from the aerial parts may also be present in the roots and released into the soil through root exudates, or decomposition of plant material, or from decaying plants used as cover mulch.

Table 2 Inhibitory activity of allelochemicals (phenolics and alkaloids) expressed as specific activity and total activity

Isolated compounds	*Content (mg/g Fw)*	*EC_{50} (ppm)*	*Specific activity (M)*	*Total activity*
Fagomine	0.22	10	68.0	22.0
2-Piperidinemethanol	0.14	10	87.0	14.0
4-Piperidone	0.12	50	505.1	2.4
Gallic acid	0.4	8	47.1	50.0
(+)-Catechin	0.2	32	110.3	6.3

CONCLUSION

Buckwheat is an economically important forage grain and cover crop that has shown strong allelopathic potential in field, greenhouses and laboratory studies. It contains and releases a number of interesting and well-characterized chemicals including gallic acid, (+)-catechin and fagomine, which show significant inhibition of root and shoot growth of lettuce and other tested weeds. From the data presented in this chapter it is concluded that gallic acid, which is present in enormously high concentrations, might play a key role and act synergistically with other minor allelochemicals to accentuate the allelopathic activity of buckwheat. It is therefore possible that in the future, buckwheat could be used as ground cover crop. This will allow long-term environmentally friendly weed suppression in agro-ecosystem.

References

Barnes, J.P., Putnam, A.R. (1981) Weed suppression with rye cover crops in vegetable cropping system. *In* the Proceedings of North Central Weed Control Conf. p. 64.

Barnes, J.P., Putnam, A.R. (1983) Rye residues contribute weed suppression in no-tillage cropping system. J. Chem. Ecol. 9: 1045-1057

Barnes, J.P., Putnam, A.R., Burke, A. (1986). Allelopathic activity of rye (*Secale Cereale* L). *In* The Science of Allelopathy, A.R. Putnam, C.S. Tang (eds.). John Wiley & Sons, NY, USA. pp. 271-286.

Chou, C.H. (1999) Roles of allelopathy in plant diversity and sustainable agriculture. Crit. Rev. Plant Sci. 18: 609-636.

DeFrank, J., Putnam. A.R. (1978) Weed and crop response to allelopathic crop residues. *In* the Proceedings of North Central Weed Control Conf. pp. 33-34.

Einhellig, F.A. (1996) Interactions involving allelopathy in cropping system. Agronomy J. 88: 886-893.

Fay, P.K., Duke, W.B. (1977) An assessment of allelopathic potential in *Avena* germ plasm. Weed Sci. 25: 224-228.

Fujii, Y. (1999a) Allelopathy of velvetbean: Determination and identification of L-DOPA as a candidate of allelopathic substances. *In* Biologically Active Natural Products., H.G. Cutler, S.J. Cutler. (eds). CRC Press. Boca Raton, Fl, USA pp. 33-48.

Fujii, Y. (1999b) Allelopathy of hairy vetch and mucuna: their application for sustainable agriculture. *In* Biodiversity and Allelopathy; From Organisms to Ecosystems in the Pacific, C.H. Chou, G.R. Wallerand, C. Reinhardt (eds.), Academia Sinica Taipei, Taiwan, Republic of China. pp. 289-300.

Guenzi, W.D., McCalla, T.M. (1962) Inhibition of germination and seedling development by crop residues. Soil Sci. Soc. Am. Proc. 26: 456-458.

Harborne, J.B. (1980) Plant phenolics *In* Secondary Plant Products. Encyclopedia of Plant Physiology; New series, E.A. Bell, B.V. Charlwood (eds). Springer-Verlag, NY, USA 8: 329-402.

Hiradate, S., Morita, S., Sugie, H., Fujii, Y. (2004) Pytotoxic *cis*-cinnmoyl glucosides from *Spiraea thunbergii*. Phytochemistry. 65: 731-739.

Irons, S.M., Burnside, O.C. (1982) Competitive and allelopathic effects of sunflower (*Helianthus annuus*). Weed Sci. 30: 372-377.

Isojima, S., Iqbal, Z., Koizumi, A., Fujii, Y. (2000) Allelopathy of *Fagopyrum esculentum*: Analysis of allelochemicals. J. of Weed Sci. Tech. (supplements in Japanese) 45: 92-93.

Iqbal, Z., Hiradate, S., Noda A., Isojima S., Fujii, Y. (2002) Allelopathy of buckwheat: assessment of allelopathic potential of extract of aerial parts and identification of fagomine and other related alkaloids as allelochemicals. Weed Biol. and Manage. 2: 110-115.

Iqbal, Z., Hiradate, S., Noda, A., Isojima, S., Fujii, Y. (2003) Allelopathic activity of buckwheat: Isolation and characterization of phenolics. Weed Sci. 51: 657-662

Kato-Noguchi, H., Kosemura, S., Yamamura, S., Mizutani, J., Koji, H. (1994) Allelopathy of Oats. I Assessment of allelopathic potential of extract of Oat shoots and identification of an allelochemical. J. Chem. Ecol. 20: 309-314.

Kushima, M., Kakuta, H., Kosemusa, S., Yamamura, S., Yamada, K., Yokotani-Tomita K., Hasegawa, K. (1998) An allelopathic substance exuded from germinating watermelon seeds. Plant Growth Regul. 25: 1-4.

Leather, G.R. (1982) Weed control using allelopathic crop plants. J. Chem. Ecol. 9: 983-989

Leather, G.R. (1983) Sunflowers (*Helianthus annuus*) are allelopathic to weeds. Weed Sci. 31: 37-42.

Lockerman, R.H., Putnam, A.R. (1979) Evaluation of allelopathic cucumbers (*Cucumis sativus*) as an aid to weed control. Weed Sci. 27: 54-57.

Lockerman, R.H., Putnam, A.R. (1981) Growth inhibitors in cucumber plants and seeds. J. Am. Soc. Hort. Sci. 106: 418-422.

Lovett, J.V., Levitt, J., Duffield, A.M., Simth, N.G. (1981) Allelopathic peotential of *Datura stramonium* L. (thorn-apple). Weed Res. 21: 165-170.

McCalla, T., M. Norstadt, F.A. (1974) Toxicity problems in mulch tillage. Agric. Environ. 1: 153-174.

Macias, F.A., Galindo, J.C.G., Massanet, G.M., Rodrigez-Luis, F., Zubia, E. (1993) Allelochemicals from *Pilcarpus goudotianus* leaves. J. Chem. Ecol. 23: 1781-1803.

Overland, L. (1966) The role of allelopathic substances in the "smoother crop" Barley. Am. J. Bot. 53: 423-432.

Putnam, A.R., DeFrank, J. (1979) Use of allelopathic cover crops to inhibit weeds. Science. 36: 58-582.

Putnam, A.R., DeFrank, J. (1983) Use of phytotoxic plant residues for selective weed control. Crop Protection. 2: 173-181.

Qasem, J.R., Foy, C.L. (2001) Weed allelopathy, its ecological impacts and future prospects. *In* Allelopathy in Agroecosystems, R.K. Kohli, H.P. Singh, and D.R. Batish (eds). Haworth Press, NY, USA. pp. 43-119

Rice, E. L. (1974) Role of allelopathy in patterning of vegetation and creation of bare areas. *In* Allelopathy, Academic press. NY, USA. pp. 126-173.

Rice, E.L. (1979) Allelopathy. An update. Bot. Rev. 45: 15-109.

Rice, E.L. (1984) Factors effecting quantities of inhibitors. *In* Allelopathy, 2nd ed. Academic Press, NY, USA. pp. 343-345.

Shilling, D.G., Libel, R.A., Worsham, A.D. (1985) Biochemical interaction among plants. *In* The Chemistry of Allelopathy, A.C. Thompson (ed.). ACS symposium series 268. American Chemical Society, Washington DC, USA. pp. 243-271.

Swain, T. (1977) Secondary compounds as protective agents. Annu. Rev. Plant Physiol. 28: 479-501.

Teasdale, J.R., Beste, C.E., Potts, W.E. (1991) Response of weed to tillage and cover crop residues. Weed Sci. 39: 195-199.

Tominaga, T., Uezu, T. (1995) Current advances in buckwheat research. 693-697.

Tsuzuki, E., Yamamoto, Y., Shimizu, T. (1987a) Fatty acids in buckwheat are growth inhibitor. Ann. Bot. (London) 60: 69-70.

Tsuzuki, E., Yamamoto, Y. (1987b) Isolation and identification of phenolic substances from wild perennial buckwheat (*F. cymosum* M.) Miyazaki Daigaku Nogakubu Kenkyu Hokoku. 34: 289-95.

Weston, L.A. (1996) Utilization of allelopathy for weed management in agro ecosystems. Agronomy J. 88: 860-866.

Weston, L.A., Duke, S.O. (2003) Weed and crop allelopathy. Crit. Rev. Plant Sci. 23: 367-389.

White, R.H., Worsham, A.D., Blum, U. (1989) Allelopathic potential of legume debris and aqueous extracts. Weed Sci. 37: 674-679.

Wink, M., Twardowski, T. (1992) Allelochemical properties of alkaloids. Effects on plants, bacteria and protein biosynthesis *In* Allelopathy: Basic and Applied Aspects, S.J.H. Rizvi, V. Rizvi (eds.). Chapman & Hall, London, UK. pp. 129-150.

Wink, M. (1993) Inhibition of seed germination by quinolizidine alkaloids. Planta. 158: 365-368.

Yamada, K., Anai, T., Hasegawa, K. (1995) Lepidimoide, an allelopathic substance in exudates from germinating seeds. Phytochemistry. 39: 1031-1032.

12

Sunflower - desired Allelopathic Crop for Sustainable and Organic Agriculture?

Helena Gawronska, Dorota Ciarka, Waldemar Bernat and Stanislaw W. Gawronski

Laboratory of Basic Sciences in Horticulture,
Faculty of Horticulture and Landscape Architecture
Warsaw Agricultural University
Nowoursynowska 159, 02-776 Warsaw, Poland

Sunflower, listed as a crop species of high allelopathic activity with the potential of its exploitation as complementary to other practices for weed managements, has received recreantly increased attention. The aim of this chapter is to describe how genotypic variation, growing conditions, and various plant organs, such as leaves, stems and inflorescences may influence the allelopathic activity of sunflower. The allelopathic effects of sunflower can be both stimulatory and inhibitory and is manifested both against weeds and crops including sunflower – autotoxicity. Sunflower shows inhibitory effect on a relatively wide range of weeds and crops as assayed under laboratory, greenhouse and field conditions. It is believed that dicotyledonous weeds are more affected than grasses though, it is not always the case as four species of *Poaceae* family are quite susceptible. Sunflower allelopathy can be exploited in several ways, among others as biologically active mulch. It effectively reduces weeds infestation in the field with some being almost completely eliminated, including the troublesome in cereals of Central Europe – *Apera spica-venti*. However, it may also adversely affect crops. It is too early to propose a ready- to- use strategy for weed management employing allelopathy and more research activities are needed. Based on available in literature data and own research, some recommendations for overcoming the negative impact on crops are discussed here and future research lines are provided.

Keywords: *allelopathic potential, bioactive mulch, genotype, Helianthus annuus* L., *natural herbicides, organic farming, Low input agriculture*

INTRODUCTION

There are many papers focusing on the effects of allelopathy on agro-ecosystems and on their exploitation in agriculture (Putman and Duke, 1978; Putman, 1985; Putman and Weston, 1986; Einhellig and Leather, 1988; Anaya, 1999; Barberi, 2002; Bhowmik and Inderjit, 2003).

In sustainable (especially organic farming) agriculture, the major threat and extremely challenging task is weed control. Weed management is an important part of the following: reducing yield, crop seeds contamination with weed seeds as well as limiting the build-up of the weed seed bank in the soil (Beveridge and Naylor, 1999; Davies and Welsch, 2002). It has recently been suggested in several papers, that allelopathy holds great prospects for finding an alternative strategy for weed management, whereby the reliance on traditional herbicides in crop production can be reduced (Weston, 1996; An *et al.*, 1998; Wu *et al.*, 1999; Kruse *et al.*, 2000; Bárberi, 2002; Bhowmik and Inderjit, 2003). Weston (1996) provides examples of using crops (rye, buckwheat, black mustard or sorghum-sudangrass hybrid) for satisfactory weed suppression with minimal or no herbicide use and Leather (1987) showed that weed suppression by sunflower was as effective as using herbicides S-ethyl dipropyl carbamothioate (EPTC) in sunflower fields. Allelopathy as a complementary element of weed management is getting increasing attention not only for organic farming but also for low chemical input agriculture. This is due to (i) the cost of high technology agriculture not acceptable everywhere, (ii) the appearance of herbicide resistant biotypes, (iii) increasing awareness of the environmental pollution by pesticides. Allelopathic-mediated weed control is considered to be: 1) cheaper (in case of sunflower or other cash-crops of allelopathic activity might be even without extra cost), 2) friendly to the environment and 3) socially acceptable to be applied in practice.

Sunflower (*Helianthus annuus* L.) is native to North America and due to its high adaptive ability (relatively tolerant to drought, can grow in quite wide temperature regimes, insensitive to photoperiod) and qualitative traits (high oil and protein content) has been growing in many parts of the world. It is cultivated for several purposes i.e. (i) eatable oil, (ii) as confecionery for human consumption, (iii) ornamental plant, (iv) green manure in mixture with legumes, and (v) for oil as a source of renewable energy (biofuel) - a newly emerging form of sunflower utilization. Sunflower is listed among species that exhibit strong allelopathic activity towards both weeds and crops (Schon and Einhellig, 1982; Leather, 1983a, b; 1987, Purvis, 1990; Purvis and Jones, 1990; Morris and Parrish, 1992; Narwal, 1999; Sedimey, 1999; Waller and Einhellig, 1999; Batish *et al.*, 2002; Ciarka *et al.*, 2002a, b, 2004b; Gawronska *et al.*, 2002, 2003, 2004; Gawronski *et al.*, 2002, 2004; Om *et al.*, 2002; Azania *et al.*, 2003; Bernat *et al.*, 2004a, b). The economic importance of sunflower and its increased acreage in many countries, has made it an important element of the local flora. Since allelopathy regulates the vegetation composition of any ecosystem (Einhellig, 1985; Putman and Tang, 1986; Chou, 1999) it would be an important factor influencing biodiversity, most often by its' reducing. However, if sunflower were a part of a rotation system in organic farming it would increase the local biodiversity as a member of the *Asteraceae* family. The increase of biodiversity, at any level of biological organization, as pinpointed by Bárberi (2002) usually results in better, healthier, more stable and a productive agro-ecosystem. Therefore, the allelopathic effect of sunflower cannot be neglected from the agro-ecosystem standpoint both in detrimental and beneficial terms. Knowledge of these traits can be managed to the benefit of farming by introducing them to weed management. On the other hand, we need to be aware that sunflower can create a problem as a weedy component of the ecosystem including agro-ecosystem if seeds are lost during harvest.

In addition to the above there are other points that make the sunflower a highly desirable candidate for weed management:

1. It is well known that fields in successive years after sunflower cultivation are substantially less weedy (Leather, 1983a, b; Semidey, 1999; Gawronski *et al.*, 2002, 2003, 2004 and in preparation).
2. Among evaluated genotypes, great differences in biological yield were recorded (Ciarka *et al.*, submitted). Those producing high biological yield, are preferred independent whether used as green manure, biologically active mulch or if left as residues after grains were harvested. In any of the above, the biomass will serve as a source of allelochemicals - a natural herbicide released from the biomass.
3. Sunflower genotypes also greatly differ in biomass partitioning patterns. The knowledge of this plays a crucial role if, looking for cultivar(s) of high allelopathic activity or of prolonged periods of allelochemicals release. In the first case *cv(s).* with high contribution of the leaves in the total biomass will be desired. The leaves, as shown by several researchers (Irons and Burnside, 1982; Leather, 1983a; Morris and Parrish, 1992; Ciarka *et al.*, submitted) possess the highest allelopathic activity. Thus, *cv(s).* with high proportion of biomass in leaves should ensure more effective allelopathic impact on weeds. If the release of allelochemicals is more important over a

longer period of time or of lower allelopathic activity, *cv(s).* with higher proportion of biomass partitioned to stem(s) will be preferred more. The sunflower stem as organ of slower than others decomposition, ensure prolonged release of allelocompounds resulting in longer allelopathic effect on weeds, what as shown in our study was observed still in during spring vegetation of the successive year.

4. Leather (1983a), Macias *et al.*, (1999) and Ciarka *et al.*, (submitted) showed great genotypic differences in allelopathic activity among 13, 26 and 37 genotypes of sunflower respectively.
5. It can be assumed that much bigger differences, in the three characters mentioned above, will be found if screening covers greater number of accessions. Thus, the potential exists to breed sunflower cultivars of higher allelopathic activity.
6. The results of some studies show that cereals are less affected than other tested species especially dicotyledonous, by allelochemicals derived from sunflower (Leather, 1983a; Narwal *et al.*, 1999a; Azania *et al.*, 2003) and as demonstrated by Gawronska *et al.*, (2003, 2004) winter wheat *cv.* Zyta possesses higher tolerance than mustard to sunflower allelochemicals both at seed germination, early seedling growth (heterotrophy) and at stages with fully developed photosynthetic apparatus (autotrophy).
7. Sunflower is suppressive against quite a broad range of weeds (Leather, 1983a; Macias *et al.*, 1999; Azania *et al.*, 2003; Ciarka *et al.*, submitted) which is a desired character for application in practice.
8. Sunflower might be a good choice to cultivate in organic farming in those parts of the world where invasion of the of parasitic weed *Orobanche cermua* Loefl. is a serious threat. According to Serghini *et al.*, (2001) there is a genotype of sunflower resistant to *Orobanche cermua* Loefl., which when infected accumulates and exudes coumarines in greater than susceptible ones (showing involvement of allelopathy). Introducing resistant genotype to cultivation would slow down the invasion of *Orobanche cermua* Loefl.
9. Sunflower green manure or mulch increase nutrients available for crop, protect water loss, improve soil structure and, as shown by Alsaadawi (1988), might suppress nitrification. These together with direct sowing with disc seeder, which is necessary when sowing into mulch, would also act against erosion.
10. Sunflower is known from synthesis of allelochemicals of great variation in chemical structure and activities, which might serve as templates for "natural herbicides" synthesis differing in chemical structure and mode of action (Macias *et al.*, 1999 and references therein).

In conclusion, sunflower, as a fast growing, producing high biomass with differences in partitioning patterns and differing in allelopathic potential is a good candidate for cultivation as a cash crop with residues left on the ground, for green manure or bio-chemically active mulch. All three serving as a source of allelochemicals as well as for isolation compounds for spray or as templates for synthesis of "natural herbicides." Therefore, it is reasonable to consider the use of sunflower allelopathy, as complementary to other practices, in weed management. It appears to be a promising tool in Low input and organic agriculture. However, intensive research activities are still required before being applicable in practical agriculture.

VARIATION IN ALLELOPATHIC ACTIVITY OF SUNFLOWER – BIOASSAYS

The allelopathic effects of any donor plant can vary depending on genotype, organ, amount of released allelocompounds, environmental conditions that occur on donor plants during growth and on the acceptor plants.

Genotypic Variation

The genetic variation for sunflower was first reported by Leather (1983a, 1983b) who evaluated responses of 17 weed species to native and 13 cultivars and hybrids of sunflower and showed that these cultivars significantly differed in the activity, as based on effects of water extracts used in bioassays. However, these differences were not so evident in field evaluation. He also showed that the native sunflower exerted significantly lower effects both in the laboratory and in the field conditions. Macias *et al.* (1999 and references therein) while looking for allelochemicals suitable to serve as templates for "natural herbicides" evaluated 26 cultivars of sunflower cultivated in the Andalusian region in Spain and found that they greatly differ in the effects exerted on lettuce germination and seedling growth. In our recent study the effect of aqueous leaf extracts of 37 cultivars and breeding lines of sunflower on mustard germinability was tested. The results revealed that allelochemicals contained in the extracts of all tested genotypes delayed and reduced germination in a dose-dependent manner and significant differences were found between the genotypes and extract concentration. On an average for four used concentrations (2.5, 5.0, 7.5 and 10% DM w/v) between 31.7% (*Albena, Mikaflor*) and 89.4% (*Goldener Neger*) of sown seeds germinated (in control 97.6% sown seeds germinated) (Table 1). All these show that there are great genotypic differences in sunflower for allelopathic traits and that there is potential to breed cultivars of high allelopathic activity.

Table 1 Germination of mustard (*Sinapis alba* L.) *cv. Nakielska* in the presence of allelochemicals contained in aqueous extracts from various organs of sunflower (*Helianthus annuus* L.) evaluated on 14th day of culture. Data, expressed as % of control, are mean of 4 extract concentrations (6 replicates with 50 seeds in each). In control (H_2O) in average for about 70 plates 97.0% of sown seeds germinated. Bold values refer to highest and to lowest values. Differences between combinations were estimated by HSD at $\alpha = 0.05$ of the Tukey test.

Genotype	*Extract from*			*Extract*
	Leaves	*Stem(s)*	*Inflorescence(s)*	*from roots*
Albena	**31.7**	80.7	84.3	96.2
Alphasol	35.3	83.4	85.5	nt*
Borowski Prazkowany	62.9	83.2	69.0	nt
Borowski Ulepszony	55.7	45.2	84.8	nt
Einfache Gemitsch	62.9	83.3	54.0	nt
Evening Sun	61.3	92.0	79.3	nt
Floristan	37.5	69.3	83.8	nt
Frankasol	71.1	82.7	75.5	nt
Full Sun	57.9	86.2	92.8	nt
Giganta	47.7	92.7	90.5	94.4
Goldener Neger	**89.4**	90.6	93.0	nt
Hallo	54.9	89.6	56.0	nt
Herbstschonheit	58.4	90.5	84.8	nt
Lech	43.6	83.9	88.8	94.8
Mikaflor	31.7	54.7	88.0	nt
Ogrodowy	77.4	89.4	94.0	**96.5**
Olbrzym Borowski	82.2	89.4	92.0	nt
Optisol	64.2	78.6	89.5	94.6
Pardisol	48.2	71.0	**95.0**	**87.4**
Pastewny	47.1	91.3	91.3	94.3
Pelny	80.0	83.9	91.0	nt
Polizie 98-43	78.3	89.2	**95.0**	94.9
Printasol	33.2	**44.7**	82.0	90.3
Schnittgold	65.6	**93.5**	76.0	nt
Sion	42.5	83.3	91.8	nt
Soraya	50.1	73.6	**45.5**	nt
Sunriche Orange	36.3	81.0	60.5	nt
Ukr Nar 96-291	85.2	91.1	86.0	nt
Valentin	59.1	83.5	87.5	nt
Wielkopolski	43.2	66.2	80.8	93.1
4449	41.7	84.4	91.3	nt
57953	57.8	47.1	88.5	nt
57956	48.4	85.2	86.8	nt
57957	53.4	82.4	65.5	nt
57958	38.7	63.9	64.3	nt
57959	45.4	73.6	71.3	nt
57960	56.4	87.2	78.3	nt

nt *– not tested
$HSD_{0.05}$ cultivars = 6.09
$HSD_{0.05}$ organs of above ground parts = 1.05
$HSD_{0.05}$ roots = 3.98

It should be pointed out that both pH and osmotic potential of the extracts (determined for sunflower cv. Ogrodowy) used in the germinating test presented in Table 1 were lowered along with increased extract concentration from 7.52 to 6.41 (pH) and from - 0.22 to -0.70 MPa. These changes however, are not the major cause for reported delay and reduction of germination since mustard delay and reduction germination in PEG solutions with respective osmotic potentials were much less affected (data not shown).

Organ-specificity

The plant parts (composed of few organs) or particular organs differ in the allelopathic activity and for sunflower it has been demonstrated in several studies (Irons and Burnside, 1982; Leather, 1983a, b; Morris and Parrish, 1992; Narwal *et al.*, 1999a, b) but results obtained are not corresponding and sometimes are even contradictory. For example, Irons and Burnside (1982), Leather (1983a) recorded higher leaves than stems with branches phytotoxicity, while Leather in other work (1983b) and Maruthi and Sankaran (2001) reported the opposite i.e. the highest stem activity followed by leaves and roots. In the study of Narwal *et al.* (1999b) addition of sunflower roots to the soil led to greater seedlings mortality of tested species when compared to soil amended with full sunflower plant biomass. In the case of Morris and Parrish (1992), leaves had higher inhibitory activity than stems against weeds but only at the highest extract concentrations while at lower concentrations no differences were recorded. In the case of wheat both leaf and stem extracts had stimulatory effects. Taking into account these discrepancies we conducted a systematic study, under the same conditions, examining allelopathic activity of aqueous extracts (at four concentrations) made of leaves, stem(s) inflorescence(s) for all 37 genotypes and of leaf petioles, roots and roots exudates, and of leaves volatiles for either 2, 10 *cvs.* or 1 *cv.* respectively. These effects significantly differed among genotypes and plant organs tested and, in almost all cases, showed dose-dependent manner. Contrary to some of the studies mentioned above, allelochemicals extracted from leaves of all sunflower genotypes were the most effective followed by stems and inflorescences in reducing mustard germination (Table 1), whereas for roots the inhibitory effect, if any, was substantially little. It is worth mentioning that leaf extracts at concentrations of 7.5 and 10.0% of most genotypes completely or strongly inhibited mustard germination. The allelopathic activity of sunflower leaf petioles (checked only for two genotypes) was similar to stems and inflorescences extracts. It is interesting that, as mentioned above, sunflower roots, contrary to all other organs, possessed very litle or no allelopathic activity. However, growth of mustard plants was significantly inhibited and biomass accumulation

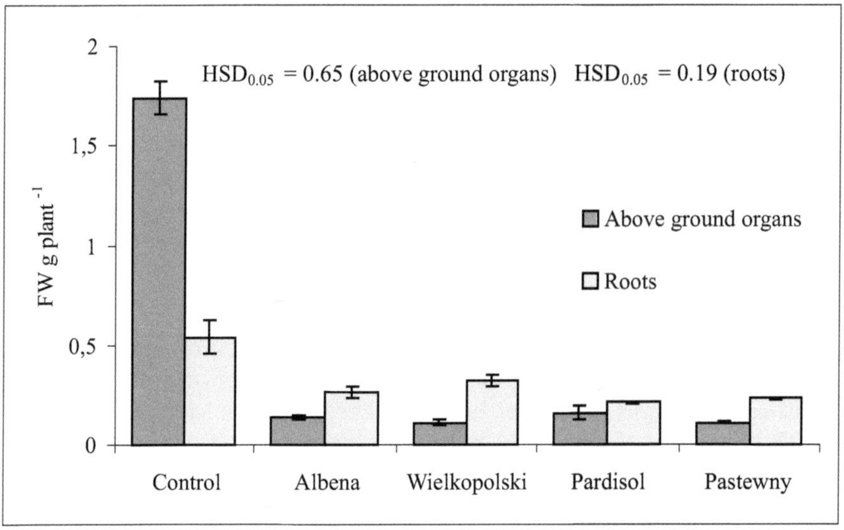

Source: Ciarka *et al.*, 2004b, with the permission of The Polish Academy of Sciences, Division of Agricultural, Forestry and Veterinary Sciences.

Fig. 1 Fresh weight of mustard (*Sinapis alba* L.) *cv.* Nakielska plants grown for 28 d in the soil previously used for cultivation of 4 *cvs.* of sunflower (*Helianthus annuus* L.). Soil, before usage, was deprived of any visible sunflower roots. Data are mean ± SE, n = 10

reduced (up to 85.1% of control plants) by allelocompounds contained in soil collected after sunflower cultivation and deprived from any visible roots (Fig. 1).

This corresponds with the results of Irons and Burnside (1982) for sorghum, soybean and sunflower. This, most probably, was an effect of the presence in the soil of (i) tiny, not possible to collect roots including roots hair and (ii) the compounds that were released by roots via exudation, which is, according to Bertin *et al.* (2003), the major way of releasing compounds into the soil environment. There are studies on sunflower root exudates, and in general sunflower root exudates are less effective than leaf and stem leachates (Irons and Burnside, 1982; Leather, 1983a, b; Azania *et al.*, 2003). Ciarka *et al.* (2004b) demonstrated inhibitory effects of sunflower roots exudates on mustard seedling in the experiment, in which mustard seedlings were co-cultured on Petri dishes together with more advanced in development sunflower seedlings (Fig. 2).

Also Ohno *et al.* (2001) showed that germinating seeds of sunflower exude substance that displayed allelopathic activity.

In response to allelocompounds the delay in mustard germination was noted for all genotypes and all tested organs except for roots, extracts of which, as in the case of reduction in germination, had no effect. Also in this

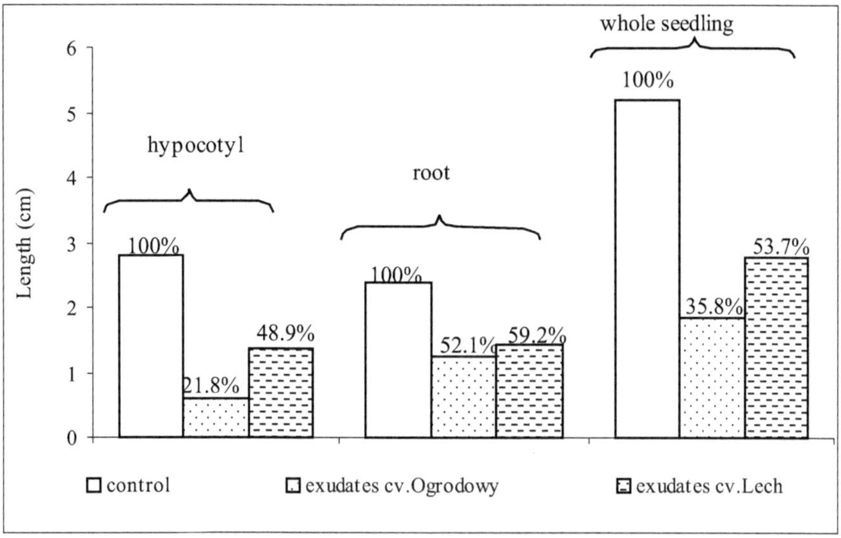

Source: Ciarka *et al.*, 2004b, with permission from The Polish Academy of Sciences, Division of Agricultural, Forestry and Veterinary Sciences.

Fig. 2 Length of mustard (*Sinapis alba* L.) *cv.* Nakielska seedlings as influenced by allelochemicals exuded by seedlings of 2 *cvs.* of sunflower (*Helianthus annuus* L.). Values above the columns are expressed as % of control. Data are mean of 8 replicates.

parameter, leaf extracts were the most active of all other organs (Ciarka *et al.*, submitted) and the greatest delay (by 6.7 d when compared with the control) caused extract made from Polize-98-43 *cv.* at the highest concentration.

The allelopathic effect of volatiles is known for other species (Qasem, 1999; Pickett *et al.*, 2003) but to the best of our knowledge, data on this for sunflower are not available yet. The results of our recent study showed that in the case of sunflower the volatiles participate in sunflower allelopathy manifestation. Allelocompounds released from fresh sunflower leaves had inhibitory effects on mustard leading to strong inhibition of seed germination and loss of vigor: seedlings developed in the presence of volatiles were weaker, shorter, stunted and their fresh weight was lower. The allelopathic effect was always greater for chopped leaves compared to intact ones and for higher amounts of biomass, from which the volatiles were released (Ciarka *et al.*, in submitted).

Effects of Environmental Conditions

The allelopathic effects of the same genotype change depending on environmental conditions, which the donor plants experienced during growth (Einhellig, 1984, 1987, 1996, 1999, 2004; Anaya, 1999; Inderjit and

Keating, 1999; Reigosa *et al.*, 1999, 2002; Inderjit and Nayyar, 2002; Gawronska and Golisz, 2005). It is commonly known that both abiotic (nutrients deficiency, elevated temperatures, drought stress, flooding, oxygen deficiency, irradiation) (Einhellig, 1987, 1996; Chou, 1999), and biotic stress factors (pathogens, insects, parasites infestations, competition, allelopathy and autoxicity (Inderjit and Dakshini, 1992; Einhellig, 1996; Singh *et al.*, 1999; Sutherland *et al.*, 1999; Yu, 1999; Kong *et al.*, 2002; Gawronska and Golisz, 2005) would elicit defense mechanism(s) against the stress, which among others might increase synthesis of some secondary metabolites. Many of these metabolites are of multifunction mode of action including allelopathic activity. Thus, enhanced levels of such metabolites might lead to increased allelopathic activity though, as demonstrated by Kong *et al.* (2002) it is not always the case. On the other hand, Einhellig (1987, 1999), Chou (1999), Inderjit and Nayyar (2002), and Reigosa *et al.* (2002) stated that allelopathy is more pronounced and could be important when acceptor plants are affected by environmental stresses. In addition, there are claims that plants under stress(es), are less vigorous, and are likely to be less tolerant to other stressful conditions including allelopathy stress (Patrick *et al.*, 1964; Reigosa *et al.*, 2002 and references therein).

For sunflower, the alteration in allelopathic activity of plants exposed to nutrient deficiency reported by Hall *et al.* (1982, 1983) is based on increased levels of phenolic compounds (expressed in chlorogenic acid equivalent). The authors demonstrated depressive effects on *Amaranthus retroflexus* seed germination by debris of sunflower plants exposed to various nutrient stress (Hall *et al.*, 1982). The increased allelopathic effect was attributed to phenolic compounds but only when released from the sunflower debris. Pure chlorogenic acid added to germinating medium did not exert such a response suggesting that other compounds and/or soil microorganisms operate in concerted action on additive/synergistic manner. The negative impact of stressed sunflower plants was also observed for biomass accumulation by *A. retroflexus* seedlings (Hall *et al.*, 1983). In our study we also observed higher allelopathic activity when field grown sunflower plants *cv.* Ogrodowy were sprayed with NaCl or $CO(NH_2)_2$ but not when mechanically injured (Bernat and Ciarka, personal communication). The higher accumulation of coumarins and greater root exudation in resistant as compared to susceptible genotype of sunflower was observed by Serghini *et al.* (2001) under biotic stress caused by plant infestation with *Orobanche cernua* Loefl.

Allelopathic Activity Against Most Common Weeds

The inhibitory effect of sunflower on weeds is quite well documented in the literature (Leather, 1983a, b, 1987; Morris and Parrish, 1992; Macias *et al.*, 1999; Nanjappa *et al.*, 1999; Azania *et al.*, 2003) though with relatively limited number of species and at differing testing conditions. In our study we tested, under the same conditions, 33 weed species, commonly occurring among cereals in Central Europe. The studies showed that in response to allelochemicals derived from sunflower leaves, germination, with few exceptions, was delayed and reduced and the response was dose-dependent. Out of 33 weed species, on average for five concentrations (1.25, 2.5, 5.0, 7.5 and 10.0% DM w/v) of aqueous extracts 18 species germinated below 25%, when compared with germination under control conditions, and further 13 between 25% and 50%. Two species *Vicia cracca* L. and *Echinochloa crus-galli* L. were hardly affected by allelochemicals of sunflower origin as they germinated in 82% of seeds germinated in H_2O (Fig. 3). The greatest reduction was noted for *Papaver rhoeas* L. (full inhibition already at 5.0% extract concentration) but complete inhibition was also noted for other weed species though, at higher concentrations (13 and 19 species at 7.5% and 10.0%, respectively) (Ciarka *et al.*, submitted). As in the case of mustard, stem extracts were less effective than leaves in almost all cases and in some at lowest concentrations stimulation was also recorded. This corresponds well to results of Leather (1983a), which also show that stem extract was not only less harmful but it also stimulated germination. On the other hand however, our results do not support the statements of Leather (1983b) that seeds of grass weeds were not affected by sunflower as well as of Azania *et al.* (2003) who also pointed out that sunflower is less harmful to species of *Graminae* (presently *Poaceae*) family. In our study (on average for five concentrations) seed germination of four species from the Poaceae family: *Avena fatuna* L., *Setaria glauca* L., *Apera spica-venti*. and *Elymus repens* L., was inhibited by between 70.6% and 78.3% when compared to seeds germinated in H_2O. As for the delay in germination, with few exceptions, for all tested weed species and both with leaf and stems extracts, the greatest effect was in case of *Amaranthus retroflexus* L. as one seed to germinate in the presence of leaf extract needed 4.4 d more than in water (Ciarka *et al.*, submitted).

The above data suggest that allelopathic activity of sunflower is effective against quite a broad spectrum of weed species, which if also true in field conditions, would be a very desired character.

Allelopathic Effects of Sunflower on Crops

The allelopathic effects of sunflower on crops including sunflower have been studied in several laboratories and most of the results show negative effects. The significant reduction in sorghum seedlings growth by sunflower aqueous extracts was found by Schon and Einhellig (1982). Irons and Burnside (1982) also noted reduced germination and seedling

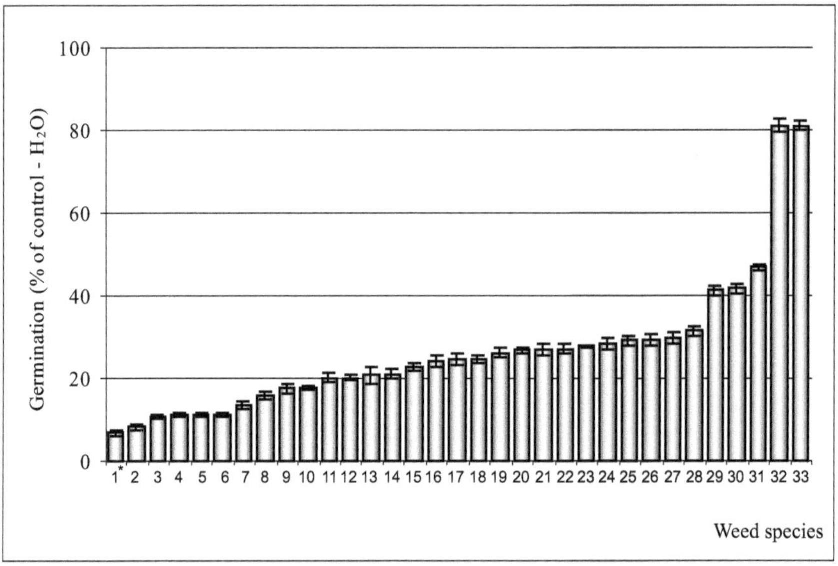

*:1-*Papaver rhoeas*, 2-*Erodium cicutarium*, 3-*Galium aparine*, 4-*Rumex acetosella*, 5-*Erigeron canadensis*, 6-*Lamium purpureum*, 7-*Polygonum persicaria*, 8-*Achillea millefolium*, 9-*Polygonum convolvulus*, 10-*Thlaspi arvense*, 11-*Viola arvensis*, 12-*Cirsium arvense*, 13-*Aegopodium podagraria*, 14-*Avena fatua*, 15-*Chenopodium album*, 16-*Convolvulus arvensis*, 17-*Amaranthus retroflexus*, 18-*Daucus carota*, 19-*Raphanus raphanistrum*, 20-*Setaria glauca*, 21-*Apera spica-venti*, 22-*Capsella bursa-pastoris*, 23-*Galeopsis tetrahit*, 24-*Stelaria media*, 25-*Matricaria inodora*, 26-*Elymus repens*, 27-*Taraxacum officinale*, 28-*Anthemis arvensis*, 29-*Delphinium consolida*, 30-*Centaurea cyanus*, 31-*Sinapis arvensis*, 32-*Vicia cracca*, 33-*Echinochloa crus-galli*.

Fig. 3 Germination of 33 weed species in the presence of allelochemicals contained in aqueous extracts of sunflower leaves (*Helianthus annuus* L.) cv. Ogrodowy evaluated after 14 d of culture. Data (average for five concentrations, mean of six replicates with 100 seeds in each), due to natural uneven weeds germination, are expressed in percentage of control (H_2O).

growth of soybean, sorghum and sunflower as well as Narwal *et al.* (1999b) in case of sunflower, three cereals and of three legume species. In our studies, with eight crop species (wheat, rye, barley, oat, tomato, lettuce, radish and mustard) germination was reduced to evidently lower degrees than weeds as on average for all four used concentrations between 66% and 95% of sown seeds germinated (Ciarka *et al.*, 2004a). The greatest reduction was recorded for radish and mustard while germination of wheat and cucumber was almost not affected. In the case of wheat we tested allelopathic activity of 10 sunflower cultivars using both leaf and stem extracts, and all of them did not reduce wheat germination (Ciarka *et al.*, in preparation). The results in regard to wheat are in agreement with data obtained by Morris and Parrish (1992) who also did not record any reduction in wheat germination. This is contrary to findings of Purvis and

Jones, (1990) who showed significant differences between 12 *cvs.* of sunflower with 10 reducing wheat germination even up to 38.8% of control. As already mentioned, the effect of sunflower was mostly phytotoxic but in few cases stimulation also took place, and this was especially evident in the work of Morris and Parrish (1992). Although Bernat *et al.* (2004b) also did not recorded reduction in wheat germination but the negative impact on growth and development of wheat seedling was evident (Fig. 4). However, it should be pointed out that this effect on wheat was substantially less when compared with mustard. At the highest concentration, mustard growth was completely inhibited while wheat, although less vigorously, continued to grow. It is interesting to note that in wheat, which develops secondary, tertiary and so on adventitious roots, along with an increase of the extract concentration, the number of roots increased too. This leads to considerable differences in the total root lengths between these two species and might be of some importance under the allelopathy stress (Bernat *et al.*, 2004b).

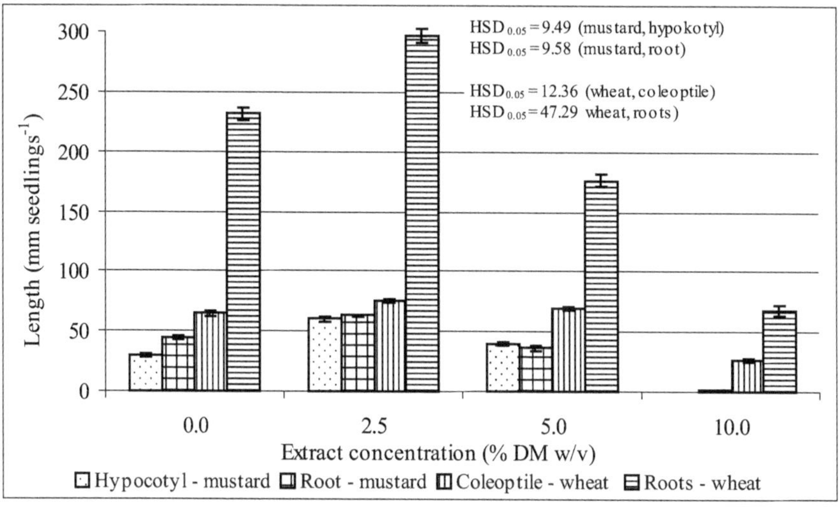

Source: Bernat, *et al.*, 2004b, with the permission of The Polish Academy of Sciences, Division of Agricultural, Forestry and Veterinary Sciences

Fig. 4 Length of winter wheat (*Triticum aestivum* L.) *cv.* Zyta and mustard (*Sinapis alba* L.) *cv.* Nakielska seedlings as influenced by allelochemicals derived from sunflower. Data are mean ± SE, *n* = 4.

The results of comparative studies on wheat and mustard (model plants in our study for cereals and dicotyledonous weeds respectively) showed that, also at the autotrophy growth stages, wheat was more tolerant than mustard to allelopathy stress generated by sunflower especially in processes related to plant water status (Gawronska *et al.*, 2004).

PERFORMANCE OF ALLELOPATHIC ACTIVITY OF SUNFLOWER IN THE FIELD

It has been implied, by several authors, that sunflower is allelopathically active under laboratory, greenhouse and field conditions and that is most often manifested in its inhibitory effects both on weeds and crops including sunflower which also exhibits autoxicity. In field conditions the allelopathic interference of sunflower with weeds and crops might perform in relation to subsequent crops both in the form of decomposing residues when sunflower is grown in rotation as well as when sunflower is cultivated as a pre-crop for mulch or green manure.

Allelopathic Activity Against Most Common Weeds

The allelopathic potential of sunflower against weeds under field conditions was clearly demonstrated in a five-years field study in which weed density and ground cover were reduced on plots with sunflower treatments when compared to control plots (Leather, 1983a, b). Moreover, weed control by sunflower treatment was equal to treatment with herbicide EPTC (S-ethyl dipropyl carbamothioate) (Leather, 1987). According to Semidey (1999) in sunflower-soybean cropping system, residues of mature sunflower plants incorporated into the soil in the summer of the year preceding soybean planting, resulted in significant reduction of weed density and biomass. When immature sunflower plants were incorporated into the soil, just before soybean planting, the suppressive effect on weeds was less evident though, still significant (Semidey, 1999).

The reduction by sunflower in germination and growth of *Amaranthus retroflexus* L. was reported by Hall *et al.* (1983). When sunflower was used as green manure in wheat-rice cropping system the population of *Phalaris minor*, a troublesome weed species in some parts of the world, was significantly reduced during two growing seasons (Om *et al.*, 2002).

The effect of sunflower mulch (in range 40-90t FW ha^{-1}, depending on the year) on weed density and community in winter wheat, produced in an organic system, was studied by Gawronski *et al.* (in preparation). In each of the three years, suppressive effect of mulch on weeds was recorded. Depending on t he year, the number of weeds was reduced by 62%-92% in autumn and by 37%-52% during spring-summer vegetation. The sunflower mulch reduced many weeds species very effectively with some being almost fully eliminated (*Papaver rhoeas* and *Viola arvensis,* and the

Table 2 Density of weed and winter wheat (*Triticum aestivum* L.) *cv.* Zyta, head number and crop yield as influenced by different load of sunflower mulch. Data are from 2003/04 vegetation season.

Combi-nation	*Weed density N° m^{-2}*			*Crop density N° m^{-2}*		*Head N°m^{-2}*	*Yield t ha^{-1}*
	EC 11-14	*EC 21-25*	*EC 37-40*	*EC 11-14*	*EC 21-25*		
Control 1	**228.5a**	**195.5a**	**195.0a**	**471.5ab**	**285.5**	**520a**	**6.80a**
Control 2	163.0ab	161.5ab	194.0a	485.0a	280.5	494a	6.01ab
Mulch 0%	140.0abc	145.0ab	147.5b	455.0ab	265.0	380c	4.70b
Mulch 50%	105.5bc	137.0ab	144.5b	417.0bc	272.5	411bc	5.00b
Mulch 100%	89.5bc	123.0ab	147.5b	411.5bc	274.0	468ab	5.86ab
Mulch 200%	45.0c	94.0b	134.0b	360.5c	243.5	456abc	5.73ab

Control 1: direct sowing, Control 2: conventional sowing

troublesome in cereal production in Central Europe *Apera spica – venti and Matricaria inodora).* However, on others such as *Echinochloa crus-galli* sunflower mulch had little effect. These confirmed the results of bioassays (see above).

The results from the trials with different mulch loads showed that significant weed suppression (*vs.* controls) was observed already in the combination of 0% mulch (only roots left in the soil) so, there is no doubt that allelochemicals released from sunflowers play a key role in weed suppression proving that this interference is of allelopathic nature. The field observations indicated that physical barriers, differences in soil moisture, soil temperatures, created by the mulch may play a role in suppressing weeds, but only if the mulch input is higher (i.e. 200% instead of 100% normal amount), though this may not be practical in normal farm management (Table 2). Also changes in soil microbial populations due to mulch deposited on soil surface might affect weed suppression. All these, to our knowledge, have not been studied for sunflower mulch as yet. Contrary to the above, results of Nanjappa *et al.* (1999) have shown a rather slight effect of sunflower mulch (10 t ha^{-1}) on weeds in succeeding crops. Germination and vigor index in Lady's Finger and seedling vigor and dry weight in Finger millet were reduced by 27%-35% while in the case of soybean, pea and cucumber slight stimulation took place.

Allelopathic Activity Towards Crops

The suppressive effect of sunflower allelopathy on weeds, beneficial to farmers in the way of weed management, unfortunately sometimes is manifested against the crop, too. However, positive effects of sunflower allelopathy on the succeeding crop are also reported. For example Semidey (1999) studied the effect of sunflower residues on soybean grown in sunflower-soybean cropping system with varying levels of residues that were incorporated either shortly or eight months before soybean planting. The results showed that in spite of reducing soybean stand in some combinations, sunflower residues, no matter when incorporated and what was the level, increased soybean yield when compared with plots free of sunflower residues (Semidey, 1999). These results correspond well and to some extend might be explained by the data of Mallik and Tesfai (1988) where the following were reported: significant stimulation of nodulation in soybean when exposed to water extracts or residues of sunflower despite the fact that soybean germination and seedling growth were reduced under the same treatments with sunflower.

Purvis (1990) showed both inhibitory and stimulatory effect on wheat, grown after sunflower stubbles were incorporated into the soil. In the first year, when wheat was grown on plots with incorporated unweathered sunflower stubble, crop germination, dry weight, tiller number, leaf area and grain yield were reduced. Contrary to this, in next year (wetter conditions) when incorporated sunflower biomass was partly decomposed, the tiller and head numbers, and grain yield of wheat were greater showing a stimulatory effect (Purvis, 1990).

The stimulatory effect on wheat germination, biomass production and grain yield (in combination sunflower residues + no-tillage) was also demonstrated by Morris and Parrish (1992). When sunflower was grown for green manure it also had a stimulatory effect on wheat, the yield of which was significantly higher over fallow land (Om *et al.*, 2002). Similarly, Gawronski *et al.* (2002) reported higher, by about 28% yield of winter wheat on mulched plots, due to greater productive tillering that compensate the reduced crop stand by mulch. In the following two years however, wheat stand, and head number were reduced by sunflower mulch leading to reduced grain yield (depending on experiments by 37%-51% in 2002/03 and by 14%-28% in 2003/04, (Table 2). The lowest yield, recorded in combination 0% mulch, was probably due to both phytotoxic effect of allelocompounds and to nutrient deficiency (these plants showed some symptoms of nutrient starvation). In these plots all sunflower biomass, which when decomposed would be a source of some nutrients (besides allelochemicals), was taken out (Gawronski *et al.*, in preparation). To check this hypothesis another experiment with sunflower + legume mulch was conducted.

The negative impact of sunflower residues on successive crop germination, height, dry matter, and yield of three cereals, three legume species and on cotton, with cereals being least affected are reported by Narwal *et al.* (1999a). Similarly, Batish *et al.* (2002) also reported phytotoxic effects of sunflower on four successive crops (*Zea mays, Pennisetum americanum, Cyamopsis tetragonoloba* and *Sorghum vulgare*). This effect was observed both when sunflower residues were left on surface as well as when they were taken out from the field but with more pronounced effects when residues were left to decompose. Emergence, crop stand, dry matter accumulation and yield of rice, soybean, pigeonpea and maize + soybean intercrop were reduced when cultivated after sunflower (Pal and Sand, 2004). The highest reduction was noted in maize-soybean followed by soybean and lowest in rice.

Autotoxicity

Several researchers reported that sunflower exerts autotoxicty and it was demonstrated for the first time by Wilson and Rice (1968). However, Irons and Burnside (1982) have found that sunflower while being phytoxic, is also affected in a stimulatory manner. Sunflower emergence, plant height fresh and dry weights were increased when plants were grown in soil amended with sunflower leaf or stem powder or in nutrient solutions containing roots exudates. At higher concentrations values for all these parameters were reduced and leaf showed higher than stem activity (Irons and Burnside, 1982). Both stimulatory and inhibitory effect of sunflower residues added to nutrient solution or soil on sunflower shoot and root growth are also reported by Malik and Tesfai (1988). Leather (1987) studying the effect of sunflower on weeds and sunflower crop over a five-year period observed that in part of second-year plots (vs. first-year plots) in addition to greater competitive weed pressure, sunflower emergence was poor, developing plants were less vigorous and head weight was reduced. The author concluded that this was the effect of autotoxicity from the preceding year (Leather, 1987). No weed reduction, even slight increase in weed DM (by 7%) in sunflower plots with sunflower mulch (10 t ha^{-1}) was reported by Nanjappa *et al.* (1999). In this work, the authors recorded drastically reduced crop stand at harvest (2.8 fold), as well as stalk (22%), seed and oil yield by 5.2 fold. The negative impact of sunflower residues on succeeding sunflower crops manifested by reduced plant height, dry matter at harvest and grain yield ranging depending on combinations of 15%-85%, 58%-94% and between 78%-82% less then in fallow lands are also reported by Narwal *et al.* (1999a).

The results presented above show that observed weeds/crops response to allelochemicals of sunflower origin even under controlled conditions often are not unidirectional and sometimes even controversial. This is most probably due to variation in the genotypic background of the tested donor and acceptor of these compounds. Also different experimental conditions, methods and time of allelochemicals applications would contribute to these differences. In the field, additionally, interferences such as competition for resources (light, water and nutrients) plant - microorganisms interactions, physical pressure of residues with strong dependence upon environmental conditions clearly underline the complexity and at least to some extent might be responsible for even controversial results.

Due to the complexity of allelopathy phenomenon and its strong dependence upon environmental conditions together with methodological difficulties to distinguish allelopathy from other interferences, skepticism often arises as to whether or not allelopathy operates in nature. In the light of the above cited results it can be said that sunflower allelopathy operates in field conditions though, other sources of interferences are not excluded. In the case of using sunflower as green manure, biologically active mulching or residues of sunflower (cultivated as a main crop) left in the field, the competition between plants for resources can be ruled out as a factor of the negative impact on succeeding plants (both weeds and crops) observed in many studies. This is because the allelochemicals are not coming from living plants. Moreover, green manure, mulch and residues serve also as a source of "nutrients" when released while decomposing. To some extent, however, there might be competition between plants and microorganisms bio-transforming the biomass though, it seems to be of little importance compared to the competition while growing plants simultaneously under competitive conditions. In the case of residue left on soil or mulch, besides allelopathy, other components of interference such as physical barriers and shading (important for seeds requiring light signal for germination) might play a role. But again, this also would not be the only, if at all, cause since as described above, weed suppression and some negative impact on wheat by sunflower were observed even when no residues were on soil surface - just roots were left in the soil.

PROSPECTS FOR EXPLOITATION OF SUNFLOWER ALLELOPATHY IN PRACTICE

Sunflower, due to its high allelopathic activity, is often considered as a crop to be utilized for weed prevention and control. Although, due to methodological difficulties in separating allelopathy from other interferences, there is still controversy on whether allelopathy operates in nature or not. In a view of the large amount of the new literature on allelopathy, in our opinion, the question is not "yes" or "no", rather the following should be posed:

- what is the contribution of allelopathy among other interferences into the observed in field responses?

- which form of using sunflower allelopathy in practice is promising and reasonable?
- where should scienific efforts be directed in order to exploit allelopathy in the near future?

Theoretically several ways of utilizing sunflower allelopathic potential can be considered: (i) cultivation as pre-crop for green manure (Om *et al.*, 2002), (ii) cultivation as pre-crop for production of biologically active mulch (Morris and Parrish, 1992; Nanjappa *et al.*, 1999; Batish, 2002; Gawronski *et al.*, 2002, 2004; Azania *et al.*, 2003; Gawronski, 2003;), (iii) cultivation of sunflower as a main crop in rotation with other species (residues after yield harvest left as a source of allelocompounds) (Leather, 1983a, b, 1987; Purvis, 1990; Purvis and Jones, 1990; Narwal *et al.*, 1999a). (iv) cultivation for isolation of allelocompounds to be used for spray, and (v) isolated from sunflower allelocompounds may serve as templates for synthesis of "natural herbicides" with new and varying modes of action (Macias *et al.*, 1999 and references therein).

Inspite of the fact that more than 30 years have past since scientists became interested in sunflower allelopathy and its knowledge has increased, we are not yet at a point to propose to farmers ready-to-use technology for sunflower allelopathy-mediated weed prevention and control. In fact, there does not appear to be a universal strategy for weed management based solely on allelopathy. Rather, there is an outline of combinations of elements recommended for a given site, since some agricultural problems are local. Therefore, recommendations would probably be of local value covering adaptation to specific climate and soil conditions together with cultural practices preferred by farmers at a given site. Moreover, they cannot be a status-quo type of recommendation and some readjustments would be necessary based on the current changes of weather and other local agro-ecological conditions.

Nevertheless, the results available in literature and our four-year-field as well as greenhouse and laboratory studies on sunflower allelopathy, suggest that there is a chance for exploitation of the sunflower allelopathy, as complementary to other cultural practices, in the weed management of the cereals.

As already mentioned, germination of winter wheat was hardly affected except for a slight delay (Morris and Parrish, 1992; Ciarka *et al.*, submitted). Also at early seedling growth (heterotrophy) and at autotrophy growth stages wheat was much more tolerant to allelocompounds of sunflower origin than mustard (Gawronska, *et al.*, 2004). Moreover, some research has found sunflower to be stimulatory to wheat germination, biomass accumulation and yield (Purvis, 1990; Morris and Parrish, 1992; Gawronski, data not published). However, in the field, depending on the specific environmental (especially weather) conditions of a given site, the allelopathic effects on crop can be adverse (Purvis, 1990; Gawronski *et al.*, in preparation).

To circumvent and counterbalance the negative impact of sunflower on the crop, based on literature data and on our own results and observations, specific adaptive changes in cultural practices can be proposed for consideration:

- only high grade, good quality and calibrated seeds should be used,
- in organic farming, it is recommended to sow winter cereals as late as possible in order to avoid weeds germination and vegetation in autumn. This might be risky if winter is early especially in coincidence with biotic stress generated by sunflower mulch. Therefore, if sunflower mulch is being used, earlier sowing time would be beneficial for good crop seedling establishment before winter. Due to allelopathic activity of the mulch, autumn weeds will be suppressed,
- since wheat density and the number of productive tillers on mulched plots were reduced increasing the sowing rate from 400 to 500-550 seeds m^{-2} is also recommended for mulched field,
- growing of the one-stem *cvs.* (recently again a direction of breeding especially for organic farming), with increased sowing rate might be a way to overcome reduced tillering by sunflower mulch. Additionally these *cvs.* as fast growing together with high sowing rate ensure high competitiveness against weed due to early ground cover and dense soil penetrations by roots,
- when sunflower is considered for allelopathy-mediated weed control, choosing a crop cultivar that will be more tolerant to, or even stimulated by sunflower allelochemicals together with some adaptation in cultural practices might benefit the crop performance and yield,
- irrigation, if possible, might be a good option both for ensuring high sunflower biomass production and/or partially decomposing of the biomass which during low rainfalls, might be a limiting step.

As already mentioned, at present there is no ready-to-use strategy for weed management that includes sunflower allelopathy. However, if in organic farms very high infestations with troublesome weeds that are well suppressed by sunflower (as for example *Apera spica-venti, Marticaria inodora*) is expected, utilization of sunflower allelopathy can be a solution.

To conclude it can be said that sunflower used as pre-crop for green manure, for biologically active mulch or as a cash crop in rotations with residues serving a source of naturally released allelochemicals as well as templates for synthesis of new "natural herbicides" can be an effective complementary tool in weed management. However, we are still faced with the dilemma that laboratory assays show only the potential, which cannot be directly extrapolated to the field. Also, at present, due to the strong dependence of allelopathic manifestation on specific climate and local conditions, consistent weed suppression and without or with low effect on the crop is not achievable yet. Moreover, as the current level of knowledge on the complexity of interactions in the field (Anaya, 1999; Chou, 1999; Barberi, 2002; Bhowmik and Inderjit, 2003) is still in the stage of infancy it requires further efforts, and close collaboration between scientists of various disciplines is necessary.

FUTURE STUDY ON SUNFLOWER ALLELOPATHY

As can be seen from the previous sections inspite of extensive work done in many laboratories on sunflower allelopathy and the large amount of literature available there is a demand for further studies both on basic mechanisms of sunflower allelopathy as well as on practically oriented aspects.

Future research on sunflower allelopathy should focus on:

- continuous assessment of the collection available in germplasms, sunflower genotypes for selecting of high allelopathic activity,
- screening sunflower genotypes with stimulatory effects on receiver crops,
- screening crop *cvs.* for higher tolerance to negative impact of sunflower allelochemicals,
- in-depth study on the interactions between released allelochemicals and all components of the underground parts (physical and chemical properties of the soil, nutrients availability and microbial population and activity) contributing to allelochemicals transformation,
- determining growing conditions for enhanced allelopathic activity (stress induced),
- elaborating cultural practices in weed management with sunflower allelopathy for specific sites and crops to be applied, (for example shallow incorporation vs. layering on surface, sunflower in mixture with legumes, using cultivars of high relative growth ratio to reduce length of pre-crop cultivation, etc),
- assessment of the ecological effects of sunflower allelopathy both on the agro-ecosystem and on neighboring ecosystem(s) in both short - (crop life cycle) and long-term (between life cycles) in order

to examine effects on all components of the environment operating in joint action,

- examining the role of sunflower volatiles, if any, in bio-regulatory signaling between plants-plants, plants-microrganisms (Petterson *et. al.*, 1996; Pickett *et al.*, 2003; Bais *et al.*, 2004; Vaughn *et al.*, 2005) and between plants and the second and third trophic levels - a novel avenue for exploitation of allelopathy (Petterson *et. al.*, 1996; Röse *et al.*, 1996; Glinwood *et al.*, 2003) unexplored for sunflower yet,
- exploration of anti-pathogens and anti-insects activities of sunflower allelochemicals for their usefulness in biocontrol,
- searching for gene(s) involved in biosynthesis of allelocompounds to be used for constructing GMO for enhanced production of allelochemicals,
- using of biotechnology tools for constructing GMO with genes of interest and possibly employing sunflower plants as bioreactors for allelochemicals production.

Acknowledgement

Authors studies were financed by grants awarded to S.W. Gawronski: 1/ 5 FP EU # QLK5 CT - 2000 – 01418 (in part for Polish Partner) and 2/ SC for SR # 117/E-985 SPUB-M/5 PR UE/DZ172/2001-2004.

References

Alsaadawi, I.S. (1988) Biological suppression of nitrification by selected cultivars of (*Helianthus annuus* L.). J. Chem. Ecol. 14: 733 –741.

An, M., Pratley, J., Haig, T. (1998) Allelopathy: from concept to reality. *In* the Proceedings of the 9th Australian Agronomy Conference, Wagga Wagga, Australia p. 5. http://www.regional.org.au/au/au/asa/1998/6/314/an.htm

Anaya, A.L. (1999) Allelopathy as a tool in the management of biotic resources in agroecosystems. Crit. Rev. Plant Sci. 18: 697-739.

Azania, A.A.P.M., Azania, C.A.M., Alves, P.L.C.A., Palaniraj, R., Kadian, H.S., Sati, S.C., Rawat, L.S., Dahiya, D.S., Narwal, S.S. (2003) Allelopathic plants. 7. Sunflower (*Helianthus annuus* L.). Allelopathy J. 11: 1-20.

Bais, H.P., Park, S.W., Weir, T.L., Callaway, R.M., Vivanco, J.M. (2004) How plants communicate using the underground information superhighway. Trends Plant Sci. 9: 26-32.

Bárberi, P. (2002) Weed management in organic agriculture: are we addressing the right issues? Weed Res. 42: 177-193.

Batish, D.R., Tung, P., Singh, H.P., Kohli, R.K. (2002) Phytotoxicity of sunflower residues against some summer season crops. J. Agron. Crop. Sci.188: 19-24.

Bernat, W., Gawronska, H., Gawronski, S.W. (2004a) Physiological effects of allelopathic activity of sunflower on mustard. Advances of Agricultural Sciences - Problem Issues [Zeszyty Problemowe Postêpów Nauk Rolniczych] [English with Polish abstract]. 496: 275 - 287.

Bernat, W., Gawronska, H., Janowiak, F., Gawronski, S.W. (2004b) The effect of sunflower allelopathic on germination and seedling vigour of winter wheat and mustard. Advances of Agricultural Sciences - Problem Issues [Zeszyty Problemowe Postêpów Nauk Rolniczych]. 496: 289 - 299. [English with Polish abstract].

Bertin, C., Yang, X., Weston, L.A. (2003) The role of root exudates and allelochemicals in the rhizosphere. Plant Soil. 256: 67-83.

Beveridge, L.E., Naylor, R.E.L. (1999) Options for organic weed control - what farmers do. Proceedings 1999 Brighton Crop Protection Conference - Weeds. pp. 939-944.

Bhowmik, P.C., Inderjit. (2003) Challenges and opportunities in implementing allelopathy for natural weed management. Crop Prot. 22: 661-671.

Chou, C-H. (1999) Roles of allelopathy in plant biodiversity and sustainable agriculture. Crit. Rev. Plant Sci. 18: 609-636.

Ciarka, D., Gawronska, H., Gawronski, S.W. (2002a) Weed species reaction to sunflower allelopathics. Abstracts of Third Word Congress on Allelopathy "Challenge for the New Millennium", Tsukuba, Japan, 2002, 26-30 August. p. 162.

Ciarka, D., Prawicz, U., Gawronska, H., Gawronski, S.W. (2002b) Genotypical differences in allelopathic potential of sunflower. Abstracts of Third Word Congress on Allelopathy "Challenge for the New Millennium", Tsukuba, Japan, 2002, 26-30 August. p. 78.

Ciarka, D., Gawronska, H., Gawronski, S.W. (2004a) Crops species reaction to sunflower allelopathics. Proceedings of Second European Allelopathy Symposium "Allelopathy - from understanding to application", Pulawy, Poland, 3-5 June 2004. p. 123.

Ciarka, D., Gawronska, H., Malecka, M., Gawronski, S.W. (2004b) Allelopathic potential of sunflower roots and root exudates. Advances of Agricultural Sciences - Problem Issues [Zeszyty Problemowe Postêpów Nauk Rolniczych]. 496: 301-313. [English with Polish abstract].

Ciarka, D., Gawronska, H., Gawronski, S.W. (2006). Allelopathic potential of sunflower. I. Genotypical and organ specific differences in allelopathic activity of sunflower. Allelopathy J. (submitted).

Ciarka, D., Gawronska, H., Gawronski, S.W. (2006). Allelopathic potential of sunflower. II. Sunflower allelo compounds are released *via* different mechanisms. Submitted in Allelopathy J.

Ciarka, D., Gawronska, H., Gawronski, S.W. (2006). Allelopathic potential of sunflower. III. Reaction of 33 weed species to allelochemicals of sunflower origin. Submitted in Allelopathy J.

Davies, D.H.K., Welsh, J.P. (2002) Weed control in organic cereals and pulses. *In* Organic Cereals and Pulses, D. Younie, B.R. Taylor, J.M. Wilkinson (eds). Chalcombe Publications, Lincoln, pp. 77-114.

Einhellig, F.A. (1984) Allelopathy - a natural protection, allelochemicals. *In* Handbook of Natural Pesticides: Vol. 1. Methods, H.B. Mandava (ed.). CRC Press, Inc. pp. 161-200.

Einhellig, F.A. (1985). Effects of allelopathic chemicals on crop productivity. *In* Bioregulators for Pest Control. P. Hedin (ed) ACS Symposium, Series no. 276 Washington DC, USA pp. 109-129.

Einhellig, F.A. (1987) Interactions among allelochemicals and other stress factors of the plant environment. *In* Allelochemicals: Role in Agriculture and Forestry, G.R. Waller (ed) ACS Symposium Series, 330. American Chemical Society, Washington, DC, USA. pp. 343-357.

Einhellig, FA. (1996) Interactions involving allelopathy in cropping systems. Agron. J. 88: 886-893.

Einhellig, F.A. (1999) An integrated view of allelochemicals amid multiple stresses. *In* Principles and Practices in Plant Ecology Allelochemical Interactions. S. Inderjit, K.M.M. Dakshini, C.L. Foy (eds.). CRC Press LLC. Boca Raton, Fl, USA. pp. 479-494.

Einhellig, F.A. (2004) Mode of allelochemical action of phenolic compounds. *In* Allelopathy Chemistry and Mode of Action of Allelochemicals. F.A. Macias, J.C.G. Galindo, J.M.G. Molinillo, H.G. Cutler (eds.). CRC Press LLC.Boca Raton, Fl, USA. pp. 217-238

Einhellig, F.A., Leather, G.R. (1988) Potential for exploiting allelopathy to enhance crop production. J. Chem. Ecol. 14: 1829-1844.

Gawronska, H., Bernat, W., Ciurzynska, M., Gawronski, S.W. (2002) Photosynthesis and water status of mustard plants as influenced by sunflower allelopathics. Abstracts of Third Word Congress on Allelopathy "Challenge for the New Millennium", Tsukuba, Japan, 2002, 26-30 August. p. 65.

Gawronska, H., Bernat, W., Gawronski, S.W. (2003) Physiological effects of allelopathy stress generated by sunflower. Abstract of Fifth International Conference "Ecophysiological aspects of plant responses to stress factors", Cracow, Poland, Acta Physiol. Plant. 25 (Supplement). pp. 24-25.

Gawronska, H., Bernat, W., Janowiak, F., Gawronski, S.W. (2004) Comparative studies on wheat and mustard responses to allelochemicals of sunflower origin. *In* the Proceedings of Second European Allelopathy Symposium "Allelopathy - from understanding to application", Pulawy, Poland, 2004, 3-5 June. p. 28.

Gawronska, H., Golisz, A. (2005) Allelopathy and biotic stress. *In* Allelopathy: A Physiological Process with Ecological Implications. M.J. Reigosa, N. Pedrol, L. Goncalez (eds.). Springer, The Netherlands, pp. 211-227.

Gawronski, S.W., Bernat, W., Gawronska, H. (2002) Allelopathic potential of sunflower mulch in weed control. Abstracts of Third Word Congress on Allelopathy: "Challenge for the New Millennium", Tsukuba, Japan, 2002, 26-30 August. p. 160.

Gawronski, S.W. (2003) Allelopathy as a strategy for weed control in organic farming. Abstract of Fifth International Conference "Ecophysiological aspects of plant responses to stress factors", Cracow, Poland, Acta Physiol. Plant. 25 (Supplement). p. 25.

Gawronski, S.W., Bernat, W., Gawronska, H. (2004) Precrop mulch of allelopathic activity for weeds management in organic winter wheat production. Abstracts of 4th International Weed Science Congress, Durban, South Africa, 2004, 20–24 June. p. 105.

Glinwood, R., Pettersson, J., Ahmed, E., Ninkovic, V., Birkett, M., Pickett, J. (2003) Change in acceptability of barley plants to aphids after exposure to allelochemicals from couch-grass (*Elytrigia repens*). J. Chem. Ecol. 29: 261-274.

Hall, A.B., Blum, U., Fites, R.C. (1982) Stress modification of allelopathy of *Helianthus annuus* L. debris on seed germination. Am. J. Bot. 69: 776-783.

Hall, A.B., Blum, U., Fites, R.C. (1983) Stress modification of allelopathy of *Helianthus annuus* L. debris on seedling biomass production of *Amaranthus retroflexus* L. J. Chem. Ecol. 9: 1213-1221.

Inderjit, Dakshini, K.M.M. (1992) Interference potential of *Pluchea lanceolata* (Asteraceae): growth and physiological responses of asparagus bean, *Vigna unguiculata* var. *sesquipedalis*. Am. J. Bot. 79: 977-981.

Inderjit, Keating, K.I. (1999) Allelopathy: Principles, Procedures, Processes and Promises for Biological Control. Adv. Agron. 67: 142-206.

Inderjit, Nayyar, H. (2002) Shift in allelochemical functioning with selected abiotic stress factors. *In* Chemical Ecology of Plants: Allelopathy in Aquatic and Terrestrial Ecosystems. Inderjit, A.U. Mallik (eds.). pp. 199-2

Irons, S.M., Burnside, O.C. (1982) Competitive and allelopathic effects of sunflower (*Helianthus annuus*). Weed Sci. 30:372-377.

Kong, Ch., Hu, F., Xu, X. (2002) Allelopathic potential and chemical constituents of volatiles from *Ageratum conyzoides* under stress. J. Chem. Ecol. 28: 1173-1182.

Kruse, M., Strandberg, M., Strandberg, B. (2000) Ecological effects of allelopathic plants - a review. NERI Technical Report No. 315 National Environmental Research Institute, Silkeborg, Denmark. G.R. Leather (1983a) Weed control using allelopathic crop plants. J. Chem. Ecol. 9: 983-989.

Leather, G.R. (1983b) Sunflowers (*Helianthus annuus*) are allelopathic to weeds. Weed Sci. 31: 37-42.

Leather, G.R. (1987). Weed control using allelopathic sunflowers and herbicide. Plant Soil. 98: 17-23.

Macias, F.A., Varela, R.M., Torres, A., Molinillo, J.M.G. (1999) Potentiality of cultivar sunflowers (*Helianthus annuus* L.) as a source of natural herbicide templates. *In* Principles and Practices in Plant Ecology: Allelochemical Interactions. Inderjit, K.M.M. Dakshini, C.L. Foy (eds.). CRC Press LLC. Boca Raton, Fl, USA pp. 531-550.

Mallik, M.A.B., Tesfai, K. (1988) Allelopathic effect of common weeds on soybean growth and soybean-Bradyrhizobium symbiosis. Plant Soil. 112: 177-182.

Maruthi, V., Sankaran, N. (2001) Allelopathic effects of sunflower (*Helianthus* spp.). Agric. Rev. 22: 57-60.

Morris, P.J., Parrish, D.J. (1992) Effects of sunflower residues and tillage on winter wheat. Field Crops Res. 29: 317-327.

Nanjappa, H.V., Poonguzhalan, R., Ramachandrappa, B.K. (1999) Influence of sunflower on subsequent crops. Allelopathy J. 6: 69-74.

Narwal, S.S (1999) Allelopathy in weed management. *In* Allelopathy Update, Vol.2, Basic and Applied Aspects. S.S. Narwal (ed.). Science Publishers, Inc. pp. 203-254

Narwal, S.S., Singh, T., Hooda, J.S., Kathuria, M.K. (1999a) Allelopathic effects of sunflower on succeeding summer crops. I. Field studies and bioassays. Allelopathy J. 6: 35-48.

Narwal, S.S., Yadava, S., Gupta, S. (1999b) Allelopathic effects of sunflower on succeeding summer crops. 2. Pot culture and biomass decomposition. Allelopathy J. 6: 209-226.

Ohno, S., Tomita-Yokotani, K., Kosemura, S., Node, M., Suzuki, T., Amano, M., Yasui, K., Goto, T., Yamamura, S., Hasegawa, K. (2001) A species-selective allelopathic susbstance from germinating sunflower (*Helianthus annuus* L.) seeds. Phytochemistry. 56: 577-581.

Om, H., Dhiman, S.D., Kumar, S., Kumar, H. (2002) Allelopathic response of *Phalaris minor* to crop and weed plants in rice-wheat system. Crop Prot. 21: 699-705.

Qasem, J.R. (1999) Biological activity of corn buttercup (*Ranunculus arvensis* L.). *In* Recent Advances in Allelopathy Vol. 1. A Science for the Future. F.A. Macias, J.C.G. Galindo, J.M.G. Molinillo, H.G. Cutler (eds.). Servicio De Publicaciones-Universidad de Cadiz. Spain pp. 287-300.

Patrick, Z.A., Toussoun, T.A., Koch, L.W. (1964) Effect of crop-residue decomposition products on plant roots. Ann. Rev Phytopathol. 2: 267-292.

Pettersson, J., Quiroz, A., Fahad, A.E. (1996) Aphid antixenosis mediated by volatiles in cereals. Acta Agric. Scan., Section B, Soil Plant Sci. 46: 135-140.

Pickett, J.A., Rasmussen, H.B., Woodcock, C.M., Matthest, M., Napiert, J.A. (2003) Plant stress signaling: understanding and exploiting plant-plant interactions. Biochem. Soc. Trans. 31: 123-127.

Purvis, C.E. (1990) Differential response of wheat to retained crop stubbles. I. Effect of stubble type and degree of decomposition. Aust. J. Agric. Res. 41: 225-242.

Purvis, C.E., Jones, G.P.D. (1990) Differential response of wheat to retained crop stubbles. II. Other factors influencing allelopathic potential; intraspecific variation, soil type and stubble quantity. Aust. J. Agric. Res. 41: 243-251

Putman, A.R., Duke, W.B. (1978) Allelopathy in agroecosystems. Ann. Rev. Phytopathol. 16: 431-451.

Putman, A.R. (1985). Weed allelopathy. *In* Weed Physiology, Vol. I. Reproduction and Ecophysiology. S.O. Duke (ed.). CRC Press Inc. pp. 131-155

Putman, A.R., Tang, C.S. (1986) The Science of Allelopathy. John Wiley & Sons. NY, USA. pp. 1-317

Putman, A.R., Weston, L.A. (1986). Adverse impacts of allelopathy in agricultural systems. *In* The Science of Allelopathy. A.R. Putman, C.S. Tang (eds.). John Wiley & Sons NY, USA. pp. 43-56.

Reigosa, M.J., Sánchez-Moreiras, A., González, L. (1999) Ecophysiological approach in allelopathy. Crit. Rev. Plant Sci. 18: 577-608.

Reigosa, M.J., Pedrol, N., Sánchez-Moreiras, A.M., González, L. (2002) Stress and allelopathy. *In* Allelopathy from Molecules to Ecosystems. M.J Reigosa, and N. Pedrol, N. (eds.). Science Publishers, Inc. pp. 231-256.

Röse, U.S.R., Manukian, A., Heath, R.R., Tumlinson, J.H. (1996) Volatile semiochemicals released from undamaged cotton leaves. Plant Physiol. 111: 487-495.

Schon, M.K., Einhellig, F.A. (1982) Allelopathic effects of cultivated sunflower on grain sorghum. Bot. Gaz. 143: 505-510.

Semidey, N. (1999) Allelopathic crops for weed management in cropping systems. *In* Allelopathy Update, Vol. 2, Basic and Applied Aspects. S.S. Narwal (ed.). Science Publishers, Inc. pp. 271- 281.

Serghini, K., Pérez de Luque, A., Castejón-Muñoz, M., Garcia-Torres, L., Jorrin, J.V. (2001) Sunflower (*Helianthus annuus* L.) response to broomrape (*Orobanche cernua* Loefl.) parasitism: induced synthesis and excretion of 7-hydroxylated simple coumarins. J. Exp. Bot. 52: 2227-2234.

Singh, H.P, Batish, D.R., Kohli, R.K. (1999) Autotoxicity: concept, organisms, and ecological significance. Crit. Rev. Plant Sci. 18: 757-772.

Sutherland, B.L., Hume, D.E., Tapper, B.A. (1999) Allelopathic effects of endophyte-infected perennial ryegrass extracts on white clover seedlings. New Zealand J. Agric. Res. 42: 19-26.

Vaughn, S.F., Isbell T.A., Weisleder, D. (2005) Biofumigant compounds released by field pennycress (*Thlaspi arvense*) seedmeal. J. Chem. Ecol. 31:166-177.

Waller, G.R., Einhelling, F.A. (1999) Overview of allelopathy in Agriculture, Forestry and Ecology. *In* Biodiversity and Allelopathy: From Organisms to Ecosystems. C-H. Chou, - G.R. Waller, C. Reinhardt (eds.). Pacific Academia Sinica, Taipei. Republic of China pp. 221-245.

Weston, L.A. (1996) Utilization of allelopathy for weed management in agroecosystems. Agron. J. 88: 860-866.

Wilson, R.E., Rice, E.L. (1968) Allelopathy as expressed by *Helianthus annuus* and its role in old field succession. Bull. Tor. Bot. Club. 95: 432-448.

Wu, H., Pratley, J., Lemerle, D., Haig, T. (1999) Crop cultivars with allelopathic capability. Weed Res. 39: 171-180.

Yu, J.Q. (1999) Autotoxic potential of vegetable crops. *In* Basic and Applied Aspects Allelopathy Update, Vol. 2. S.S. Narwal (ed.). Science Publishers Inc. pp.149-162.

13

The Potential for Allelopathy During Decomposition of Hairy Vetch Residue

John R. Teasdale[1], Aref A. Abdul-Baki[1], Yong Bong Park[2] and Richard C. Rosecrance[3]

[1]*USDA-ARS Sustainable Agricultural Systems Laboratory, Building 001 Room 245, Beltsville, MD 20705 USA*

[2]*Faculty of Horticultural Life Science, Cheju National University, Jeju, Korea*

[3]*Chico State University, Chico, CA 95926 USA*

Residue of leguminous cover crops such as hairy vetch (*Vicia villosa* Roth) on the soil surface in minimum tillage cropping systems contributes to integrated weed management and provides nitrogen for subsequent crops. Research was conducted to determine the role of allelopathy on suppression of weed emergence by hairy vetch residue as the residue decomposes throughout the season. Residue was removed from the field at various times before and after the vetch was killed in spring and assayed for suppression of emergence and growth of selected species in greenhouse assays or Petri dish assays. Results showed that hairy vetch residue that was allowed to decompose in the field or that was subjected to leaching in the laboratory was less suppressive of emergence and/or growth than was residue that was fresh. These results could be explained by the presence of aqueous soluble phytotoxins in fresh hairy vetch residue that were lost during the decomposition process in the field or the leaching process in the laboratory. Aqueous extracts of the upper leaf tissue were more inhibitory of seedling root and hypocotyl growth than the lower leaf or stem tissue. This suggests that phytotoxins are present primarily in the metabolically active upper leaves of hairy vetch and that allelopathy contributes to weed suppression for a relatively short time following cover crop kill until leaf tissue has decomposed.

Keywords: *cover crop, no-tillage, germination, emergence, synergism*

INTRODUCTION

Cover crops are non-crop species introduced into crop rotations to provide beneficial services to the agro-ecosystem (Sustainable Agriculture Network, 1998). These services include influences on the physical environment such as soil protection from erosion, modification of the soil temperature and moisture regime, and improvement in soil tilth. They also include influences on the chemical environment such as capture and recycling of soil nutrients, fixation of nitrogen by legumes, and release of potentially allelopathic secondary compounds. Finally, cover crops can influence the biology of the soil by increasing biological diversity, enhancing beneficial organisms, and suppressing weed and pest populations.

Within the past two decades, there has been an upsurge of interest by weed scientists to explore the potential for using cover crops to enhance weed suppression in sustainable production systems. This attention has been encouraged by the growing interest in reduced tillage cropping systems that offers the opportunity for using cover crop vegetation and/or residue on the soil surface to interfere with weed establishment and growth. Numerous researchers have demonstrated weed suppression in addition to enhanced production of both agronomic and horticultural crops using cover crops in no-tillage production systems throughout the world (Araki and Ito, 1999; Morse, 1999; Akobundu *et al.*, 2000; Caamal-Maldonado, 2001; Cheema and Khaliq, 2000; Hutchinson and McGiffen, 2000). However, in most instances, weed suppression by cover crops is limited and inconsistent, often occurring only for selected species, in selected seasons, and during the early portion of the production season. Part of the problem is incomplete coverage of the soil by cover crops. For example, even seemingly uniform soil coverage by cover crop residue provides a highly heterogeneous environment for reception of light at the soil surface that can facilitate germination and growth of weeds (Teasdale and Mohler, 1993). Research on the physical mechanisms of weed suppression by cover crop residue has demonstrated that cover crops both 1) attenuate environmental cues such as light and soil temperature amplitude that trigger germination and 2) interfere directly with the upward movement of the emerging seedling as well as with the downward movement of light to seedlings (Teasdale and Mohler 1993, 2000). Although release of allelopathic phytotoxins has been hypothesized as a mechanism for cover crop effects for many years, this has been difficult to document in natural conditions where the numerous effects that cover crops have on the physical, biological, and chemical environment are confounded. Nonetheless, extensive literature describing the toxicity of plant extracts and the isolation of toxic compounds (Inderjit and Keating, 1999) suggest that allelopathy can be an important factor in natural systems despite the difficulty in detecting its presence.

Hairy vetch is a winter annual legume that has become an important cover crop for sustainable production systems. It is usually planted in late summer or early fall after summer crops are harvested in temperate climates. Abundant lateral growth by vines in spring allows hairy vetch to provide almost complete ground cover by the time summer crops are planted. Hairy vetch is usually killed either mechanically or chemically and left as a mulch on the soil surface before crops are planted in spring (Abdul-Baki and Teasdale, 1997). Due to high biomass production and high nitrogen content, a hairy vetch cover crop can contribute most of the nitrogen needs of high nitrogen-requiring crops such as corn (Decker *et al.*, 1994) or tomato (Abdul-Baki *et al.*, 1997). Although this rapid decomposition and release of nitrogen from vetch residue can benefit crop production (Wagger, 1989), it poses several problems for weed management. These problems include the loss of weed suppressing residue late in the season (Mohler and Teasdale, 1993) and the stimulation of germination of selected weed species by the release of nitrogenous compounds from decomposing residue (Teasdale and Pillai, 2005). The mixture of legume and grass cover crops has been one successful solution for delaying residue decomposition and enhancing weed suppression (Teasdale and Abdul-Baki, 1998).

Less is known about the production of allelopathic compounds by hairy vetch than about its nutrient content and contributions to soil fertility (perhaps the strong growth-enhancing effects of nitrogenous compounds have confounded efforts to detect inhibitory effects). Extracts and surface residues have been shown to both inhibit and stimulate growth of crop and weed species (White *et al.*, 1989). Cyanamide was identified as being responsible for the allelopathic activity of 9-d hairy vetch seedlings (Kamo *et al.*, 2003) while volatile ketones emitted by legume cover crops accounted for inhibition of cotton (Bradow, 1993). Ammonia released by hairy vetch into the soil atmosphere explained suppression of the pathogen *Thielaviopsis basicola* (Candole and Rothrock, 1997).

Although rapid decomposition is a recognized characteristic of hairy vetch residue, there has been little research to document the change in allelopathic potential of this residue during decompostion. In this chapter, we present a series of experiments that provide insight into the loss of allelopathic potential in hairy vetch residue during the period following operations that kill hairy vetch for crop production in spring. We then discuss the potential contribution of allelopathy and the physical factors toward weed suppression by hairy vetch residue.

MATERIALS AND METHODS

Rhizobium inoculated common hairy vetch seed were planted at the Beltsville Agricultural Research Center, Beltsville, MD, USA, in September and sampled in the following spring under various experimental conditions described below.

Hairy vetch above-ground vegetation was harvested from six sites in the field at three times and brought to the greenhouse for assay. Harvests were made in spring 1 d before desiccation with paraquat, 5 d after desiccation, and one month after desiccation. Flats with an area of 0.09 m^2 containing a sand-greenhouse soil mix (81% sand, 10% silt, 9% clay and 2.2% organic matter) were planted with 50 seeds each of velvetleaf (*Abutilon theophrasti* Medic.), green foxtail (*Setaria viridis* (L.) Beauv.), and smooth pigweed (*Amaranthus hybridus* L.) in rows across the width of each flat. Hairy vetch residue was placed on top of flats in amounts that represented 2X, 1X, 0.5X and 0.25X of the field rate at each site harvested. Emergence of weeds was determined 3 wk after planting. Greenhouse temperatures were maintained at relatively uniform conditions during assays following each harvest; averaging 31, 30 and 32°C maximum daily temperature and 19, 20 and 20°C minimum daily temperature for the pre-desiccation, 5 d post-desiccation, and one-month post-desiccation sampling, respectively. Emergence as a function of dry weight of hairy vetch residue was modeled using the log-logistic dose-response equation Y = 1/(1+(X/c)**b) where Y is emergence expressed as a fraction of the control, X is residue dry weight, c is a parameter defining the I50 value, and b is a parameter defining curve steepness.

Above-ground and root samples were taken from the field five times at 3 wk intervals before the vetch was killed mechanically. Three additional above-ground residue samplings were taken at 3, 6 and 9 ws after killing. Aqueous crude extracts from the ground shoot and root tissues of the vetch were prepared following established procedures (Barnes and Putnam, 1986; Burgos and Talbert, 2000) with some modification. Samples of dry, ground tissue (12 g) were incubated in 180 ml deionized water (1:15 wt/v) in 500-ml Erlenmeyer flasks. The flasks were sealed with a parafilm layer, placed on a shaker (Gyrotory G-2, Edison NJ), and incubated for 24 h in the dark at 4°C with constant shaking at 100 rpm. The slurry was filtered through six layers of cheesecloth and centrifuged at 2°C for 10 min at 3,046 x *g*. The supernatant solution (≈134±2 ml) that represented the crude aqueous extract was held in ice and used in bioassays. The undiluted extract and dilutions of 1:1, 1:3 and 1:7 were used in bioassays to determine the suppressive activity of the vetch shoot or root tissue on growth of hypocotyls or roots of 'Great Lakes' lettuce (*Lactuca sativa* L.). Lettuce was chosen because it is a standard test plant for determining allelopathic activity (Fujii *et al.*, 2004). Seeds of lettuce (50 per dish) were germinated in 100 by 15-mm Petri dishes over a Whatman No. 1 filter paper using 2.5 ml of extract per dish. The dishes were sealed with parafilm, placed on trays, and kept in the dark in a constant-temperature growth chamber at 26°C. The trays were positioned at 45° angle to encourage geotropic growth to facilitate hypocotyl and root measurements after 3 d (Burgos and Talbert, 2000). All experiments were repeated twice for each sampling period using five replications for each extract and dilution.

Fresh hairy vetch was collected from the field in spring and used intact (unleached) or was leached in a container overnight with a circulating pump (350 g residue per 50 L water). Flats were planted with velvetleaf, giant foxtail (*Setaria faberi* Herrm.), smooth pigweed, and common lambsquarters (*Chenopodium album* L.) as described above. Leached or unleached hairy vetch residue was placed on top of flats at the rates of 30 or 60 g per flat (approximately 3,300 or 6,700 kg/ha, respectively). Flats with residue were watered before placing residue on the surface and, thereafter, watered as needed using water only. Control flats without residue were watered with either water or with the leachate from the leaching process (stored under refrigeration, but brought to room temperature on the day of use). Flats were placed in the greenhouse until emerged seedlings were counted. The experiment was repeated and data were subjected to analysis of variance.

RESULTS

Emergence of each weed species in the greenhouse assay was reduced as a function of hairy vetch residue (Fig. 1). The log-logistic dose-response function fit data well since R-squared values ranged from 0.74 to 0.91 (Table 1). Less hairy vetch residue harvested before desiccation was required to suppress emergence of smooth pigweed (I50 = 3,932 kg/ha) than was required to suppress green foxtail (I50 = 5,014 kg/ha) or velvetleaf (I50 = 5,588 kg/ha).

The 95% confidence intervals for the I50 values overlapped when the vetch was harvested 1 d before or 5 d after desiccation for all weed species (Table 1). This suggests there was no significant difference between vegetative or recently killed vetch in suppressing weeds when compared at equivalent biomass. However, vetch harvested just before or just after

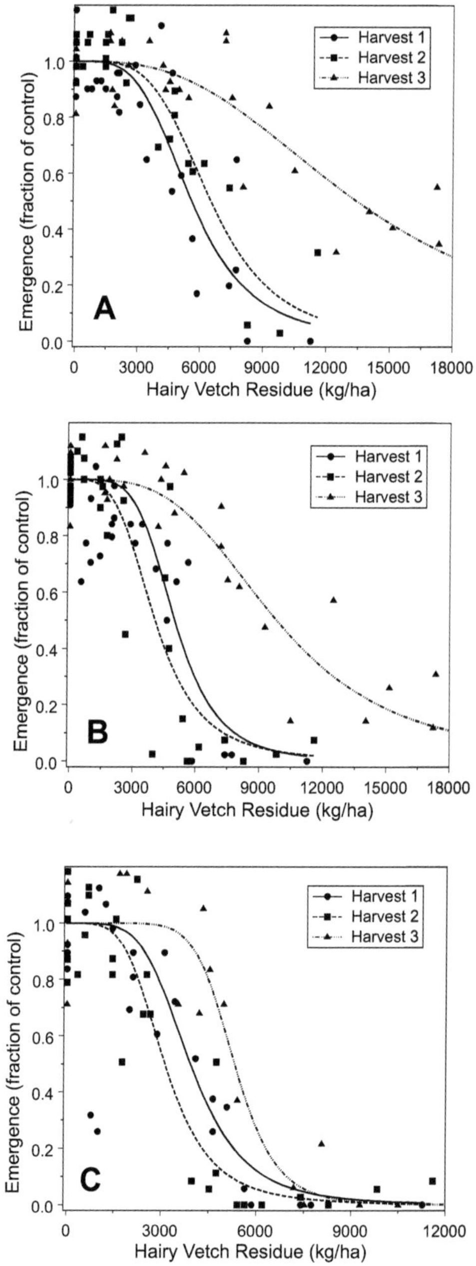

Fig. 1 Emergence of A) velvetleaf, B) green foxtail, and C) smooth pigweed as a function of dry biomass of hairy vetch residue on the soil surface of flats in the greenhouse assay. Hairy vetch was harvested from the field either 1 d before desiccation (Harvest 1), 5 d after desiccation (Harvest 2), or one month after desiccation (Harvest 3). Model details are shown in Table 1.

desiccation suppressed all three species more than the vetch residue that remained in the field for one month (Fig. 1). The I50 values were approximately doubled when residue remained in the field one month and 95% confidence intervals did not overlap (Table 1). The amount of hairy vetch residue required for 50% reduction of emergence was 13,449 versus 6,511, 9,830 versus 4,075, and 5,322 versus 3,127 kg/ha for one-month versus 5 d desiccated residue for velvetleaf, green foxtail, and smooth pigweed, respectively. This means that it required approximately twice as much desiccated residue to suppress weed emergence by 50% when decomposed for one month versus when recently killed.

Table 1 Coefficients for the model Y = 1/(1+(X/c)**b) where Y is weed emergence as a fraction of control, X is the dry weight of hairy vetch residue on the soil surface (kg/ha), and b and c are parameters. The parameter c is the residue level giving 50% emergence reduction (I50 value). I50 values within species followed by the same letter are not significantly different according to overlap of 95% confidence intervals.

Species	*Harvest* [a]	*b*	*c (I50)*	*R-square*
Velvetleaf	1 d before	3.85	5588 b	0.77
	5 d after	4.11	6511 b	0.86
	One month after	2.87	13449 a	0.80
Green foxtail	1 d before	4.88	5014 b	0.79
	5 d after	3.73	4075 b	0.79
	One month after	3.48	9830 a	0.89
Smooth pigweed	1 d before	4.65	3932 b	0.74
	5 d after	4.31	3127 b	0.87
	One month after	8.12	5322 a	0.91

[a] Time of vetch residue harvest before or after hairy vetch desiccation

Aqueous extracts of hairy vetch shoots were more inhibitory than extracts of root tissue (lettuce root length was inhibited by 80% by hairy vetch shoots versus 44% inhibition by vetch roots averaged across all rates and harvest dates before dessication). There was little difference in the activity of extracts from hairy vetch shoots harvested at the five dates before killing the cover crop (activity ranged from 73 to 86% inhibition of lettuce root length). Extracts of living hairy vetch harvested before the cover crop was killed were more toxic to lettuce roots than extracts of dead hairy vetch residue (Fig. 2). Three week after kill, extracts of hairy vetch residue had lost almost all activity at the one-eighth dilution and 23 to 51% of activity in the other dilutions relative to the activity of extracts from live hairy vetch tissue. Six to nine week after kill, extracts of hairy vetch residue had little activity left.

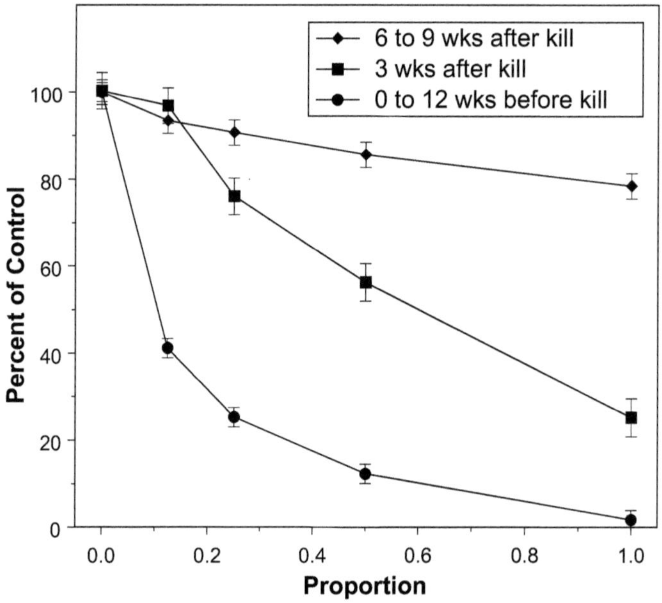

Fig. 2 Lettuce root growth in response to proportions of full strength extracts from hairy vetch above-ground tissue. Hairy vetch was harvested at five dates ranging from 0 to 12 wk before kill (grouped together in curve designated by circles), at 3 wk after kill (squares), and at 6 to 9 wk after kill (grouped together in curve designated by diamonds). Error bars represent the standard error of the mean.

Extracts of the upper leaves from hairy vetch plants were more phytotoxic to lettuce growth than extracts of the lower leaves or stem tissue (Table 2). Lower leaves, which were senescing at the time of harvest, in turn, had higher activity than that of stem tissue. Generally, these data show more activity in shoot than root tissue and demonstrate that the more actively growing, photosynthetic tissue of hairy vetch plants tend to have the most active extracts.

Table 2 Lettuce root and hypocotyl length in response to upper leaf, lower leaf and stem tissue of the hairy vetch. Values within column with the same letter are not significantly different ($P < 0.05$).

Tissue	*Percent of control*	
	Root	*Hypocotyl*
Upper leaf	11 c	26 c
Lower leaf	16 b	42 b
Stem	27 a	72 a

Unleached hairy vetch residue on the soil surface inhibited emergence of all weed species more than a comparable amount of leached residue (Table 3). The leachate derived from the leached residue had no inhibitory effect on emergence relative to the control and it stimulated emergence in the case of common lambsquarters. Stems of leached residue were similar in diameter to those of unleached residue, averaging 1.8 mm. However, unleached hairy vetch residue had intact leaves whereas leached residue had shriveled or missing leaves.

Table 3 Emergence of selected annual weeds in response to leached and unleached residue of hairy vetch. Values within species followed by the same letter are not significantly different ($P < 0.05$).

Species	*Surface residue*	*Treatment*	*Emergence (no./flat)*
Velvetleaf	Absent	None	40 a
		Leachate	37 ab
	Present	Leached	31 b
		Unleached	13 c
Giant foxtail	Absent	None	61 ab
		Leachate	69 a
	Present	Leached	53 b
		Unleached	28 c
Smooth pigweed	Absent	None	33 ab
		Leachate	45 a
	Present	Leached	22 b
		Unleached	2 c
Common lambsquarters	Absent	None	34 b
		Leachate	47 a
	Present	Leached	35 b
		Unleached	17 c

DISCUSSION

Evidence for Allelochemicals in Hairy Vetch Leaves

This research shows that hairy vetch residue that had been allowed to decompose in the field or that was subjected to leaching was less suppressive of seedling emergence/early growth than hairy vetch residue that was fresh. This result applied to residue that decomposed in the field and was assayed for suppression of weed emergence through residue placed on the soil surface in greenhouse flats (Fig. 1) or that was assayed for suppression of seedling growth by aqueous extracts in Petri dishes (Fig.

2). In addition, residue lost suppressive potential when leached overnight in water (Table 3).

All of these results could be explained by the presence of aqueous soluble allelopathic compounds in fresh hairy vetch residue that were lost during the decomposition process in the field or the leaching process in the laboratory. Alternately, suppression of weed emergence by surface residue also could be explained by the physical interference of the emergence process by mulch material (Teasdale and Mohler, 2000). However, the fact that the same amount of leached residue did not inhibit emergence as much as unleached residue suggests that allelochemicals were probably a contributing factor. Another alternative is that suppression of lettuce root length by extracts could be explained by suppressive osmotic levels of solutes in these extracts rather than by the direct toxicity of those solutes. Teasdale and Pillai (2005) showed that the toxicity of undiluted hairy vetch extracts could result from high osmotic levels. However, White *et al.* (1989) could not account for most of the toxicity of aqueous extracts of legume cover crop residue by osmotic effects. Given the consistency of results between the aqueous extract assay and the greenhouse assays, it seems more reasonable to assume that inhibition was caused by phytotoxic solutes than by inhibitory osmotic levels of solutes leached from residues into the soil.

Higher inhibition of seedling growth by aqueous extracts of upper leaves compared to lower leaves or stem tissue (Table 2) is consistent with the findings that fresh and unleached tissue were more inhibitory than decomposed and leached tissue. Generally, decomposed and leached material had lost most leaf tissue and was composed primarily of stems. This suggests that allelochemicals are present primarily in the most metabolically active upper leaves of fresh hairy vetch residue and that allelochemicals in this leaf tissue have sufficient activity to inhibit weed emergence from the soil.

Synergism between Natural Toxins and Surface Residue in Natural Systems

Allelochemicals are natural toxins that often have limited toxicity. In addition, their toxicity in the soil is limited further by potential degradation and/or adsorption processes similar to those that deactivate herbicides in the soil (Ohno, 2001; Kobayashi, 2004). The leachates from the leached hairy vetch residue had no inhibitory activity alone, even though they contained a chemical component that, when removed from residue, made that residue to be less suppressive than unleached residue (Table 3). This apparent paradox can be explained in two ways.

One possible explanation is the loss of allelochemicals from both the residue and the leachate during the leaching process. This would be possible for volatile compounds such as cyanamide identified by Kamo *et al.* (2003) or ketones identified by Bradow (1993). Kamo *et al.* (2003) also demonstrated that there was a higher concentration of cyanamide in the hairy vetch shoots than in the roots, which is consistent with the relative activity of the vetch shoot and root extracts observed in our studies. However, Kamo *et al.* (2003) also conducted their investigation with 9-d old seedlings and it remains to be determined whether cyanamide is an important constituent of mature hairy vetch tissue.

A second explanation for the lack of activity in leachates compared to unleached tissue is the potential synergism that may be operable when cover crop residue and plant toxins interact. In this type of synergistic interaction, the combination of cover crop residue and a phytotoxin would have substantially more activity than either the residue or the phytoxin would have when acting alone. An example of this type of synergism has been demonstrated by Teasdale *et al.* (2005); they have shown that hairy vetch residue synergizes the activity of very low rates of the herbicide metolachlor. For example, smooth pigweed emergence was reduced 13% by 5,000 kg/ha of hairy vetch residue, alone, and was reduced 16% by 10 g/ha of metolachlor, alone, but together they reduced smooth pigweed emergence by 86%. It was hypothesized that seedlings growing through residue would become etiolated, deprived of carbon, unable to deactivate metolachlor through glucose conjugation, and finally die before emergence. On the other hand, seedlings only emerging through low to moderate amounts of mulch would become etiolated but would survive and seedlings only exposed to low rates of metolachlor would have sufficient glucose reserves to detoxify metolachlor and survive.

Synergism could be a more widespread phenomenon defining interactions between cover crop residue and any phytotoxin with a tolerance mechanism based on metabolic deactivation. It is possible that many potential allelopathic compounds could have more activity if they were combined with a cover crop mulch. Allelochemicals are often tested in isolation from the residue they are derived from. Likewise, cover crop residue is often tested for weed suppression without attempting to optimize the release of allelochemicals. More research is needed to understand the mode of action of allelochemicals and to determine the best timing and management of cover crops for optimizing allelochemical activity.

Model for Weed Suppression by Hairy Vetch Residue and the Role of Allelopathy

This research suggests that allelopathy may play a role in weed suppression by hairy vetch mulch. The important factors that control weed suppression by residue of a legume cover crop such as hairy vetch are shown in Fig. 3. The emergence process begins when seed dormancy is broken and the germination process is initiated by environmental cues such as exposure to light, high alternating daily temperature amplitude, imbibed soil moisture, and dissolved oxygen.

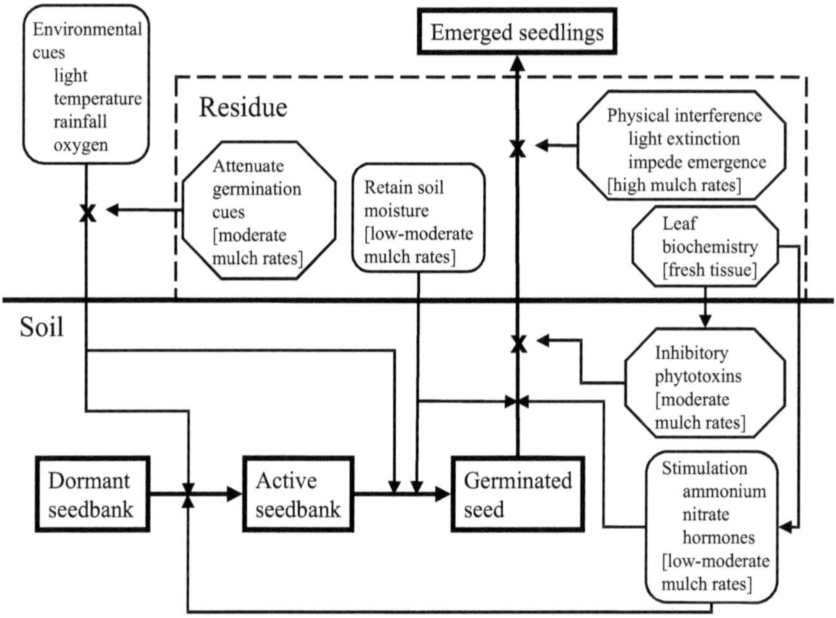

Fig. 3 Model of mechanisms for suppression or enhancement of weed emergence by legume cover crop residue. Rectangles connected by arrows represent weed states leading from dormant seed to emerged seedlings. Hexagons represent effects that inhibit emergence at the point marked with an "X". Rounded rectangles represent effects that enhance emergence at the points designated by the attached arrow. Brackets indicate residue conditions where the effect is most likely to occur.

Residue can enhance emergence by retention of soil moisture or by release of nitrogenous compounds that interact with environmental cues to break dormancy. Residue can inhibit emergence by attenuating environmental cues, physically interfering with the emergence process, and releasing phytotoxic compounds. The conditions under which these processes are most likely to occur are discussed below.

Research in the past, has shown that residue from cover crops such as hairy vetch or most other organic mulches must be present in very high amounts to provide a high level of physical suppression of annual weeds. For example, Teasdale and Mohler (2000) showed that greater than 75% inihibition of weed emergence is consistently achieved only when mulch biomass exceeds 8,000 kg/ha and mulch thickness exceeds 10 cm. The mechanism for this is primarily physical interference with the upward movement of the emerging seedling and the downward penetration of light to seedlings.

Most cover crops such as hairy vetch produce natural levels of approximately 3,000 to 5,000 kg/ha and, therefore, would not be sufficient to smother weed emergence but, rather, would operate through effects on either germination cues or allelopathy. Teasdale and Mohler (1993) showed that these intermediate rates of residue can limit light and temperature fluctuation cues that weed seeds may require for initiation of germination. In addition, release of phytotoxic compounds that inhibit weed roots and hypocotyl growth immediately following germination could prevent emergence of seedlings through natural residue levels by the kind of synergistic process discussed above. Research presented in this chapter suggests that this would most likely occur immediately following the vetch kill when residue is fresh and contains intact leaf material. A pulse of these compounds probably would be released when the first rains begin the degradation of fresh residue material. After the initial degradation process, residue would become decomposed and lose the leaf material where the highest activity resides. In addition, phytotoxicity of legume cover crops would not be expected to persist more than 2 wk in the soil (Ohno *et al.*, 2000). Subsequent weed suppression beyond the initial 2 wk following legume kill would probably result more from physical effects on germination cues and emerging seedlings than from allelopathic influences.

It is also important to recognize that release of inorganic nitrogenous compounds can stimulate germination of selected weed seeds under selected conditions when low or moderate levels of residue are present. For example, smooth pigweed germination was stimulated when a dormant seed lot was treated with hairy vetch extracts under moderate temperature and light conditions (Teasdale and Pillai, 2005). The activity of hairy vetch extracts and ammonium ions on germination was highly correlated suggesting the role of inorganic nitrogen in this process. Several researchers have shown that weed emergence can be enhanced intermittently by cover crop residue at lower rates in the field (Teasdale and Mohler, 2000; Gallagher and Cardina, 2003) and results reported in this paper showed an enhancement of emergence of common lambsquarters by hairy vetch leachate (Table 3). As a result, it may be difficult to predict weed emergence at natural levels of leguminous cover crops such as hairy vetch because released compounds could be both inhibitory and stimulatory. Specific effects of these compounds as well as effects on germination cues would also be highly dependent on local conditions and environmental parameters. Clearly, more research is needed to identify important compounds, their interactions with soil and residue, and their role in weed suppression.

References

Abdul-Baki, A.A., Teasdale, J.R. (1997) Sustainable production of fresh-market tomatoes and other summer vegetables with organic mulches. USDA-ARS Farmers' Bull. No. 2279 (Revised Aug. 1997).

Abdul-Baki, A.A., Teasdale, J.R., Korcak, R.F. (1997) Nitrogen requirements of fresh-market tomatoes on hairy vetch and black polyethylene mulch. HortSci. 32: 217-221.

Akobundu, I.O., Udensi, U.E., Chikoye, D. (2000) Velvetbean suppresses speargrass and increases maize yield. Int. J. Pest Manage. 46: 103-108.

Araki, H., Ito, M. (1999) Soil properties and vegetable production with organic mulch and no-tillage system. Jap. J. Farm Work Res. 34: 29-37.

Barnes, J.P., Putnam, A.R. (1986) Evidence for allelopathy by residues and aqueous extracts of rye (*Secale cereale* L.). Weed Sci. 34: 384-390.

Bradow, J.M. (1993) Inhibitions of cotton seedling growth by volatile ketones emitted by cover crop residues. J. Chem. Ecol. 19: 1085-1108.

Burgos, N.R., Talbert, R.E. (2000) Differential activity of allelochemicals from *Secale cereale* in seedling bioassays. Weed Sci. 48: 302-310.

Caamal-Maldonado, J.A., Jiménez-Osornio, J.J., Torres-Barragán, A., Anaya, A.L. (2001) The use of allelopathic legume cover and mulch species for weed control in cropping systems. Agron. J. 93: 27-36.

Candole, B.L., Rothrock, C.S. (1997) Characterization of the suppressiveness of hairy vetch-amended soils to *Thielaviopsis basicola*. Phytopathology. 87: 197-202.

Cheema, Z.A., Khaliq, A. (2000) Use of sorghum allelopathic properties to control weeds in irrigated wheat in a semi arid region of Punjab. Agric. Ecosyst. Environ. 79: 105-112.

Decker, A.M., Clark, A.J., Meisinger, J.J., Mulford, F.R., McIntosh, M.S. (1994) Legume cover crop contributions to no-tillage corn production. Agron. J. 86: 126S135.

Fujii, Y., Shibuya, T., Nakatani, K., Itani, T., Hiradate, S., Parvez, M.M. (2004) Assessment method for allelopathic effect from leaf litter leachates. Weed Biol. Manage. 4: 19-23.

Gallagher, R.S., Cardina, J., Loux, M. (2003) Integration of cover crops with post emergence herbicides in no-till corn and soybean. Weed Sci. 51: 995-1001.

Hutchinson, C.M., McGiffen, M.E. (2000) Cowpea cover crop mulch for weed control in desert pepper production. HortSci. 35: 196-198.

Inderjit, Keating, K.I. (1999) Allelopathy: principles, procedures, processes, and promises for biological control. Adv. Agron. 67: 141-231.

Kamo, T., Hiradate, S., Fujii, Y. (2003) First isolation of natural cyanamide as a possible allelochemical from hairy vetch (*Vicia villosa*). J. Chem. Ecol. 29: 275-283.

Kobayashi, K. (2004) Factors affecting phytotoxic activity of allelochemicals in soil. Weed Biol. Manage. 4: 1-7.

Mohler, C.L., Teasdale, J.R. (1993) Response of weed emergence to rate of *Vicia villosa* Roth and *Secale cereale* L. residue. Weed Res. 33: 487-499.

Morse, R.D. (1999) No-till vegetable production - Its time is now. HortTechnol. 9: 373-379.

Ohno, T., Doolan, K., Zibilske, L.M., Liebman, M., Gallandt, E.R., Berube, C. (2000) Phytotoxic effects of red clover amended soils on wild mustard seedling growth. Agric. Ecosyst. Environ. 78: 187-192.

Ohno, T. (2001) Oxidation of phenolic derivatives by soil and its relevance to allelopathic activity. J. Environ. Qual. 30: 1631-1635.

Sustainable Agriculture Network. (1998) Managing Cover Crops Profitably. 2nd ed. Handbook Series Book 3. Beltsville, MD, USA.

Teasdale, J.R., Abdul-Baki, A.A. (1998) Comparison of mixtures vs. monocultures of cover crops for fresh-market tomato production with and without herbicide. HortSci. 33: 1163-1166.

Teasdale, J.R., Mohler, C.L. (1993) Light transmittance, soil temperature, and soil moisture under residue of hairy vetch and rye. Agron. J. 85: 673-680.

Teasdale, J.R., Mohler, C.L. (2000) The quantitative relationship between weed emergence and the physical properties of mulches. Weed Sci. 48: 385-392.

Teasdale, J.R., Pillai, P. (2005) Contribution of ammonium to stimulation of smooth pigweed germination by extracts of hairy vetch residue. Weed Biol. Manage. 5: 19-25.

Teasdale, J.R., Pillai, P., Collins, R.T. (2005) Synergism between cover crop residue and herbicide activity on emergence and early growth of weeds. Weed Sci. 53.

Wagger, M.G. (1989) Time of desiccation effects on plant composition and subsequent nitrogen release from several winter annual cover crops. Agron. J. 81: 236-241.

White, R.H., Worsham, A.D., Blum, U. (1989) Allelopathic potential of legume debris and aqueous extracts. Weed Sci. 37: 674-679.

14

Allelopathic Effect of *Astragalus adsurgens* Pall Root Culture

Yong-qing Ma

Institute of Soil and Water Conservation, Chinese Academy of Sciences and Ministry of Water Resources, Northwest Sci-Tech University of Agriculture and Forestry, Yangling, 712100, Shaanxi Province, P. R. China

The allelopathic effect of *A. adsurgens* root culture on wheat, lettuce, millet, alfalfa and *A. adsurgens* are discussed in this Chapter. The results showed that *A. adsurgens* root culture filtrates strongly inhibited radicle elongation of the test species. The preliminary isolation of the inhibitors from the culture filtrates indicated that there are at least two groups of the chemical function. The results also indicated that *A. adsurgens* has autotoxicity.

Keywords: *allelopathy, Astragalus adsurgens Pall, root culture, Stellera chamaejasme* L.

INTRODUCTION

Astragalus adsurgens Pall (*Astragaus* spp.) is a Chinese origin legume pastures. It has been cultivated in arid areas of North China for more than 100 years because of its tolerance to drought, sandstorms and its high palatability with valuable protein. *A. adsurgens* has been recognized as an important forage legume, However, sub-optimal quality, poor seed productions as well as long growth periods have obstructed the application of this species (Lou and Jia, 1998; Luo *et al.*, 2005).

Wu *et al.* (1991) reported the content of toxin 3-nitropropanic acid (3-NPA) in *A. adsurgens* plants, which is less than 10 mg/kg. It belongs to a low toxin plant. Sheep which consumed nsumed 80% of *A. adsurgens* grew healthy and did not appear to have any poisoning symptoms. Wang *et al.*

(1992) conducted research on the determination of trace 3-NPA in *A. adsurgens* plants by gas chromatography. The results indicated that the contents of 3-NPA in *A. adsurgens* were very low; the highest content was not more than 100 mg/kg. The ranges of the contents in the stem were 0.2-37.4 mg/kg, in the leaf 0.43-37.5 mg/kg and in the flower 0.98-11.02 mg/kg. The contents of 3-NPA in different plant parts were decreased successively in order of leaf > stem > flower within one growing season.

Sheng *et al.* (1993) studied the chemical constituents of *A. adsurgens.* In their study they excluded the toxins and volatile compounds. Fifteen compounds have been isolated from *A. adsurgens* of which structures of 11 compounds were identified as known compounds. Among them (20R, 24S)-3,16-dicorbonyl-6α, 25-dihydroxy-20, 24-epoxy-9, 10-cyclpadaurgenin and (20R, 24S)-3, 16-dicorbonyl-6α, 25-dihydroxy-20, 24-epoxy-23-nitrogen-9, 19-cycloadsurgenin were two new identified compounds.

Several groups conducted research on tissue culture of *A. adsurgens.* Luo *et al.* (1991) conducted research on plant regeneration from protoplast culture of *A. adsurgens.* Protoplasts were enzymatically isolated from hypocotyls of *A. adsurgens.* The protoplasts formed calli in K & P liquid medium. Plants were regenerated from protoplasts of hypocotyls. Sun *et al.* (1992) studied the tissue culture of *A. adsurgens* in histology. The results showed that buds are exogenous and originated from the mass of meristematic cells of the callus, which have no relation with vascular tissue nodes. Luo *et al.* (1998) conducted research on plant regeneration from callus protoplasts of the forage legume *A. adsurgens.* A reproducible release of viable protoplasts was obtained from friable calli of *A.s adsurgens.* Luo *et al.* (1999a) reported again on high frequency somatic embryogenesis and plant regeneration in callus cultures of *A. adsurgens.* Their results indicated that efficient plant regeneration through somatic embryogenesis was achieved in *Astragalus adsurgens.* Embryogenesis callus was induced from hypocotyl, not root or cotyledon explants, and the maximum induction frequency (62%) was obtained on Murashige and Skoog (1962) basal medium (MS) containing 2.0 mg/L 2,4-dichlorophenoxyacetic acid (2,4-D) and 0.5 mg/L N superior6-benzylaminopurine (BA). The rooted plantlets were successfully transferred to the soil with 60-80% survival, and the plants showed normal morphological characteristics. In another report Luo *et al.* (1999) found that addition of ascorbic acid (ASA) at 200 mg/L to enzyme solution used for protoplast isolation and protoplast culture medium could diminish phenolic browning and its effect on propoplast growth, which resulted in the maintenance of high protoplast viability and enhancement of colony formation. Zhang *et al.* (1996) studied plantlet regeneration from leaf of *A.*

adsurgens. Five completes were finally obtained by the further transferring shoots onto the rooting medium.

Wang *et al*. (1999) reported selection and characteristic analysis of NaCl-tolerant cell line from *A. adsurgens* callus cultures. Their results only indicated that under NaCl treatment four new bands appeared in SDS-PAGE when compared with the control. Kita and Hongo (1998) studied the cytology of wild and cultivated forms and inter-regional hybrids of *A. adsurgens* from China and Japan. The results showed that there was a high degree of homology between the chromosomes of the diploid and tetraploid populations, which supported the hypothesis that the latter are autotetraploids. The use of the wild populations for the improvement of the cultivated form is referred to.

Stellera chamaejasme L. is a naturally occurring poisonous plant and an indicator for grassland degradation. It is mainly distributed in Ninxia, Gansu, Tibet and Inner Mongolia of China (Shi *et al*., 1997). The appearance of *S. chamaejasme* on the grassland is the result of long-term inverse succession. In other words, the grassland changed from normal to slightly degraded, then moderately and finally into severely-degraded grassland. Growing *S. chamaejasme* is not only dangerous for animals; it also competes for soil water and nutrients as well as space with other good quality grasses. Due to its prosperous growth, the other forage grasses gradually decreased. The productivity and the unit animals for the grassland will decline and the quality of the forage grasses will descend. In addition, it affects the sustainable development of animal husbandry.

The research results of our Institute have already indicated that *A. adsurgens* could inhibit *S. chamaejasme* growth. Our group is now conducting research and trying to understand the mechanisms of these inhibitions by using root culture techniques.

MATERIALS AND METHODS

Root Culture

A. adsurgens seeds were surface sterilized by 75% (V/V) ethanol for 3 min and then washed with autoclaved distilled water 3 times. The seeds were again put into 1% NaClO for 5 min and washed with autoclaved distilled water 3 times. Surface sterilized seeds were cultured into 200 mL flask containing 50 mL of MS medium with agar for 12 d in a incubator with 3,000 lux of light and temperature at 25°C. The root of the seedlings were excised and put into 200 mL flask containing 50 mL of B5 medium with 30 g of sucrose and 0.2 mg/L of NAA. The excised roots were cultured in the dark in a rotary shaker (70 rpm) at the consistent temperature at 25°C and it was sub-cultured at 3 wk intervals.

Collection of Culture Filtrate

A. adsurgens excised roots were cultured for 3 wk, there after the medium was filtered through a filter paper and stored in a freezer for bioassay.

Bioassay Procedure

Aqueous solutions were assayed directly, by applying 3 mL aliquots of the respective test solution to accepting plant seeds on 9 cm filter paper, placed in Petri dishes. For solutions containing organic solvents, aliquots (3 mL) of the test solution were applied to 9 cm filter paper. The filter paper was allowed to dry and then moistened with 3 mL of distilled water. The treated seeds were incubated at 25°C and examined for the length of radicle and coleoptiles 72 h later. Distilled water control was included where appropriate. Individual treatments were replicated four times. Data were analyzed by SAS software for the variance.

Experiment 1

A. adsurgens excised roots culture filtrates (3 wk after sub-culture) were diluted to 5-, 10-,50-, 100-fold with distilled water, respectively. Undiluted culture filtrate and distilled water were treated in the same way. The acceptor species were alfalfa (*Medicago sativa*), *A. adsurgens* and wheat.

Experiment 2

A. adsurgens excised roots culture filtrates (3 wk after sub-culture) were collected and stored in a freezer. A total of 200 mL of culture filtrate were partitioned with ethyl acetate [culture filtrates: ethyl acetate = 2:1 (v/v)] four times, each time samples were taken both from ethyl acetate and water phase. The solution was diluted to 5-, 10-, 50-, 100-fold with methanol for ethyl acetate phase and distilled water for water phase, respectively and utilized for bioassay test. Three mL of samples from ethyl acetate were added onto Petri dishes containing two layers of filter paper and the ethyl acetate was removed by air drying at 25°C and then 3 mL of distilled water was added onto the filter paper again. Three mL of samples from water phase were added onto two layers of filter paper and allowed to air dry at 25°C, the Petri dishes were weighed, after the trace ethyl acetate was removed and the same weight before air drying was maintained by adding distilled water. Undiluted solution and the mixed ethyl acetate phase as well as culture filtrates were included in the experiment and treated in the same way. The acceptor species were alfalfa, *A. adsurgens* and wheat.

Experiment 3

A. adsurgens excised roots culture filtrates (3 wk after sub-culture) were collected and stored in a freezer. A total of 6 L of culture filtrate were partitioned with ethyl acetate [culture filtrates : ethyl acetate = 2:1 (v/v)] four times and the ethyl acetate was removed by rotary evaporator at 35°C and dissolved into 200 mL of distilled water and used for bioassay. The solution was diluted to 5-, 10-, 50-, 100-fold with distilled water utilized for bioassay test. The acceptor species were alfalfa (*Medicago sativa*), *A. adsurgens*, lettuce (*Lactuca sativa*), millet (*Setaria italica*) and wheat.

Experiment 4

A. adsurgens excised roots culture filtrates (3 wk after sub-culture) were collected and used directly for bioassay. *Stellera chamaejasme* L. seeds were pre-germinated in an incubator at 25°C for 7 d. The germinated seedlings were subjected to the test solution. The radicle length was measured every day.

RESULTS

Effect of *A. adsurgens* Root Culture Filtrates on Radicle and Coleoptile Elongation of Alfalfa, *A. adsurgens* and Wheat (*Triticum aestivum* L.)

Undiluted culture filtrates strongly inhibited test crops radicals and coleoptiles elongation (Fig. 1). When alfalfa seeds were treated with undiluted culture filtrates and five-fold dilutions, the radicle elongations were significantly inhibited compared with distilled water and 10-, 50- and 100-fold dilutions. A significant inhibition was displayed for coleoptile elongation of alfalfa only by undiluted culture filtrated (Fig. 1). Effect of *A. adsurgens* root culture filtrated on radicle and coleoptile elongation of *A. adsurgens* exhibited a similar trend as for alfalfa. The only difference is that *A. adsurgens* radical was dead after 72 h incubation and the color of radicle tips became brown.

Wheat seedlings were measured for the longest radicals and the total length of radicals as well as coleoptiles were calculated. Undiluted culture filtrates strongly inhibited the longest radicle growth of wheat; the inhibitory effects declined as the dilution was increased. The total length of radicle displayed the same results (Fig. 1). This inhibitory effect was continued up to 10-fold-dilution solution and the differences were significant. In the case of coleoptiles undiluted and five-fold dilution showed a significant difference with distilled water control (Fig. 1).

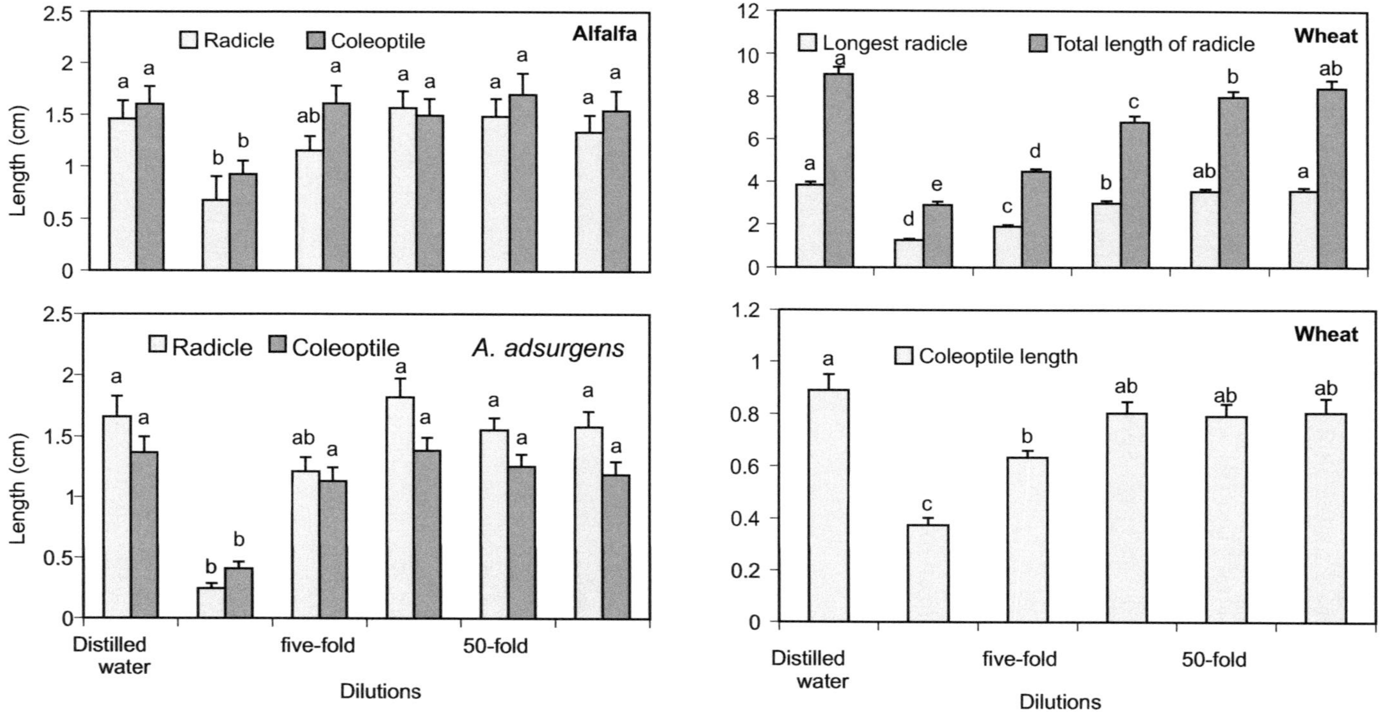

Fig. 1 Effect of different dilutions of *A. adsurgens* root culture filtrates on alfalfa, *A. adsurgens* (left side) and wheat (right side), radicle and coleoptile growth. Data are average of 30 means. With each length, bars marked by the same letter are not significantly different ($P > 0.001$).

Table 1 Ethyl acetate partitioning of *A. adergens* root culture filtrate on growth of acceptor crops

		A. adergens (Length cm ± SE)[A]		*Alfalfa* (Length cm ± SE)[A]	
	Fractions	Radicle length	Coleoptile length	Radicle length	Coleoptile length
Ethyl acetate phase	1st	1.72 (0.16)[aB]	1.20 (0.12)[a]	1.50 (0.21)[a]	1.38 (0.20)[a]
	2nd	1.71 (0.14)[a]	1.09 (0.10)[a]	1.45 (0.17)[a]	1.39 (0.16)[a]
	3rd	1.65 (0.13)[a]	1.13 (0.13)[a]	1.54 (0.20)[a]	1.27 (0.15)[a]
	4th	1.59 (0.10)[a]	1.07 (0.10)[a]	1.33 (0.16)[a]	1.12 (0.16)[a]
	Mixed	1.40 (0.13)[a]	0.89 (0.11)[a]	1.26 (0.17)[a]	1.34 (0.18)[a]
Water phase	1st	0.08 (0.03)[b]	0.18 (0.04)[b]	0.25 (0.03)[b]	0.31 (0.04)[b]
	2nd	0.09 (0.02)[b]	0.10 (0.02)[b]	0.31 (0.04)[b]	0.30 (0.04)[b]
	3rd	0.19 (0.04)[b]	0.23 (0.03)[b]	0.26 (0.04)[b]	0.29 (0.04)[b]
	4th	0.12 (0.03)[b]	0.17 (0.03)[b]	0.22 (0.04)[b]	0.26 (0.04)[b]
	Culture filtrate	0.11 (0.01)[b]	0.24 (0.02)[b]	0.21 (0.04)[b]	0.22 (0.04)[b]
	Distilled water	1.72 (0.12)[a]	1.06 (0.10)[a]	1.34 (0.19)[a]	1.34 (0.19)[a]

[A] Each mean represents 30 plants with 10 seeds per Petri dishes. [B] Data in each column marked by the same letters are not significantly different (P > 0.001), same as Table 2.

Ethyl acetate partitioning of *A. adergens* root culture filtrate on growth of *A. adergens*, alfalfa, and wheat. The inhibitory substances from *A. adergens* root culture filtrates could not be partitioned into ethyl acetate phase. From Table 1 and 2 it is clear that both radicle and coleoptile lengths in *A.adergens*, alfalfa and wheat were not affected by ethyl acetate extraction. The active materials remained in the water phase.

Ethyl acetate partitioning of *A. adergens* root culture filtrate on growth of lettuce, millet, *A.adergens*, alfalfa, and wheat. From the results shown above we are not sure whether *A. adergens* root culture filtrates containing inhibitory substance, can be partitioned into ethyl acetate. The results displayed that some of the inhibitory substances can be partitioned into ethyl acetate, but this compound(s) has a selective inhibitory effect as shown in Fig. 2. Ethyl acetate phase contained substance(s) which inhibited radicle growth of lettuce, millet, and wheat. On the other hand, radicle growth of alfalfa and *A. adsurgens* were not affected (Fig. 2).

Effect of Root Culture Filtrate on Growth of *S. chamaejasme* Radicle

Undiluted root culture filtrates significantly inhibited the elongation of *S. chamaejasme* radicle. This is the first report on inhibitory effect of *A. adsurgens* root culture filtrates on *S. chamaejasme* radicle (Fig. 3).

Table 2 Ethyl acetate partitioning of *A. adergens* root culture filtrate on growth of wheat seedling

	Wheat (T. aestivum) (length cm ± SE)[A]			
	Fractions	*Length of radicles*	*Longest radicle length*	*Coleoptile length*
Ethyl acetate phase	1st	7.39 (0.19)[a]	3.55 (0.05)[a]	0.49 (0.03)[a]
	2nd	7.35 (0.36)[a]	3.42 (0.14)[a]	0.53 (0.04)[a]
	3rd	7.80 (0.32)[a]	3.64 (0.12)[a]	0.55 (0.05)[a]
	4th	7.22 (0.21)[a]	3.42 0.10)[a]	0.55 (0.06)[a]
	Mixed	7.51 (0.36)[a]	3.55 (0.12)[a]	0.52 (0.06)[a]
Water phase	1st	3.31 (0.11)[b]	1.34 (0.05)[b]	0.26 (0.02)[b]
	2nd	2.95 (0.14)[b]	1.20 (0.05)[b]	0.24 (0.02)[b]
	3rd	2.69 (0.13)[b]	1.25 (0.05)[b]	0.21 (0.01)[b]
	4th	2.76 (0.16)[b]	1.10 (0.05)[b]	0.23 (0.02)[b]
	Culture filtrate	3.14 (0.15)[b]	1.41 (0.04)[b]	0.26 (0.03)[b]
	Distilled water	8.18 (0.37)[a]	3.80 (0.13)[a]	0.53 (0.06)[a]

[A] Each mean represents 30 plants with 10 seeds per Petri dishes

DISCUSSION

Prairie milkvetch (*A. adsurgens.*) is an important multi-year growing leguminous pasture in North China or green manure in South China. To meet the demand of feeding animals, the acreage of growing *A. adsurgens* is increasing. *S. chamaejasme* has long been recognized as a noxious weed and a great deal of work has been done to control or eradicate it. Huang and Zhou (1993) have suggested the use of biological control methods to control *S. chamaejasme*, which could be parasitized by *Aphis brachychaeta* Holman (Aphis:Aphididae) (Shi *et al.*, 1997) or infested by *Melamspora stellera* Teich (Melampsora: Melampsoraceae (Yao *et al.*, 2004). However, it seems likely that it will take a long time to develop the use of these natural enemies for the biological control of *S. chamaejasme* (Ji, 2003). On the degraded grassland, due to *S. chamaejasme* scatter sowing of some strongly competitive forage grasses by using biological competition to inhibit *S. chamaejasme* growing will not only make the coverage of *S. chamaejasme* decline but it also will leave *S. chamaejasme* on the soil surface and this will prevent soil erosion due to cultivation.

It is difficult to have a unique germination rate under laboratory conditions. Yang *et al.* (2004) reported the influence of sulfuric acid, methiophanate-methyl sulfur, GA_3, varied temperature, ultraviolet radiation treatments on germination rate of *S. chamaejasme* seeds. In addition, after overcoming the difficulties of seed germination, the

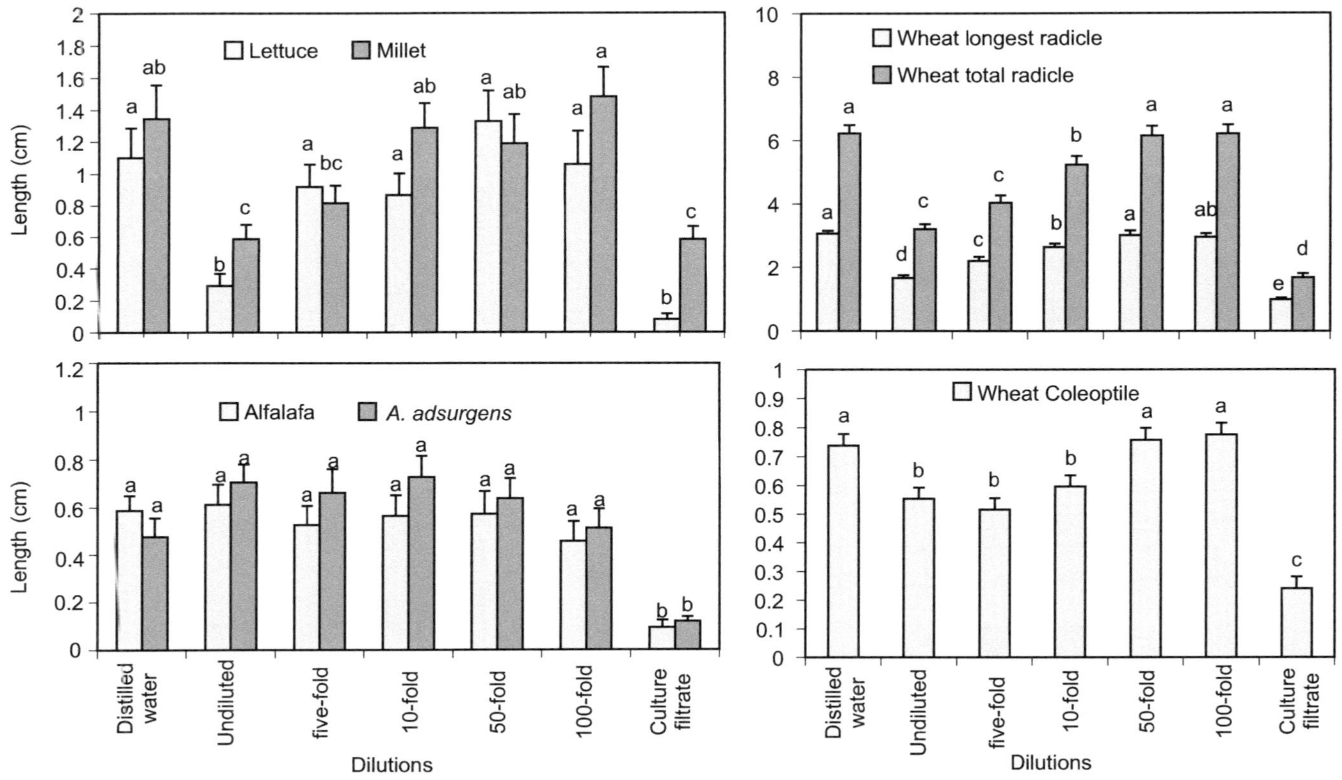

Fig. 2 Effect of ethyl acetate phase on radicle growth of lettuce, millet, alfalfa, wheat and *A. adsurgens*. Data are averages of 30 means. Bars marked by the same letter are not significantly different (P > 0.001).

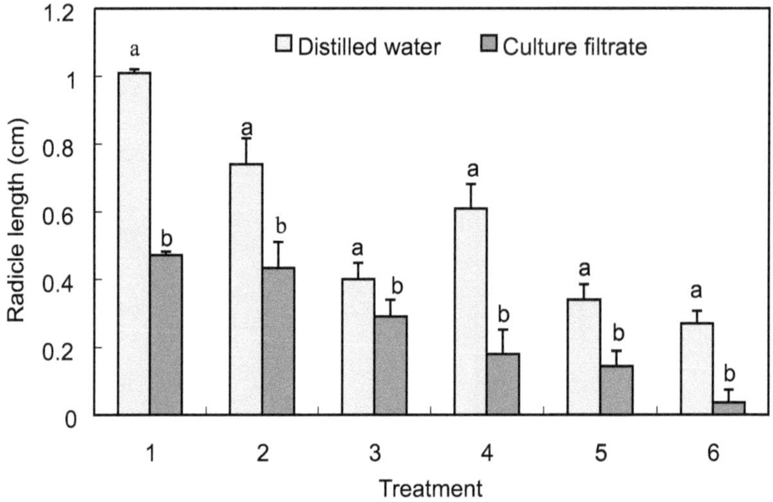

Fig. 3 Effect of ethyl acetate phase on radicle growth of *S. chamaejasme* L. Data are averages of 30 means. Treatment 1 = Differences of *S. chamaejasme* radicle between 3 d after culture filtrate treatment and before treatment; Treatment 2 = Differences of *S. chamaejasme* radicle between 2 d after culture filtrate treatment and before treatment; Treatment 3 = Differences of *S. chamaejasme* radicle between 1 d after culture filtrate treatment and before treatment; Treatment 4 = Differences of *S. chamaejasme* radicle between 3 d after culture filtrate treatment and 1 d after treatment; Treatment 5 = Differences of *S. chamaejasme* radicle between 2 d after culture filtrate treatment and 1 d after treatment; Treatment 6 = Differences of *S. chamaejasme* radicle between 3 d after culture filtrate treatment and 2 d after treatment; Bars marked by the same letter are not significantly different ($P > 0.001$).

culturing of *S. chamaejasme* seedlings faced another problem, as they would not grow well under both laboratory and field conditions. Our group started to use tissue culture techniques for obtaining and propagation of *S. chamaejasme* (Yu *et al.*, 2005). The study of *A. adsurgens* root culture filtrates on tissue cultured *S. chamaejasme* seedling is under way.

Our research results have shown that *A. adsurgens* cultured roots do exudate inhibitory substances that cause inhibition of the growth of radicle and coleoptile of wheat, millet, lettuce and *A. adsurgens*. Yu *et al.* (2005) have reported that the *A. adsurgens* culture filtrate have allelopathic effect on seedling growth of radish (*Raphanus satiivus*) and wheat (*T. aestivum*). The antitoxic effect of *A. adsurgens* is obvious. The inhibitory substance(s), which could be partitioned into ethyl acetate phase, indicated that this compound(s) has selective effects on the concerned species. The results also showed that *A. adsurgens* cultured roots exudate inhibitory compound, which can slow down the root elongation of *S. chamaejasme*. As the present study was conducted under aseptic culture, this unambiguously demonstrates that these inhibitory compound(s) are of plant origin.

Acknowledgements

We acknowledge the financial support of Innovation project from Northwest Sci-Tech University of Agricultural and Forestry, Led by Professor Shao Ming-an.

References

Kita, F., Hongo, I. (1998) Cytology of wild and cultivated forms and inter-regional hybrids of *Astragalus adsurgens* Pall. from China and Japan. SABRAO J. Bree. Gene. 30: 35-44.

Huang, Z.J., Zhou, S.Q. (1993) The important poisonous plant *Stellera chamaejasme* in grassland. Sichuan Grassland. 4: 24-27.

Ji, Y.J. (2003) The poisonous plant *Stellera chamaejasme* in Qinghai grassland and its interative control. Qinghai J. Animal Veter. Sci. 33: 35-36.

Luo, J.P., Jia, J.F. (1998) Plant regeneration from callus protoplasts of the forage legume *Astragalus adsurgens* Pall. Plant Cell Reports. 4: 313-317.

Luo, J.P., Gu, Y.H., Liu, J., Jia, J.F. (1999) High frequency somatic embryogenesis and plant regeneration in callus cultures of *Astragalus adsurgens* Pall. Plant Sci. 1: 93-99.

Luo, J.P., Gu, Y.H., Wang, S.B., Liu, J. (1999) Effect of ascorbic acid on membrane damage and activity of some related enzymes in *Astragalus adsurgens* protoplasts during their isolation and culture. Acta Phytophysiologica Sinica. 4: 343-349.

Luo, J.P., Zha, X.Q., Shi, W. (2005) Selection of ethionine-resistant variants with increased accumulation of methionine from embryogenic protoplasts of the legume *Astragalus adsurgens*. Plant Cell Tiss.Org. Cul. 82: 75-81.

Luo, X.M., Zhao, G.L., Xie, X.J., Liu, Y.Z. (1991) Plant regeneration from protoplast culture of *Astragalus adsurgens*. Acta Genetics Sinica. 5: 239-243.

Sheng, X.W., Sun, L.P., Song, Z.W., Wang, L., Zheng, S.Z. (1993) The chemical constituents of *Astraglus adsurgens*. Acta Botanica Sinica. 10: 807-812.

Shi, Z.C., Zou, K.N., Cao, G.R., Hong, Z.L. (1997) Important poisonous plants of China grassland. China Agricultural Press, Beijing.

Sun, H., Xu, A.J. (1992) Histological study of regenerative shoots of prairie milkvetch under tissue culture. Grassland of China. 6: 63-11.

Wang, J., Lei, Z.Y., Fen, X.Q. (1992) The determination of trace 3-nitropropionic acid in prairie milkvetch (*Astragalus adsurgens*) by gas chromatography. Grassland of China. 3: 69-72.

Wang, Y, Jia, J.F. (1999) Selection and characteristics analysis of NaCl-tolerant cell line from *A. adsurgens* callus cultures. Chin. J. Appl. Environ. Biol. 6: 5 47-550.

Wu, Z.L., Li, S.Q., Wang, Q. (1991) Dynamic-modification of toxin 3-NPA in *Astragalus adsurgens*. Grassland of China. 2: 59-60.

Yang, S.Y., He, J.Z., Shui, J.F., He, X.L., Ma, Y.Q. (2004) Influence if different treatments on the germination rate of *Stellera chamaejasme* L. seeds. Agri. Res. Arid Areas. 22: 66-69.

Yao, T., Kou, J.C., Liu, Y. (2004) *Stellera chamaejasme* rust diseases caused by *Melamspora stellera*. Chin. J. Biol. Contr. 20: 142-144.

Yu, F.K., Ma, Y.Q., Li, X.W. (2005) Tissue culture and rapid propagation of *Stellera chamaejasme*. Plant Physi. Communi. 41: 59.

Yu, F.K., Ma, Y.Q., Shui, J.F., Li, X.W. (2005) Effects of cultured *Astragalus adsurgens* root on other plants growth. Chin. J. Appl. Ecol. 16: 758-762.

Zhang, C.G., Liu, J.L., Feng, J. (1996) Study on callus induction and plantlet regeneration from leaf explants in *Astragalus adsurgens*. Acta Agrestia Sinica. 2: 148-154.

15

Evaluation of Weed Suppressive Effect of Allelochemicals of Ornamental Marigold Species

Nataliya P. Didyk and Svitlana P. Mashkovska
"M.M.Gryshko" National Botanical Garden of the National Academy of Sciences of Ukraine

Weed suppressive effect of allelochemicals of three ornamental species of marigold (*Tagetes erecta, T. patula, T. signata*) was investigated. It was found that in field conditions increasing abundances of plants of the *Tagetes* species reduced distribution of a number of widespread noxious weeds. The results of allelopathic assessment of volatile excretions and water extracts from fresh parts of the *Tagetes* plants showed that *T. signata* was more allelopathic than *T. erecta* and *T. patula*. The water extracts from *Tagetes* leaves and inflorescences were noticeably more phytotoxic than those from the other parts. Species of weeds showed differential sensitivity to allelochemicals of the *Tagetes* species. Treatment of the water extracts with polyamide, which is a specific sorbent for phenolic substances, showed that these substances contributed substantially to allelopathic activity of various parts, especially, the fresh leaves of *Tagetes* plants. In general, dynamics of allelopathic activity of decay products of phytomass of the three *Tagetes* species were similar and characterized by a sharp increase in phytotoxicity during first 2-4 wk of decomposition with subsequent gradual decrease and change to stimulation. However, the degree and duration of each of the mentioned above phases differed among different species of *Tagetes*. Decay products of the phytomass of *T. patula* and *T. erecta* were more effective against weeds than those of *T. signata*. The authors consider allelochemicals of both the fresh parts and their decay products of the studied *Tagetes* species to be promising for weed control in manipulated plant communities.

Keywords: *allelopathy, Tagetes sp., weed suppression*

INTRODUCTION

Allelopathic interactions between domesticated plants and weeds have been traditionally considered as detrimental for the former (Putnam and Weston, 1986). It was shown that many species of weeds produce allelochemicals inhibiting growth and development of domesticated plants (Patterson, 1986). The recent advances in allelopathic studies have shown that allelochemicals of cultivated plants could be effectively used for weed control. This approach gains importance as growers try to adopt alternative weeding techniques, which rely less on chemical herbicides.

Many of cultivated species such as barley, oat, rice, wheat, rye, canola, mustard species, buckwheat, red clover, white clover, sweet clover, hairy vetch, creeping red fescue, tall fescue and perennial ryegrass have been used effectively to suppress weeds, especially, broadleaf annuals (Liebman, 1988; Teasdale *et al.*, 1991; Fujii, 1992; Boydston and Hang, 1995; Weston, 1996; Hanano *et al.*, 1998).

Unfortunately, only a few studies have been devoted to weed suppressive allelochemicals of ornamental species despite of the fact that a number of the latter were successfully used for biological control of pests and phytopathogenic fungi. Ornamental flora is highly diverse and includes many poisonous, medicinal and aromatic i.e. potentially allelopathic plants. Thus, it presents an inexhaustible source of species perspective as donors of weed suppressive allelochemicals. Among the latter, ornamental annuals of marigold (*Tagetes* L.) genus could be considered, which were shown to produce large amounts of physiologically active secondary metabolites (flavonoids, phenolic acids, terpenoids, polythienyls) with fungicidal, bactericidal and nematocidal properties (Jimener *et al.*, 1990; Mashkovska *et al.*, 2002; Brovdy *et al.*, 2004). As these substances are phytotoxic in certain concentrations, it was suggested that *Tagetes* species could also be perspective donors of weed suppressive allelochemicals. Therefore the aim of our study was to test this suggestion by evaluating weed suppressive effect of allelochemicals of three ornamental species of marigold: *T. erecta L., T. patula L., T. signata Bartl.*

MATERIALS AND METHODS

Field observations on the effect of the abundances of the *Tagetes* species (*T. erecta, T. patula, T. signata*) on distribution of weeds were conducted on the plantations of "M.M. Gryshko" National Botanical Garden (Kyiv city).

Two hundred and fifty vegetation descriptions on plots with an area of 0.5×0.5 m^2 located along a transect crossing the studied plantations were made. The effect of abundances of the *Tagetes* species on the distribution of weeds was evaluated by the method of ceno-gradient analysis (Grodzinsky *et al.*, 1985).

The allelopathic activity of volatile excretions and water extracts of fresh plant material was studied immediately after collection at the flowering stage of the donor-species (Mashkovska, 2000) using bioassays on radicle elongation of cress (*Lepidium sativum L.*), as well as on radicle elongation, seed germination and rhizome bud sprouting of some selected weeds (Grodzinsky, 1991). Seeds of fathen (*Chenopodium album* L.), galinsoga (*Galinsoga parviflora* Cav.), amaranth (*Amaranthus retroflexus* L.) as well as rhizome fragments with a bud of quack-grass (*Elytrigia repens* (L.) Nevski) and bindweed (*Convolvulus arvensis* L.) were used as test-plants. Seeds or rhizome fragments with a bud were placed in Petri dishes on a filter paper with the appropriate test extract. The concentrations for water extracts were 0.2%, 1%, 2% and 5%. Volatile excretions were obtained from crushed fresh plant material placed into small open containers, which were put into hermetically sealed Petri dishes with test-plants (Grodzinsky, 1991). The concentrations of volatile excretions were 0.005 and 0.02 g of fresh phytomass of donor-plants per cm^3. Control dishes received water only.

Decay of phytomass of *T. signata, T. erecta, T. patula* in gray podzol soil was modeled in laboratory and plot experiments. The latter was established on the plantations of monoculture of lily (*Lilium* × *hybrida* hort.) at "M.M.Gryshko" National Botanical Garden (Kyiv city). The plot size was 2.5×2 m^2. Seeds of *Tagetes* were sown in row-spacing of lily monoculture after emerging of *Lilium×hybrida* plants in April. After reaching the budding stage (end of June) *Tagetes* plants were cut, crushed and mixed with surface soil in concentration of 3 kg of fresh phytomass per 1 m^2. Plots without *Tagetes* phytomass were used as a control. For two month of incubation of *Tagetes* phytomass in the surface soil regular weeding (once in 2 wk) with subsequent evaluation of fresh weight for each weed species were conducted. During the last weeding (end of August) shoots of lilies were also harvested and their fresh weight was measured.

The soil for laboratory experiment was collected from row-spacing of monoculture of *Lilium* × *hybrida* in early spring, before the emergence of ornamentals—dried, sifted and placed in pots with sizes of 1.5 L. Dried and finely ground (to accelerate decomposition process) phytomass of the *Tagetes* plants was mixed with the soil in concentrations of 1% to soil dry weight. Soil without addition of phytomass served as a control. The soil was moistened to 60% of the total moisture capacity. Such soil moisture level was maintained till the end of the experiment. The pots were kept in phytotron at the temperature of 20-26°C. Immediately after addition of *Tagetes* phytomass to the soil and afterwards regularly (within a period of 7 d) soil samples were collected and tested for allelopathic activity by direct bioassay method (Grodzinsky, 1991). The duration of the experiment was 140 d.

RESULTS

The analysis on the effect of marigold plants abundances on distribution of weeds on the plantations of the "M.M. Gryshko" National Botanical Garden (Kyiv city) showed that out of 28 weed species found on the plots *T. signata* reliably inhibited distribution of 9, *T. erecta* – 7, and *T. patula* – 5 (Fig. 1).

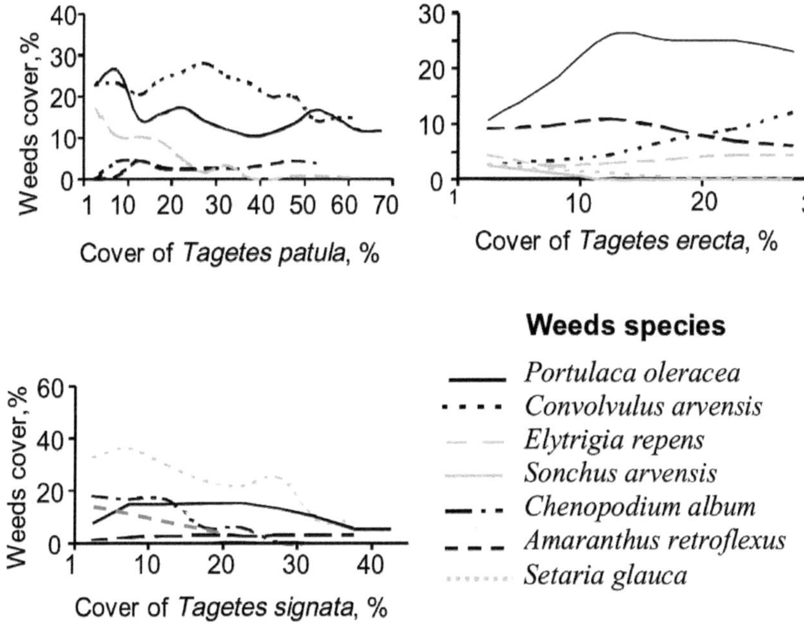

Fig. 1 The effect of the *Tagetes* species cover on the distribution of some weeds

The increase in *Tagetes* density reliably reduced distribution of such widespread noxious weeds as quack-grass (*Elytrigia repens*), amaranth (*Amaranthus retroflexus*), bristle grass (*Setaria glauca* (L.) Beauv.), fathen (*Chenopodium album*), perennial sow thistle (*Sonchus arvensis* L.). Field bindweed (*Convolvulus arvensis*) and purslane (*Portulaca oleracea* L.) were slightly affected by *Tagetes* plants even at high densities of the latter.

To elucidate the role of the allelopathic factor in the observed selective weed suppressive effect of the *Tagetes* species allelopathic assessment of their volatile excretions, water extracts of various parts and decay products have been conducted using laboratory bioassays with *Lepidium sativum* and some weeds.

Volatile excretion of the *Tagetes* plants released by their fresh phytomass (in concentrations of 0.02 g of per cm^3) inhibited radicle elongation of *L. sativum* (Fig. 2). The excretions of different plant parts showed different degrees of inhibitory effect. The highest allelopathic activity demonstrated volatile excretions of leaves and inflorescences. There was no apparent effect of volatile excretions of seeds of marigold plants on *L. sativum* radicle elongation. *T. signata* demonstrated the highest inhibitory activity of volatile excretions of all parts except for roots. While *T. erecta* demonstrated the lowest inhibitory activity of volatile excretions of all parts except for roots.

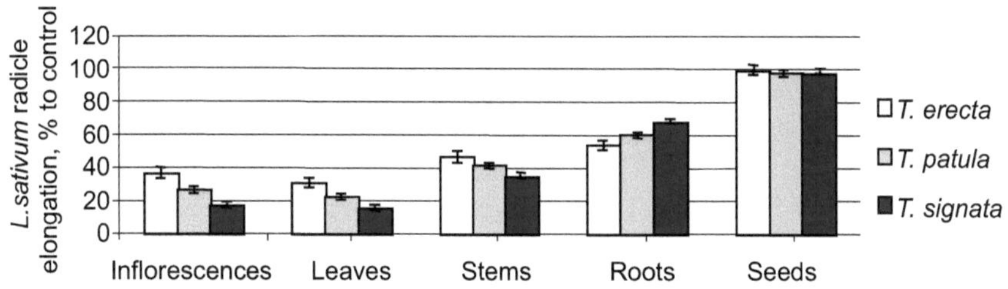

Fig. 2 Allelopathic effect of volatile excretions of parts of *Tagetes* plants (0.02 g of fresh phytomass per cm^3) on *L. sativum* radicle elongation. Bars are the averages of five replications of samples and the lines indicate standard errors.

Radicle elongation of *Amaranthus retroflexus* and *Chenopodium album* was inhibited by volatile excretions of inflorescences and leaves of *Tagetes signata* (0.005 g of fresh phytomass per cm^3) by 14% and 30%, respectively, as compared to the control. Volatile excretions of inflorescences of *T. patula* and *T. erecta* had no significant effect on these weeds. Conversely, volatile excretions of leaves of *T. patula* and *T. erecta* in the above mentioned concentration stimulated radicle elongation of *Amaranthus retroflexus* and *Chenopodium album* by 11% and 26%, respectively.

Water extracts from the *Tagetes* plants in concentrations of 2% and more inhibited radicle elongation of *L. sativum* (Fig. 3). There was no apparent effect only for water extracts of roots of *T. signata*. On the whole, allelopathic activity of water extracts obtained from different parts of the *Tagetes* plants correlated with that of volatile excretions. The highest allelopathic activity demonstrated water extracts from leaves and the lowest - from roots of *Tagetes* plants. *T. signata* demonstrated the highest inhibitory activity of water extracts obtained from all parts except for

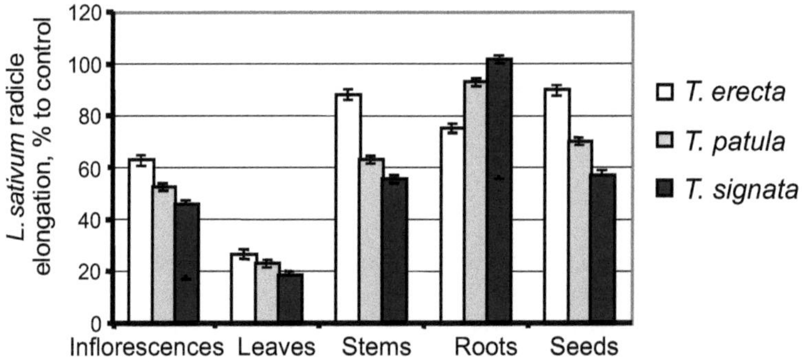

Fig. 3 Allelopathic effect of 2% water extracts from different parts of the *Tagetes* plants collected at the budding phase on *L. sativum* radicle elongation. Bars are the averages of five replications of samples and the lines indicate standard errors.

roots. While *T. erecta* demonstrated the lowest inhibitory activity of water extracts obtained from all parts except for roots.

Differential tolerance to water extracts from various parts of the *Tagetes* plants was observed among weeds, with *Amaranthus retroflexus* and *Chenopodium album* showing the most sensitivity. These weeds were inhibited by 33% and 46%, respectively, after treatment with 1% extract from leaves of *T. signata.* While *Convolvulus arvensis* showed the highest tolerance to water-soluble allelochemicals of the *Tagetes* species as evidenced by little effect of the extracts on its rhizome buds sprouting over the whole range of concentrations tested (0.2%-5%). *Galinsoga parviflora* and *Elytrigia repens* displayed moderate sensitivity to water-soluble allelochemicals of the *Tagetes* plants.

Tagetes are known to produce water-soluble phenolic allelochemicals: flavonoids and phenolic acids (Jimener *et al.*, 1990; Israilev and Abdala, 1993; Meches and Gaspar, 1993). To assess the contribution of these substances to allelopathic potential of the studied *Tagetes* species, water extracts were treated with polyamide, which is a specific sorbent for phenolics (Fig. 4).

It was found that such treatment noticeably reduced inhibitory effect of water extracts of all parts of the *Tagetes* species, especially of fresh leaves. The fractions of phenolic allelochemicals extracted from polyamide were only slightly less inhibitory than the initial extracts. This reduction of inhibitory activity could be due to irreversible absorption of some phenolic allelochemicals by polyamide.

The results of the pot experiment modeling decay of *Tagetes* phytomass in soil showed noticeable increase in phytotoxicity of the latter reaching maximum values on the 14th to 42th d (depending on a species of *Tagetes*)

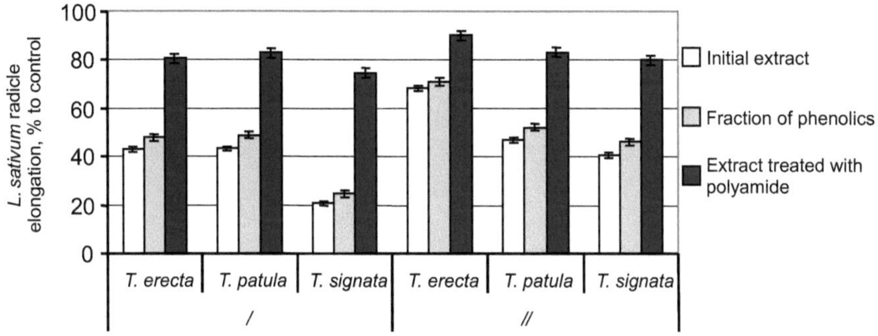

Fig. 4 Allelopathic effect of 1% water extract from leaves (*I*) and shoot residues (*II*) of the *Tagetes* species on *L. sativum* radicle elongation. Bars are the averages of five replications of samples and the lines indicate standard errors.

with subsequent gradual decrease and approaching the control level on the 48th to 77th d (Fig. 5). Afterwards the character of allelopathic activity changed to stimulation, which after reaching maximum values on 91th to 105th d began to decrease gradually approaching the control level at the end of the fourth month of experimentation. The highest phytotoxicity values were observed for decay products of phytomass of *T. patula*. While the highest stimulation values were observed for decay products of residues of *T. signata*.

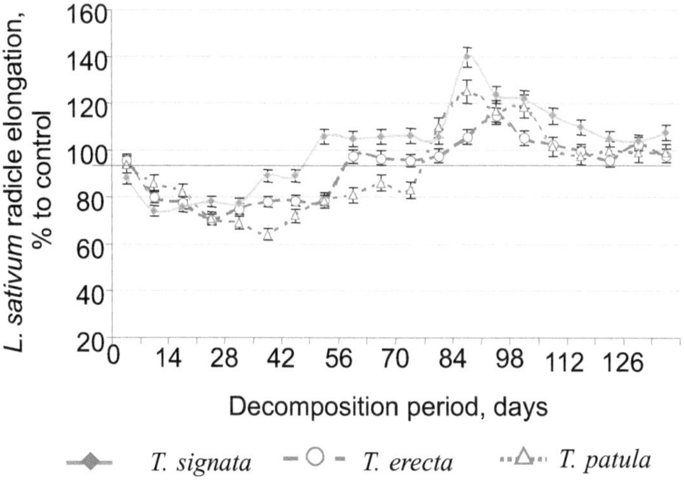

Fig. 5 Dynamics of allelopathic activity of soil mixed with decaying residues of the *Tagetes* species in concentration of 1% to soil dry weight. Points are the averages of eight replications of samples and the lines indicate standard errors.

The results of the plot experiment (Table 1) showed that incorporation of phytomass of the *Tagetes* plants into the soil between rows of lily (*Lilium* × *hybrida*) monoculture strongly reduced total phytomass of weeds during the whole period of observations (two months).

High sensitivity to marigold decay products revealed such widespread and noxious weeds as *Elytrigia repens, Convolvulus arvensis, Galinsoga parviflora, Portulaca oleracea,* shepherds purse (*Capsella bursa-pastoris* (L.) Medik.), chickweed (*Stellaria media* (L.) Vill.). While summer grass (*Digitaria sanguinalis* (L.) Scop.) and groundsel (*Senecio vulgaris* L.) were more tolerant to allelopathic excretions of decay products of *Tagetes*. The highest suppression of weeds was observed during the first observation on the 15-th d of decomposition of *Tagetes* phytomass. As the decomposition period increased weed suppressive effect of *Tagetes,* decay products decreased reaching the lowest values at the end of observation period (on the 60-th d of decomposition). The most effective against weeds were decay products of *T. patula*. Decay products of *T. signata* demonstrated the lowest weed suppressive properties in the plot experiment.

Analysis of the effect of the decay products of the *Tagetes* plants incorporated into soil between rows on phytomass accumulation of lily plants showed that the former did not affect this ornamental much. Such results could be due to the fact that lily plants passed critical phases of development (i.e. the most sensitive to stress factors) before the time of incorporation of *Tagetes* phytomass into row-spacing soil.

CONCLUSION

This study focused on weed suppressive effect of allelochemicals of three ornamental species of marigold (*T. erecta, T. patula, T. signata*). It was found that in field conditions increasing abundances of *Tagetes* plants reduced distribution of a number of widespread noxious weeds. From the studied *Tagetes* species, live plants of *T. signata* showed the most distinguished inhibiting effect on weeds. This observation could not be explained by competition as two other *Tagetes* species (*T. erecta, T. patula*) were characterized by much higher rates of growth and phytomass accumulation compared to *T. signata*. The results of allelopathic assessment of volatile excretions and water extracts from fresh parts of the *Tagetes* species showed that *T. signata* was more allelopathic than *T. erecta* and *T. patula*. Such allelopathic potential of *T. signata* could contribute to its suppressive effect on certain weeds observed in the field.

Species of weeds showed differential sensitivity to *Tagetes* allelochemicals. Among sensitive weeds were *Amaranthus retroflexus, Chenopodium album, Elytrigia repens, Galinsoga parviflora*. While *Convolvulus*

Table 1 Effect of decay products of the *Tagetes* plants incorporated into row-spacing of lily monoculture (3 kg/m^2) on fresh weight of weeds (g) depending on decomposition period. Figures are the averages of eight replications of plots.

Treatment	Decomposition period	Weeds phytomass, g										
		Convolvulus arvensis	*Portulaca oleracea*	*Galinsoga parviflora*	*Capsella bursa-pastoris*	*Stellaria media*	*Elytrigia repens*	*Digitaria sanguinalis*	*Poa annua*	*Erigeron canadensis*	*Other weeds*	*Total weed phytomass*
Control	15	104,5	96	64,5	52,5	27	15	11	8,5	10	26,5	415,5
	30	65	163,5	41,5	28,5	48	8	2,5	33,5	7	17,5	415
	45	30,5	145	15	35	110	10	15	30	10	23	423,5
	60	45	65	15	25	95	25	20	19	8	9	326
T. signata	15	15	3,25	0	0,75	0,25	0	10,75	8,75	2,5	4,5	45,75
	30	35	129,8	3,8	9,75	4,5	1,75	17	13,4	12	15,25	242,2
	45	22,5	90	0	31,5	7,25	0	15	21	5	13,25	205,5
	60	40,5	52	5,25	22,5	7,6	2,5	18	20,14	4,8	8	181,3
T. patula	15	22,5	1,5	0,5	0,5	1,25	0	7,25'	12	15	3,25	63,5
	30	30,75	56,75	2	7,5	7	0	14	11,5	3,25	17,5	150,3
	45	22,5	25	0	37,5	11,25	0	12	9	5	19	141,3
	60	35,1	38	0,75	13,6	23,75	0,75	13,6	10,26	3,6	12	164,1
T. erecta	15	12	32,5	0	0	3, 5	0	45	2,5	12	5	72
	30	5	72	4	12	28	0	12	17	0	13	163
	45	25	45	5	15	20	0	18	15	15	23,5	181,5
	60	24	40	5	10,71	14	0	14	18	19	27	171,7

arvensis was tolerant. Thus, the effect of the abundances of the *Tagetes* species on weed distributions observed in the field correlated with the results of allelopathic studies. There is evidence of participation of the allelopathic factor in the selective weed suppressive effect of *Tagetes* plants observed in the field.

The root extracts of the *Tagetes* species were noticeably less phytotoxic than the extracts from the other parts although some plants are known to release allelochemicals from their roots. A weak association between allelopathic activity of root extracts and the effect of the *Tagetes* species abundances on weeds observed in the field could be explained by the immediate release of allelochemicals after translocation from the leaves. Another possibility could be that allelochemicals of the studied *Tagetes* plants were not released from roots but were leached directly from the leaves and inflorescences.

The results of the experiments with treatment of water extracts with polyamide (which is a specific sorbent for phenolic substances) showed that these substances contributed substantially to allelopathic activity of various parts, especially the fresh leaves of the *Tagetes* species. However, the role of hydrophobic allelochemicals in allelopathic potential of the *Tagetes* species also needs to be investigated.

In general, dynamics of allelopathic activity of decay products of phytomass of the three *Tagetes* species were similar and characterized by a sharp increase of phytotoxicity during the first 2-4 wk of decomposition with subsequent gradual decrease and change to stimulation. However, the degree and duration of each of the above menioned phases differed among different species of *Tagetes*. Though allelochemicals of fresh leaves and inflorescences of *T. signata* demonstrated the highest inhibition in bioassays, decay products of this plant (since the 35-th d of decomposition till the end of the experiment) demonstrated the lowest values of phytotoxicity and the highest values of stimulation in comparison to *T. erecta* and *T. patula*. These results could be explained by the fact that decay products of *T. signata* released substances inhibiting phytopathogenic fungi as well as microorganism releasing phytotoxines (Brovdy, 2004; Mashkovska *et al.*, 2002). Another reason could be the possible lower stability of *T. signata* allelochemicals in soil.

The dynamics of weed suppressive effect of decay products of the *Tagetes* plants observed in the plot experiment correlated with the results of the laboratory study. The highest suppression of weeds was observed during the first observation on the 15 th d of decomposition of *Tagetes* phytomass, and the lowest during the last observation on the 60 th d. Most species of weeds found on the experimental plots revealed sensitivity to decay products of the *Tagetes* phytomass. Among the studied weeds comparatively tolerant to allelopathic excretions of marigold decay products were *Digitaria sanguinalis* and *Senecio vulgaris*.

In both the plot and laboratory experiments, decay products of phytomass of *T. patula* and *T. erecta* were more effective against weeds than those of *T. signata*. Besides, the first two species are characterized by quick development and almost twice as higher phytomass per plant compared to *T. signata*. Therefore, we consider *T. patula* and *T. erecta* to be more productive for utilization as weed suppressive mulches.

To summarize, the studied *Tagetes* species may be candidates for further investigation of their allelopathic effect on weeds and may be used for the development of alterative weed control strategies. Allelochemicals of both fresh plant parts and their decay products are promising as inhibitors of weeds in manipulated plant communities.

References

Boydston, R.A., Hang, A. (1995) Rapeseed (*Brassica napus*) green manure crop suppresses weeds in potato (*Solanum tuberosum*). Weed Tech. 9: 669-675.

Brovdy, V.M., Guly, V.V., Fedorenko, V.P. (2004) Biological protection of plants. 348 p.- "Svit", Kyiv (in Ukrainian).

Fujii, Y. (1992) The potential biological control of paddy weeds with allelopathy (allelopathic effect of some rice varieties). *In* Proc. Int. Symp. Biol. Control and Integrated Management of Paddy and Aquatic Weeds in Asia, Natl. Agric. Res. Cent., Tsukuba, Japan. 19-25 Oct. Food and Fert. Technol. Cent. for the Asian and Pacific Region, China. pp. 305-320.

Grodzinsky, A.M. (1991) Allelopathy of plants and soil sickness: selected works. P.1-432. - Naukova Dumka, Kyiv. (In Russian).

Grodzinsky A.M., Maryushkina V.Y.A., Hazimametov, R.M., Onishenko, L.I. (1985) The experience of statistical analysis of cenotic interactions between *Ambrosia artemisiifolia* and other weeds in the communities of the Steppe zone of Ukraine. *In* Phytocenologiya of antropogennaya rastitelnost B.G. Mirkin, L.M. Lukhina (eds.). Rotaprint, Ufa. (In Russian) pp. 40-48.

Hanano, Y., Fujii, Y., Sato, K., Osozawa, S., Fujihara, S. (1998) Weed control by hairy vetch (*Vicia villosa* LOTH.) in Shikoku area: Vegetation test and field survey in 1993 to 1997. Bull. Shikoku National Agricultural Experiment Station. 62: 45-69.

Israilev, L.R., Abdala, D.A. (1993) Flavonoides como marcadores taxonomicos interspecificos en *Tagetes* (*Compositae*). Lilloa. 38: 101.

Jimener, G.F., Amader, P.M., Montes, M.A. (1990) Caracterizacion de compuestos fitotoxicos aislados de *Tagetes signata* Bartl. (Compositae). Fyton. 51: 99-102.

Liebman, M. (1988) Organic and low herbicide approaches for weed management. *In* the Proceedings of the Bioregional Farmer/Scientist Workshop in Organic Agriculture (Maine & Maritime Provinces 1988) D. Patriquin, ed. Fredricton, New Brunswick, USA.

Mashkovska, S.P. (2000) Allelopathic peculiarities of introduced species of *Tagetes* L. Plant introduction. No. 3-4 : 56-59.

Mashkovska, S.P., Yunosheva, O.P., Vergun, O.M. (2002) Phytosanitary role of the species of *Tagetes* L. genus in the manipulated ecosystems of ornamentals. *In* Yu. D. Kleopov and Modern Botany: Proceedings of Readings Devoted to 100 Anniversary of Yu. D. Kleopov. Fitosocioceter, Kyiv. pp. 378-383. (in Ukrainian).

Meches, L.M., Gaspar, J.S. (1993) Phytotoxic effect of methanolic extracts from *Porophyllum macrocephalum* and *Tagetes erecta*. Physiotherapy. 64: 35-41.

Patterson, D.T. (1986) Allelopathy. *In* Research Methods in Weed Science, N.D. Camper, (ed.), 3rd ed., Weed Science Society, Champaign, IL, USA. pp. 111-134.

Putnam, A.R., Weston, L.A. (1986) Adverse impacts of allelopathy in agricultural systems. *In* The Science of Allelopathy, A.R. Putnam, C.H. Tang, (eds.). John Wiley and Sons, NY, USA. pp. 43-56.

Teasdale, J.R., Beste, C.E., Potts, W.E. (1991) Response of weeds to tillage and cover crop residue. Weed Science. 39: 195-199.

Weston, L.A. (1996) Utilization of allelopathy for weed management in agroecosystems. Agronomy J. 88: 860-866.

SECTION 5

Rice Allelopathy

16

Rice Allelopathy

Kaworu Ebana and Kazutoshi Okuno
Plant Genetic Resources Laboratory, Genebank,
National Institute for Agrobiological Sciences,
2-1-2 Kan-nondai, Tsukuba, Ibaraki 305-8602, Japan

Allelopathic effect of rice against lettuce was investigated with water soluble extracts from two rice (*Oryza sativa* L.) varieties, "PI312777" (highly inhibitory) and "Rexmont" (less inhibitory), which are suppressive and non-suppressive to weeds in the field. Water soluble extracts derived from rice seedlings at various stages were applied to lettuce seeds. Leaf extracts from the six-leaf stage rice seedlings were appropriate for the assessment of the difference of the two varieties. Quantitative trait loci (QTL) underlying allelopathic effect of rice were identified using RFLP markers and the water-soluble extracts method. QTL analysis was carried out using the F_2 population from the cross between PI312777 and Rexmont. Among the seven QTL detected, the QTL on chromosome 6 had the largest effect and explained 16.1% of phenotypic variation. The other six QTL explained from 9.4% to 15.1% of the variation. A multiple QTL model estimated that five QTL with LOD scores of higher than 3.0 explained 36.6% of the total phenotypic variation. Digenic interactions in five pairs between the nearest RFLP markers to each of the seven QTL detected were significant at the 1% level. QTL on the chromosome 7 was also detected as the QTL for necrosis.

Keywords: *QTL, rice, water-soluble extract*

INTRODUCTION

Asian cultivated rice (*Oryza sativa* L.) is one of the most important crops in the world. Its cultivated area was 150 million ha and the grain production was 600 million ton in 1999 (FAOSTAT Database, 2000). Rice has been cultivated for 7,000 years and has differentiated to diverse landraces, adapted to the environment, such as upland or irrigated fields and from temperate to tropical climate regions. Weeds could decrease the rice yield 19-71% (Smith, 1981; 1988), and the damage is especially serious in upland conditions. Herbicides are used to prevent weeds, but for low input agriculture, rice with the ability to inhibit the weed growth via allelopathy is an important target for breeding.

Allelopathy is a phenomenon in which plants influence other organisms and their environment by the chemical compounds they produce. These compounds are released from plants through root exudation, leaching, volatilization or decaying plant tissue and they suppress or promote the growth of neighboring plants (Rice, 1984). Allelopathy in crops may act as biological weed control agents in the agroecosystem, which may be a target of genetic improvement in breeding programs. In the 1970s, germplasm assessment was extensively undertaken to detect intraspecific variation of allelopathy, and accessions with inhibitory effect to weeds were selected in many crops (Rice, 1984). Cultivated and wild cucumbers were studied in pot and field tests (Putnam and Duke, 1974) and the accessions with inhibitory effect on weeds were used as breeding material (Lockerman and Putnam, 1979). The accessions of oat, *Avena sativa* L., exude a large amount of an allelochemical, scopoletin (Fay and Duke, 1977). Genetic analysis of the strigolactone in *Sorghum bicolor* (L.) Moench (Vogler *et al.*, 1996) suggested that allelopathic activity of crops can be improved genetically. However, breeding of highly allelopathic cultivars has not been successful. Allelochemicals are secondary metabolites whose synthesis is controlled by large numbers of genes. Marker-assisted selection based on the QTL (quantitative trait loci) analysis is necessary for breeding such complicated traits (Tanksley, 1993).

Research on allelopathy in rice has been undertaken to find a variety with high inhibitory effects to weeds, and to identify new allelochemicals. Since the landraces adapted to various environmental conditions, wild relatives have survived competition with other weeds leading to the possibility of the production of allelochemicals, which significantly suppress the weeds. Rice accessions that inhibited the growth of an aqueous weed, ducksalad, [*Heteranthera limosa* (Sw.) Willd.] in the field were found in rice germplasm (Dilday *et al.*, 1989). Field tests for rice to inhibit weed growth were assessed with some important weeds, such as *Echinochloa crus-galli* (L.) P. Beauv. or *Sagitarria* spp. (Dilday *et al.*, 1989; Lin *et al.*, 1994 a; Hassan *et al.*, 1994). USDA researchers screened rice cultivars with high allelopathic activity in the field and found active varieties against annual weeds such as, *Heteranthera limosa*, *Cyperus iria* L., *Ammannia coccinea* Rottb., *Echinochloa crus-galli* (Dilday *et al.*, 1994, 1998).

To study allelopathy in rice, simple and stable evaluation methods have been developed. Fujii (1992) screened numerous rice accessions by the plant box method in which lettuce (*Lactuca sativa* L.) seeds were placed together with rice plants in agar media, and found some rice accessions with high inhibitory effect. He also developed the Sandwich method to evaluate the allelopathy of plant residues and applied it to rice varieties. Rice varieties with allelopathic effects on barnyardgrass were analyzed using a relayed-seeding technique (Olofsdotter and Navarez, 1996). Bioassays with not only intact plants, but extracts have also been used. Aqueous extracts of decomposing rice residues in soil inhibited the growth of rice roots and lettuce seedlings (Chou and Lin, 1976). Water-extracts from rice were assessed to evaluate their allelopathy to barnyard grass (Chung *et al.*, 2001), lettuce (Ebana *et al.*, 2001a) and *Sagitarria montevidensis* (Seal *et al.*, 2004). Although significant variations were found in these studies, different methods showed different results. These observations suggest that rice has several ways to release allelochemicals.

Chemical analysis to find rice allelochemicals has been conducted by various researchers. Phenolic compounds were identified from the residue of rice straw (Chou and Lin, 1976). Six allelochemicals, phenolic acids, *p*-hydroxybenzoic, *p*-coumaric, vannilic, syringic, ferulic, and *o*-hydroxyphenyl acetic acids were isolated from decomposing rice straw and paddy soil (Chou, 1980). According to a review of rice allelopathy (Olofsdotter *et al.*, 1995), a total of 16 potential allelochemicals, including the above mentioned compounds were reported. However, these candidates have not been identified as allelochemicals, because the intraspecific differences in allelopathic effect of rice observed in the field or in laboratory experiments could not be explained. Recently momilactone B has been reported as a rice allelochemical of the Japanese cultivar, "Koshihikari" (Kato *et al.*, 2002). Momilactone is an anti-fungae compound found in rice hull, seeds (Kato *et al.*, 1973) and leaves (Cartwright *et al.*, 1977). It needs to be confirmed whether this compound is an allelochemical released from other rice varieties. Substances in rice hulls that promote the germination of *Monochoria vaginalis* was reported and it is another aspect of rice allelopathy (Kawaguchi *et al.*, 1997).

For the genetic improvement of rice allelopathy, rice germplasm with highly inhibitory effects on weeds are necessary. Understanding the genetic control of allelopathy, production of allelochemicals, is also important for allelopathy breeding. The Rice Genome Research Program (RGP) in Japan produced a high-density linkage map with more than 2,300 DNA markers on 12 rice chromosomes (Kurata *et al.*, 1994). This linkage map enables the analysis of QTL underlying allelopathic effects in rice.

Our objectives are to establish the evaluation method of rice allelopathy and to study this phenomenon with genetics. We adopted the water-soluble extract method (Ebana *et al.*, 2001a), because it is stable and appropriate to deal with a lot of samples in a short time. At first, we studied the allelopathic effects in different growth stages and tissues of rice plants. Next, QTL analysis was conducted to show the gene location on rice chromosomes associated with the inhibitory effect.

MATERIALS AND METHODS

Plant Materials

Two rice varieties, "PI312777" and "Rexmont" were used for the study of assessment condition. These two cultivars were selected as representative of suppressive and non-suppressive rice based on the earlier germplasm evaluation (Dilday *et al.*, 1994). PI312777 is a progeny line derived from the cross combination, "Taichung 65" *2 / "Taichung Native 1" ("TN 1"), and Rexmont is not a suppressive and commercial cultivar in the USA. These two cultivars were used to investigate which parts and growth stages of rice plants are the most suppressive against lettuce. Plants were grown in a greenhouse with controlled temperatures between 25 to 35°C. Roots, stems (3 cm from the basal node) and leaves were sampled from rice plants grown in the greenhouse at the two-leaf, three-leaf, four-leaf, five-leaf, six-leaf, eight-leaf, flowering and grain filling stages of PI312777 and Rexmont grown in the greenhouse. Each experiment was triplicated.

F_2 and F_3 progenies derived from the cross of PI312777 and Rexmont were used for the QTL analysis. A total of 192 F_2 individuals were grown in the field. Young green leaves were sampled from each F_2 plants to extract DNA. Thirty seeds of each F_3 line from each F_2 individual were sown in a greenhouse at 25-35°C in duplicates. They were sampled after 80% of the individuals were grown to the six-leaf stage.

Extraction of Water-soluble Components

Fresh samples were stored at –80°C and then freeze dried. The samples were ground to a powder and stirred in cold sterilized distilled water. Ten ml of sterilized and distilled water per gram of fresh sample was added to the samples. Then the mixture was kept in a refrigerator for 2 h, it was stirred on a rotary shaker for one h and then centrifuged at 1,600 rpm for 15 min. The supernatant was recovered and stored in a refrigerator until used as crude water soluble extracts (Ebana *et al.*, 2001a).

Bioassay with Lettuce

Fifty seeds of the lettuce cultivar, "Great lakes 366" were placed on a filter paper (Advantec No. 2) in a 9-cm diameter Petri dish to evaluate the allelopathic effect of water-soluble extracts. Three ml of the crude extract or sterilized distilled water, as controls, were applied to them. They were incubated at 25°C in the dark, for 3 d. The number of germinating seeds was recorded and the length of roots and hypocotyls of 10 randomly selected 10 germinating seeds was measured. Each extract was quintupled. Data were processed with GLM procedure of SAS (SAS, 1990).

For the QTL analysis, the percentage of the root length of lettuce to the control was used as the phenotypic value and the average of the replications were used for the analysis. The frequency of necrotic roots was also evaluated as eight classes, from level; most of the seeds could not germinate and the germinated seeds showed necrotic roots, to level seven; all of the seeds showed no necrotic roots.

DNA Analysis

Total DNA was extracted following the CTAB method (Murray and Thompson, 1980). DNA was digested with eight restriction enzymes, *Apa*I, *Bam*HI, *Bgl*II, *Dra*I, *Eco*RI, *Eco*RV, *Hind*III and *Kpn*I and transferred to the nylon membrane after electrophoresis. Southern hybridization and signal detection were conducted with ECL direct nucleic acid labeling and detection kit (Amersham Pharmacia Biotech, Uppsala). A total of 125 RFLP probes which showed polymorphism between the parental cultivar were used.

QTL Analysis

The program MAPMAKER/EXP ver. 3.0b (Lander *et al.*, 1987) based on the Kosambi function was used to establish an RFLP linkage map. MAPMAKER /QTL ver. 1.0 (Lander and Botstein, 1989) was used to detect QTL and also to obtain estimates of the percentage of phenotypic variance explained by each QTL. PROC GLM in the computer program SAS (SAS Institute, 1989) was used to determine the relationship between RFLP markers and the allelopathic effect. Putative QTL for the allelopathic effect were identified using the LOD threshold of 2.0 in MAPMAKER/QTL and 5% significant level in PROC GLM analysis. PROC GLM was also used to analyse the possible digenic interaction between the nearest marker loci to each QTL.

RESULTS

Inhibitory Effect Detected in the Extracts from Rice Plants at Different Growth Stages

Germination rate of lettuce was 87 ± 6.2% with the water-soluble extract of PI312777 and 87 ± 6.0% with Rexmont, whereas 92 ± 3.3% with the sterilized and distilled water. All of the extracts reduced the germination of lettuce seeds, but no significant difference was observed between the two rice cultivars. The effect of extracts on the lettuce hypocotyl elongation was variable among replications (data not shown). When applying the leaf extract of PI312777, the average length of hypocotyl was 4.2 ± 4.15 mm. The length of lettuce roots was used as a standard to assess allelopathic activity of rice, because the effect on the root elongation was stable among the replication.

All extracts showed allelopathic inhibitory effect on root elongation of lettuce (Fig. 1). Inhibitory effects of the water-soluble extract from the leaves of PI312777 changed according to the growth stage. The extracts from the two-leaf stage showed strong inhibitory effects. It decreased at the four-leaf stage and then increased until the six-leaf stage. After that, the effect gradually decreased and was weakest in the grain filling period. Extracts from Rexmont also showed strong inhibitory effect at the two-leaf stage. However, after that the inhibitory effects decreased until the four-leaf stage. Then, the effects gradually increased until the eight-leaf stage. In the grain filling stage, the effect was also weak.

Leaf extracts from the six-leaf stage rice plants inhibited lettuce root growth more than extracts from other tissues and stages. The extracts caused necrosis on the tip of lettuce roots and most of the seeds with necrotic root tips could not develop normal hypocotyls. The suppressive effects of extracts from the plants during the grain filling period were low. Stem and root extracts were less effective compared to leaf extracts. Those extracts of Rexmont showed small age specificity as well as tissue specificity on lettuce growth, compared with PI312777 (Fig. 1). The inhibitory effects of the water-soluble extracts from rice were different among the different growing stages and the tissues (Fig. 1). Leaf extracts showed stronger inhibitory effect than stem or root extracts. Differences between the cultivars were more clear with leaf extracts at the fourth to sixth-leaf stage than at other growth stages.

QTL Analysis

The allelopathic effect of each F3 line was calculated by the inhibition of lettuce root length treated with water-soluble leaf extract from rice plants at the six-leaf stage. The frequency distribution of the allelopathic effects in the F3 lines was continuous (Fig. 2), ranging from 3.3% to 39.5% of the control and the mean value of the F3 lines was 15.4%. Root lengths of lettuce treated with extract from PI312777 and Rexmont was 5.8% and 28.7% of the control, respectively. A map of 15 linkage groups constructed covers a genetic distance of 1,336.2 cM (Ebana *et al.*, 2001b). The analysis by MAPMAKER/QTL showed seven QTL for the allelopathic effects with

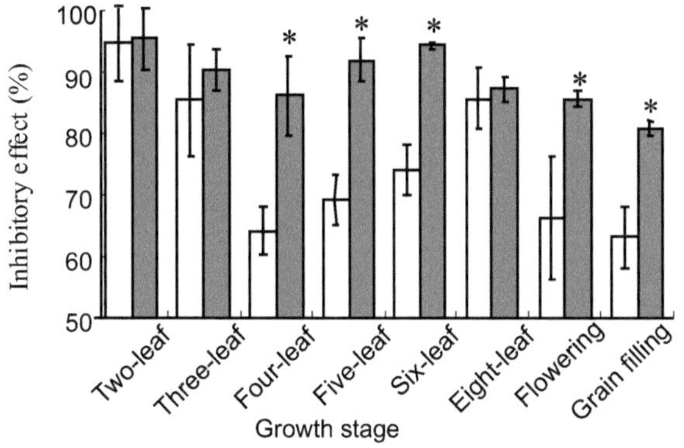

a. Inhibitory effect of leaf extracts

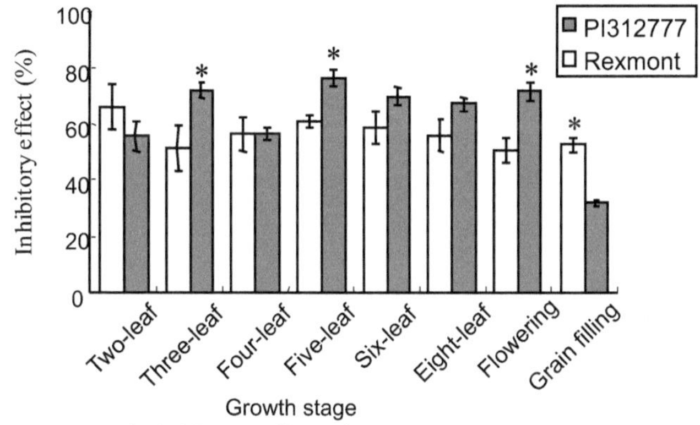

b. Inhibitory effect of stem extracts

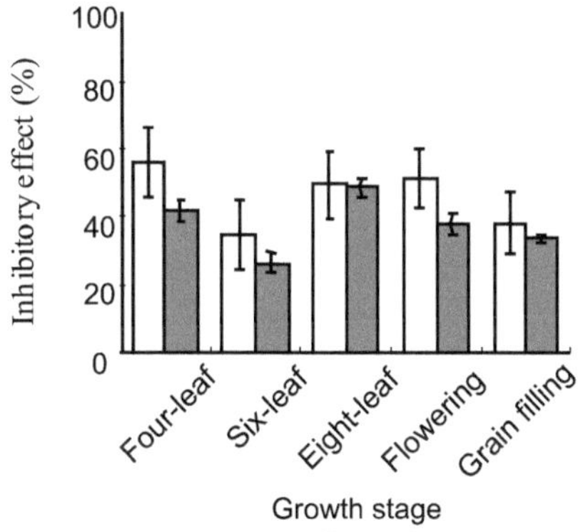

c. Inhibitory effect of root extracts

Fig. 1 Inhibitory effect of water-soluble extracts to lettuce root elongation

*: significant difference between the two cultivars

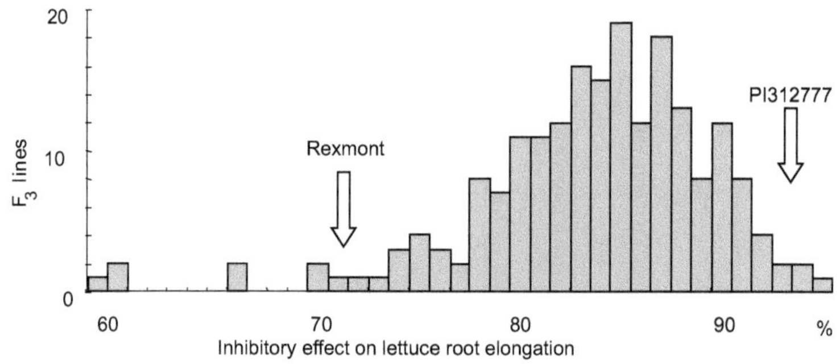

a. Effect of rice extracts on lettuce root evaluated by the inhi bitory effect of root elongation

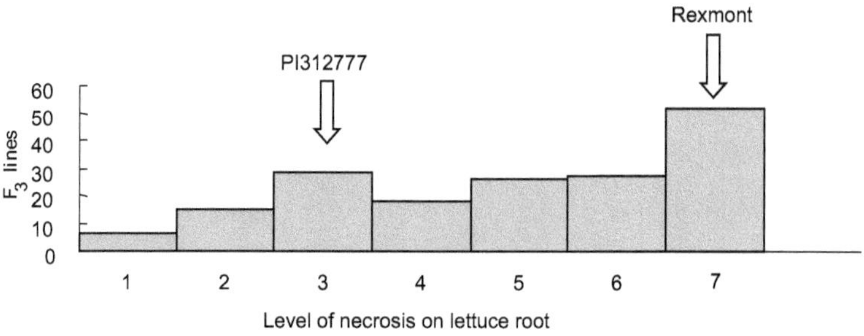

b. Effect of rice extracts on lettuce root evaluated by the necrosis frequency of root tip

Fig. 2 Inhibitory effect of extracts from F_3 lines of PI312777 and Rexmont

LOD scores higher than 2.0 located on seven chromosomes, chromosomes 1, 3, 5, 6, 7, 11 and 12 (Table 1). Each of 7 QTL explained 16.1% (chr.6), 15.1% (chr.7), 13.2% (chr.5), 12.5% (chr.12), 10.4% (chr.3), 9.6% (chr.1) and 9.4% (chr.11) of phenotypic variation, respectively. The nearest RFLP markers to QTL for the allelopathic effects were significantly associated with this trait at the 5% level based on SAS/GLM. A multiple QTL model estimated that five QTL with LOD score higher than 3.0% explained 36.6% of the total phenotypic variation. The PI3127777 alleles inhibited the lettuce roots more than those of Rexmont in all QTL loci. Dominance effect was found in the QTL linked to RFLP markers R1167, R830, R1709 and R2159 and suggested that the PI312777 allele in these QTL was dominant. Digenic interaction between the nearest RFLP markers to each QTL was analyzed with SAS/GLM PROC (Table 2). Interactions in five combinations between R1167 on chr.6 and R2401 on chr.7, between R2401 and R830 on chr.5, between R830 and R1925 on chr.3, between R1709 on chr.12 and R1925 and between R1925 and G257 on chr.11 were significant at the 1% level. The other four combinations between R2401 and G257,

R1709 and R2159, R1925 and R2159, R1709 and G257 were significant at the 5% level.

With the same materials, inhibitory effect was also assessed by the frequency of necrosis observed on the lettuce roots. The level of necrosis was divided into eight categories, from zero to PI312777 was at level three and Rexmont at level seven. QTL analysis of necrosis detected three QTL and they explained the 10.4% (chr.2). 10.3% (chr.7) and 6.9% (chr.3) of the phenotypic variation, respectively. A multiple QTL model estimated that three QTL explained 21.9% of the total phenotypic variation. The QTL on chromosome 7 was common to the former detected QTL with the index of root length growth inhibition.

Table 1 QTLs of inhibitory effect of water-soluble extract detected in the F_2 population of PI312777 and Rexmont

Chromosome	*Nearest marker locus*	*LOD value*	*Contribution to phenotypic variation (%)*	*PI312777 allele*		*t-test* [1]
				Additive effect	*Dominance effect*	
6	R1167	4.03	16.1	-2.72	-3.61	**
7	R2401	3.99	15.1	-2.78	-2.00	**
5	R830	4.01	13.2	-2.25	-2.78	*
12	R1709	3.14	12.5	-2.59	-2.37	**
3	R1925	3.74	10.4	-2.37	-0.23	**
1	R2159	2.58	9.6	-1.72	-2.60	*
11	G257	2.74	9.4	-2.30	-1.26	**

[1]** and * means that the correlation between the genotype and phenotype was significant at 1% and 5%, respectively

Table 2 Significant interaction detected among the RFLP marker loci where the QTL of allelopathy was detected.

The nearest marker locus to QTLs	*Chromosome*	*R1167*	*R2401*	*R830*	*R1709*	*R1925*	*R2159*	*G257*
R1167	6	-						
R2401	7	**[1]	-					
R830	5	ns	**	-				
R1709	12	ns	ns	ns	-			
R1925	3	ns	ns	**	**	-		
R2159	1	ns	ns	ns	*	*	-	
G257	11	ns	*	ns	*	**	ns	-

[1]** and * means that the interaction between the genetic loci was significant at 1% and 5%, respectively. ns means non-significant.

CONCLUSION

In this study, a bioassay of allelopathic activity using water-soluble extracts was established. Water-soluble extracts from the two rice varieties at various growth stages were compared. Leaf extract from the six-leaf stage plants was appropriate for the assessment of the difference between the two varieties, PI312777 and Rexmont, reflecting the weed suppressive effect in the field. This bioassay, with water-soluble extracts using lettuce as a test plant, was easier than the field test for the allelopathic effects, which needs several years replication.

Inhibitory effect of water-soluble extract from the leaves has changed according to the growth stage and the pattern was different between PI312777 and Rexmont. Extracts from the two-leaf-stage plant did not show the difference in inhibitory effect between the varieties, and extracts from later stages showed a different effect. Rice seedlings before four-leaf stage were dependent on the endosperm as a nutrient source (Matsuo *et al.*, 1995), whereas after the four-leaf stage, they depend on their own photosynthetic activities. It is assumed that different metabolism resulted in the difference in the quality and/or quantity of allelochemicals. One hypothesis is that during the young seedling stage, common compounds such as momilactone included in the seed, are effective and after the independent nutrient stage the metabolites produced are used as allelochemicals.

The results of the present study showed that root extracts of rice were less effective than other extracts, although plants generally release allelochemicals from roots (Inderjit, 1996). Weak association of the effects of root extracts with allelopathic effect in the field can be explained by immediate release after translocation from the leaves. Another possibility is that rice allelochemicals are not released from roots but by leaching directly from the leaves, such as a napthol derivative and scopoletin identified in the leachate of sunflower (Wilson and Rice, 1984).

In this study, QTL analysis was conducted with water-soluble extracts bioassay to find the chromosomal region controlling the allelopathic activity. Seven QTL were detected in the population of PI312777 and Rexmont, and the PI312777 alleles inhibited the growth of lettuce seedlings more than the Rexmont allele at all the QTL loci. Among the possible combinations of digenic interaction between the QTL, about half were significant. This suggested that the allelopathic effects of rice was controlled by a complex genetic mechanism.

Compounds associated with allelopathic effect in crops are synthesized by the secondary metabolism pathway (Rice, 1984). In rice, candidates have been reported, but they have not been proved to be allelochemicals because of the discrepancy between their abundance and the allelopathic activity of the plant (Olofstdotter *et al.*, 1995). Red rice varieties inhibited the growth of lettuce more frequently than white rice varieties (Fujii, 1992). Red rice varieties possess the *Rc* gene (Nagao, 1951) on the middle of chromosome 7 (Lin *et al.*, 1994b). This gene is responsible for the production of catechol (Nagao *et al.*, 1957). This compound is known to inhibit the growth of plants (Rice, 1984). In the chromosomal region near the *Rc* locus on the middle of chromosome 7, a major QTL was detected as well as the QTL for necrosis. Severe necrosis on the root made its length short. Consequently this chromosomal region seems important to cause necrosis on the lettuce root by some compound(s). The involvement of *Rc* with allelopathy should be studied by fine genetic study. Recently the rice genome was completely sequenced and functional analysis of candidate gene approach will help to identify relevant genes for allelopathy.

This study is the first step in identifying the QTL underlying allelopathic effects in rice. Closely linked DNA markers to the QTL detected here could be used for the marker-assisted selection to limit the candidate chromosomal region. Recombinant inbred lines are useful for analyzing allelopathy, which is controlled by complicated genetic system. They will be useful in identifying allelochemicals or to study the physiological aspect of allelopathy. Our final goal is germplasm enhancement focusing on biological weed control in rice fields.

References

Cartwright, D., Langcake, P., Pryce, R.J., Leworthy, D.P., Ride, J.P. (1977) Chemical activation of host defense mechanisms as a basis for crop protection. Nature. 267: 511-513.

Chou, C.H., Lin, H.J. (1976) Autointoxication mechanism of *Oryza sativa* L. I. Phytotoxic effects of decomposing rice residues in soil. J. Chem. Ecol. 2: 353-367.

Chou, C.H. (1980) Allopathic researches in the subtropical vegetation in Taiwan. Comparative Physiol. Ecol. 5: 222-234.

Chung, I.M., Ahn, J. K., Yun, S.J. (2001) Assessment of allelopathic potential of barnyard grass (*Echinochloa crus-galli*) on rice (*Oryza sativa* L.) cultivars. Crop Protection. 20: 921-928.

Dilday, R.H., Nastasi, P., Smith, R.J. (1989) Allelopathic observation in rice (Oryza sativa L.) to ducksalad (*Heteranthera limosa*). *In* the Proceedings of Arkansas Academy of Sci. 43: 21-22.

Dilday, R.H., Lin, J., Yan, W. (1994) Identification of allelopathy in the USDA-ARS rice germplasm collection. Aust. J. Expt. Agric. 34: 901-910.

Dilday, R.H., Yan, W.G., Moldenhauer, K.A.K., Gravois, K.A. (1998) Allelopathic activity in rice for controlling major aquatic weeds. *In* Allelopathy in Rice. M. Olofsdotter (ed) IRRI, Manila. Phillipines. pp. 7-26.

Ebana, K., Yan, W.G., Dilday, R.H., Namai, H., Okuno, K. (2001a) Variation in allelopathic effect of rice (*Oryza sativa* L.) with water-soluble extracts. Agronomy J. 93: 12-16.

Ebana, K., Yan, W.G., Dilday, R.H., Namai, H., Okuno, K. (2001b) Analysis of QTL underlying the allelopathic effect of rice using water-soluble extracts. Breeding Sci. 51: 47-59.

FAOSTAT Database (2000) Food and Agriculture Organization of United Nations, http://apps.fao.org/.

Fay, P.K., Duke, W.B. (1977) An assessment of allelopathic potential in *Avena* germplasm. Weed Sci. 25: 224-228.

Fujii, Y. (1992) The potential biological control of paddy weeds with allelopathy (allelopathic effect of some rice varieties). *In* the Proceedings of International Symposium of Biological Control and Integrated Management of Paddy and Aquatic Weeds in Asia. pp. 305-320.

Hassan, S.M., Rao, A.N., Bastawis, A.O., Aidy, I.R. (1994) Weed management in broadcast seeded rice in Egypt. Proceedings from an International Workshop on Constraints, Opportunities and Innovations for Wet-seeded Rice. Bangkok, Thailand. p. 205.

Inderjit (1996) Plant phenolics in allelopathy. The Bot. Rev. 62: 186-202.

Kato, T., Kabuto, C., Sasaki, N., Tsunegawa, M., Aizawa, H., Fujita, K., Kato, Y., Kitahara, Y., Takahashi, N. (1973) Momilactones, growth inhibitors from rice *Oryza sativa* L. Tetrahedron Lett. 39: 3861-3864.

Kato-Noguchi, H., Ino, T., Sata, N., Yamamura, S. (2002) Isolation and identification of a potent allelopathic substance in rice root exudates. Physiol. Plant. 115: 401-405.

Kawaguchi, S., Takeuchi, Y., Ogasawara, M., Yoneyama, K., Konnai, M. (1997) Allelopathic potential of rice seed (*Oryza sativa* L.) on seed germination of *Monochoria vaginalis* var. plantaginea. Weed Res., Japan. 42: 262-267.

Kurata N., Nagamura, Y., Yamamoto, K., Harushima, Y., Sue, N., Wu, J., Antonio, B.A., Shomura, A., Shimizu, T., Lin, S.Y., Inoue, T., Fukuda, A., Shimano, T., Kuboki, Y., Toyama, T., Miyamoto, Y., Kirihata, T., Hayasaka, K., Miyao, A., Monna, L., Zhong, H.S., Tamura, Y., Wang, Z.X., Momma, T., Umehara, Y., Sasaki, T., Minobe, Y. (1994) A 330 kilobase interval genetic map of rice including 883 expressed sequences. Nature Genetics. 8: 365-372.

Lander, E.S., Green, O., Abrahanson, J. (1987) MAPMAKER An interactive computer package for constructing primary genetic linkage maps of experimental and natural populations. Genomics. 1: 173-381.

Lander, E.S., Botstein, D. (1989) Mapping Mendelian factor underlying quantitative traits using RFLP maps. Genetics. 121: 185-199.

Lin, J., Smith, R., Dilday, R.H. (1994 a) Biological control of weeds. Weed Sci. Society Abst. 34: 48.

Lin, S.Y., Nagamura, Y., Kurata, N., Yano, M., Minobe, Y., Sasaki, T. (1994 b) DNA markers tightly linked to genes, *Ph*, *alk* and *Rc*. Rice Genetics Newsletter 11: 108-109.

Lockerman, R.H., Putnam, A.R. (1979) Evaluation of allelopathic cucumber (*Cucumis sativus*) as an aid to weed control. Weed Sci. 27: 54-57.

Matsuo, T., Kumazawa, K., Ishii, R., Ishihara, K., Hirata, H. (eds) (1995) Science of the rice plant. Vol. 2. Physiology Food and Agriculture Policy Research Center, Tokyo, Japan.

Murray, M.G., Thompson, W.F. (1980) Rapid isolation of high molecular weight plant DNA. Nucleic Acid Res. 8: 4321-4325.

Nagao, S. (1951) Genic analysis and linkage relationship of characters in rice. Advances in Genetics. 4: 181-212.

Nagao, S., Takahashi M., Miyamoto, T. (1957) Genetical studies on rice plant XXI. Biochemical studies on red rice pigmentation. Japanese J. Genetics. 32: 123-328.

Olofsdotter, M., Navarez, D. (1996) Allelopathic rice for *Echinochloa crus-galli* control. *In* Proceedings of the Second International Weed Control Congress. H. Brown (ed)

Department of Weed Control and Pesticide Ecology, Copenhagen,Denmark. pp. 1175-1181.

Olofsdotter, M., Navarez, D., Moody, K. (1995) Allelopathic potential in rice (*Oryza sativa* L.) germplasm. Annals of Appl. Biol. 127: 543-560.

Putnam, A.R., Duke, W.B. (1974) Biological suppression of weeds: evidence for allelopathy in accessions of cucumber. Science. 185: 370-372.

Rice, E.L. (1984) Allelopathy. 2nd ed. Academic Press Inc., London, UK.

SAS Institute (1989) SAS/STAT users guide: version 6.03, 4th ed. SAS Institute Inc., Cary, North Carolina, USA.

Seal, A.N., Pratley, J.E., Haig, T., Lewin, L.G., (2004) Screening of rice varieties for allelopathic potential against arrowhead (*Sagittaria montevidensis*), an aquatic weed infesting Australian Riverina rice crops. Aust. J. Agric. Res. 55: 673-680.

Smith, R.J. (1981) Control of red rice (*Oryza sativa*) in water-seeded rice (Oryza sativa). Weed Sci. 29: 663-666.

Smith, R.J. (1988) Weed threshold in southern US rice *Oryza sativa*. Weed Tech. 2: 232-241.

Tanksley, S.D. (1993) Mapping polygenes. Annual Rev. Genetics. 27: 205-233.

Vogler, R.K., Ejeta, G., Butler, L.G. (1996) Inheritance of low production of *Striga* germination stimulant in sorghum. Crop Sci. 36: 1185-1191.

Wilson, R.E., Rice, E.L. (1984) Allelopathy as expressed by *Helianthus annuus* and its role in old field succession. Bull. Torrey Bot. Club. 95: 432 448.

17

Allelochemicals Involved in Rice Allelopathy

Chuihua Kong
Institute of Applied Ecology, Chinese Academy of Sciences, Shenyang 110016, China and State Key Laboratory of Elemento-organic Chemistry, Nankai University, Tianjin 300071, China

Allelochemicals play an important role in natural and managed ecosystems. Food crop *Oryza sativa* L. can produce and release allelochemicals participating in its defense against weeds. However, it remains obscure which allelochemicals in rice are predominantly involved in defense mechanisms against weeds. In this study, many kinds of compounds were systematically isolated and identified from allelopathic rice PI312777 seedlings. Among them, alkaloids, alkylresorcinols, cyclohexenone and urea derivatives, flavonoids and their glucosides, diterpenoids and triterpenes had inhibitory activities on the growth of *Echinochloa crus-galli, Cyperus difformis* and *Leptochloa chinensis* weeds associated with rice. However, the presence of these compounds in allelopathic rice seedlings does not necessarily mean that they can be released into the environment to demonstrate their allelopathic effects under natural field conditions. Accordingly, allelochemicals released from selected rice seedlings into the soil were determined and analyzed by LC/MS and HPLC. As a result, phenolic acids, flavonoids, diterpenoids and cyclohexenone and urea derivatives were found in soil- grown allelopathic rice seedlings. Particularly, 3-isopropyl-5-acetoxy-cyclohexene-2-one-1, momilactone B and 5,7,4'-trihydroxy-3,5-dimethoxyflavone occurred in the soil at day 15 after rice seedlings emergence, and reached the concentration initiated for inhibition on associated weeds at day 30 after rice emergence, while other compounds were not detected or they were present as trace only in the soil. These results suggest that 3-isopropyl-5-acetoxycyclohexene-2-one-1, momilactone B and 5,7,4'-trihydroxy-3,5-dimethoxyflavone are allelochemicals involved in rice allelopathy and allelopathic rice seedlings can establish their own weed defense mechanism through production and release of these allelochemicals.

Keywords: *allelochemicals, allelopathy, isolation and identification, Oryza sativa L., secondary metabolite, soil*

INTRODUCTION

Allelopathy is an interference mechanism in which living or dead plants release allelochemicals exerting an effect (mostly negative) on associated plants (Rice, 1984). Accordingly, allelopathy is a phytotoxic interference in most circumstances (Romeo, 2000; Weidenhamer *et al.*, 1989). In-depth understanding of the allelopathic mechanism in natural and managed ecosystems can develop new environmentally safe, biological control methods or agents for sustainable agriculture (Kim and Shin, 2000).

Rice (*Oryza sativa* L.) is one of the principal food crops in the world. Its allelopathy has received a great deal of attention and an increasing number of studies since 1991 (Dilday, 1994). Rice allelopathy, which results from the allelochemicals synthesized and released from rice itself, can potentially be used to improve weed management in rice production. Therefore, searching for allelochemicals from rice has been extensively pursued in recent years (Chung *et al.*, 2001; Rimando *et al.*, 2001; Kato-Noguchi *et al.*, 2002, 2003; Kim and Kim, 2002; Kong *et al.*, 2004a, b; Seal *et al.*, 2004a, b). Phenolic acids, flavonoids, diterpenoids and other types of compounds, were identified as potent allelochemicals from rice tissues and root exudates. However, it still remains obscure which allelochemicals in rice are predominantly involved in defense mechanisms against weeds. Recent studies have shown that phenolic acids are unlikely to account for allelopathy in rice since their concentrations in paddy never reach phytotoxic levels (Olofsdotter *et al.*, 2002b), and a few putative allelochemicals were not detected in allelopathic rice root exudates (Seal *et al.*, 2004a, b). The objectives of this study were thus to systematically isolate and identify secondary metabolites present in allelopathic rice accession (PI312777), and then to evaluate which are the allelochemicals involved in rice allelopathy.

MATERIALS AND METHODS

Instruments and Allelopathic Rice Seedlings

The NMR spectra were measured in deuterated dimethyl sulfoxide or chloroform with a Brücker AC-P300Q spectrometer (400 MHz for ^{1}H,

100 MHz for ^{13}C). All chemical shifts are reported as δ values relative to TMS. High-resolution mass experiments were carried out with a JEOL JMS-01SG-2 mass spectrometer (EI: 70 eV). Low resolution mass spectra were obtained with a Finigan TSQ-7000 LC/MS, using a direct inlet system.

The allelopathic rice accession PI 312777 was selected, and cultivated in a rice field with 17.7±0.5 g/kg organic matter, 80.1±0.5 mg/kg available N, 80.1±0.9 g/kg available P, and 102.5±0.05 mg/kg available K.

Isolation and Identification of Secondary Rice Metabolites

PI312777 plants were randomly collected from a rice field at the six-leaf stage, and their above-ground parts were freeze-dried and ground. The powder was extracted with 70% methanol over 24 h. The filtrate was concentrated *in vacuo,* and the concentrated aqueous extract was partitioned three times with EtOAc. The EtOAc phases were combined and subsequently evaporated under nitrogen gas, and then subjected to silica gel column chromatography with various organic solvents or their mixtures, thereby affording several fractions. The solvent in most fractions was removed to give individual compounds. The compounds in a few fractions were further purified and isolated by preparative TLC (silica gel) with various organic solvents. The above aqueous phase was concentrated *in vacuo,* and then applied to a Sephadex column and eluted with methanol. The eluted fractions were further purified by precoated preparative TLC plates and finally purified by preparative HPLC with a C_{18} reversed phase column.

The molecular formulae of all isolated compounds from the allelopathic rice plants were determined by high-resolution mass spectral analysis and then their structures were further identified from analysis of NMR spectra. Furthermore, low resolution mass spectra of the compounds were conducted with EI at 70 eV.

Analysis of Allelochemicals in Soil

Five pre-germinated PI312777 seeds were sown in each pot (5 × 5 cm) containing 100 g of soil collected from a rice field. Pots without PI312777 seeds served as controls. All pots were placed in a greenhouse, where night and daytime temperatures ranged roughly from 20-30°C, and watered once a day. The pots were individually taken out from the greenhouse at different day intervals and the soil in each pot extracted with 100 mL methanol (agitated for 48 h at 25°C, then centrifuged at 1,200 G for 30 min). The extracts were concentrated in a vacuum at 40°C to give an aqueous residue. The residues were respectively dissolved in 50% aqueous methanol (v/v, 2 mL) and loaded onto reversed phase C_{18} Sep-Pak cartridges (Waters, Co.). The cartridges were eluted with 50% aqueous methanol (5 × 3 mL) and then methanol (3 × 3 mL), and the methanol fraction then concentrated by nitrogen gas blow-down to obtain 100 μL concentrate for LC/MS and HPLC analysis (Kong *et al.*, 2004a, b).

Qualitative analysis of the allelochemicals released from allelopathic PI312777 into the soil was determined using a liquid chromatography/ mass spectral (LC/MS) technique. A Finigan TSQ-7000 instrument equipped with a C_{18} reversed phase column (Hypersil 125 mm × 4.0 mm, 5 μm) was used. Elution with MeOH-H_2O (v:v, 7:3) was carried out at a flow rate of 0.5 mL/min, and detected at 228 nm. The injection volume of the samples was 10 μL. The mass spectrometer was equipped with an atmospheric pressure chemical ionization (APCI) inlet system in a positive-ion mode, and nitrogen served as the collision gas. A multiple reaction monitoring (MRM) mode was used to monitor the parent and product ions. The dwell time was set at 500 ms, and the duration time was 10 min. The allelochemicals were detected and identified from a combination of their molecular and specific ions and from pure reference compounds isolated from PI312777 plants described above.

Quantitative analysis of allelochemicals in soil-grown allelopathic rice seedlings was carried out with an HPLC Hitachii L 7100 equipped with a C_{18} reversed phase column (Hypersil 125 mm × 4.0 mm, 5 μm). HPLC determination conditions: mobile phase was a mixture of 75% methanol and 25% water, eluted at a flow rate of 2.0 mL/min, and detected at 265 nm. The injection volume of samples was 10 μL. All samples were filtered through a 0.25 μm nylon syringe pre-filter before analysis. Three allelochemicals involved in rice allelopathy, 3-isopropyl-5-acetoxycyclohexene-2-one-1, momilactone B and 5,7,4'-trihydroxy-3,5-dimethoxyflavone in soil, were each quantified by interpolating the peak height on the chromatograms from HPLC to a standard curve constructed by the peak height of pure 3-isopropyl-5-acetoxycyclohexene-2-one-1, momilactone B and 5,7,4' -trihydroxy-3',5'-dimethoxyflavone isolated from PI312777 seedlings described above.

Bioassays

Echinochloa crus-galli, Cyperus difformis and *Leptochloa chinensis* are major weeds associated with rice in China. Inhibitory activity of the compounds isolated from the rice plants on the growth of these weeds was evaluated using a pot-culture method (Weidenhamer *et al.*, 1989). At least 50 seeds were sown in each 5 × 5 cm pot containing 150 g soil. After emergence, the seedlings were thinned to 10 plants per pot, and then the tested compounds at 100 μg/mL concentration were applied to each of the treated pots, respectively. The control pots received water only. All pots were placed in a controlled environmental chamber (3 m^3) with a 12 h day length and approximately 350 μmol/m^2/s light intensity, 25-28°C daytime temperature, and 70% relative humidity. Pots were watered and randomized once a week. The seedlings were harvested after 4 wk. Shoots were clipped at the point of first root, dried for at least 48 h at 80°C, and their dry weights were determined. Percentage inhibition was obtained from the comparison of weed dry weights between the treated and control pots. The same manipulation was conducted three times for each determination under identical conditions.

RESULTS

More than one hundred low-molecular-mass secondary metabolites, including alkaloids, alkylresorcinols, cyclohexenone and urea derivatives, flavonoids and their glucosides, diterpenoids and triterpenes, were isolated and identified from allelopathic rice PI312777 seedlings. Many of these compounds were originally found and isolated in rice plants as phytoalexins and potent allelochemicals, a few have never been reported in rice plants. Besides the compounds mentioned above, a range of phenolic acids, volatile terpenes and other types of compounds were also identified from the rice seedlings. They have been well documented in the literature (Kim and Kim, 2000; Chung *et al.*, 2001; Rimando *et al.*, 2001; Kim and Kim, 2002; Seal *et al.*, 2004a, b).

Alkaloids (Fig. 1) and alkylresorcinols (Fig. 2) were identified from etiolated rice seedlings as antifungal agents (Suzuki *et al.*, 1996; 1998). Several alkylresorcinols were detected in the rice root exudates (Bouilant *et*

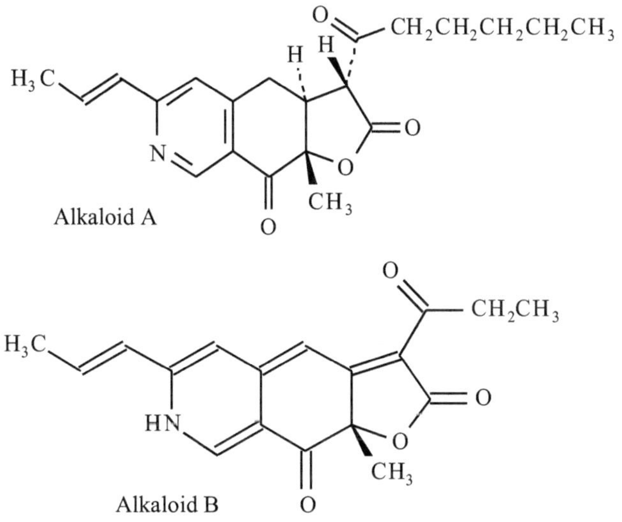

Fig. 1 Alkaloids identified from etiolated rice seedlings

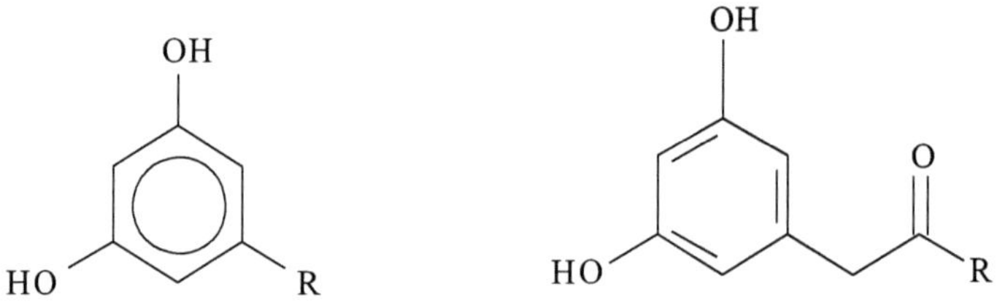

R = $-(CH_2)_{12}CH_3$, $-(CH_2)_{14}CH_3$, $-(CH_2)_{16}CH_3$, $-(CH_2)_7CH = CH(CH_2)_5CH_3$,
$-(CH_2)_7CH = CH(CH_2)_7CH_3$, $-(CH_2)_7CH = CHCH_2CH = CH(CH_2)_4CH_3$

Fig. 2 Alkylresorcinols isolated from PI312777 rice seedlings

al., 1994; Kong *et al.*, 2002b) and their inhibitory activities on the growth of *E. crus-galli*, *C. difformis* and *L. chinensis* were related to their structures. All percentage weed inhibitions were measured at a test-compound concentration of 100 μg/mL (Table 1).

Table 1 Inhibitory effects of representative compounds isolated from allelopathic rice seedlings PI312777 on the growth of major weeds.

Compounds	*E. crus-galli*	*C. difformis*	*L. chinensis*
Alkaloid A	39.8±2.1[a]	39.7±7.5[a]	35.3±5.0[a]
Alkylresorcinol[*1]	22.5±7.6[b]	35.9±8.1[a]	39.9±5.8[a]
Cyclohexenone[*2]	71.2±6.3[c]	76.3±6.5[b]	35.6±7.1[a]
Urea derivative[*3]	55.6±12.3[d]	60.5±8.7[c]	59.6±7.1[b]
Flavonoid[*4]	77.7±11.0[c]	63.2±9.2[c]	60.2±7.3[b]
Sakuranetin	27.5±7.9[b]	20.5±7.2[d]	38.9±4.5[a]
Oryzalexin A	37.3±6.1[a]	37.2±6.2[a]	40.2±5.3[a]
Oryzalexin E	28.6±8.0[b]	35.3±6.6[a]	37.7±4.1[a]
Oryzalexin S	6.1±1.1[e]	0.9±0.3[e]	1.8±0.6[c]
Momilactone A	53.5±6.4[d]	37.6±5.4[a]	60.8±9.9[b]
Momilactone B	72.1±6.5[c]	60.6±8.6[c]	57.9±10.1[b]
Phytocassane A	52.1±6.5[d]	58.3±3.2[c]	48.3±3.2[d]
Triterpene glucoside	5.0±0.9[e]	1.6±0.3[e]	7.2±1.6[e]

*1, 5-(12-heptadecenyl)-resorcinol; *2, 3-isopropyl-5-acetoxycyclohexene-2-one-1; *3, ethyl-4-(*o*-nitrophenyl)-3-allophanate; *4, 5,7,4'-trihydroxy-3',5'-dimethoxyflavone. Test concentration of all compounds was 100 μg/mL. All data are inhibitory percentage and mean of 3 replicates with standard error. Data in a column not followed by the same letter are significantly different, p = 0.05, ANOVA with Duncan's multiple range test.

Cyclohexenone and urea derivatives are two types of compounds rarely found in rice plants before. In our studies, some of them, such as 3-isopropyl-5-acetoxycyclohexene-2-one-1, ethyl-4-(*o*-nitrophenyl)-3-thio-allophanate and ethyl-4-(*o*-nitrophenyl)-3-allophanate (Fig. 3), were

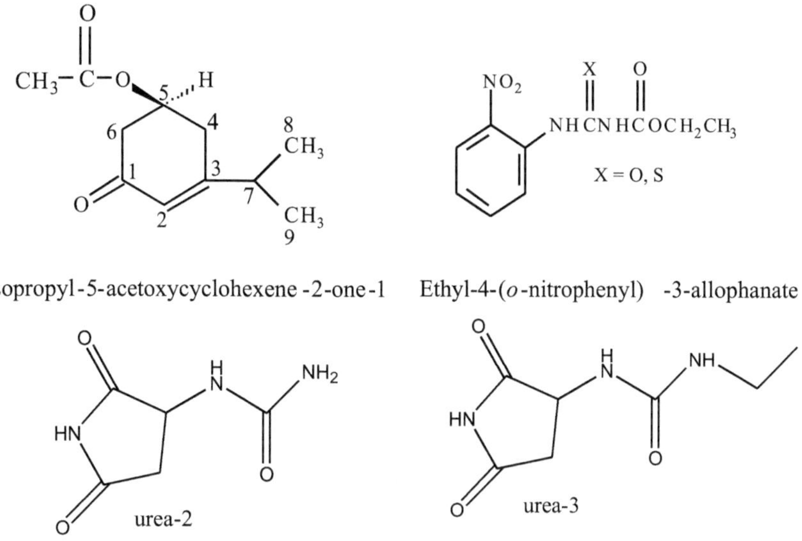

Fig. 3 Cyclohexene and urea derivatives isolated from PI312777

isolated and identified from allelopathic rice PI312777 seedlings. Both cyclohexenone and urea derivatives significantly inhibited the growth of *E. crus-galli*, *C. difformis* and *L. chinensis* (Table 1).

Several flavonoids and their glucosides were isolated from allelopathic rice PI312777 seedlings. Among them, sakuranetin (5,4'-dihydroxy-7-methoxyflavanone) and 5,7,4'-trihydroxy-3',5'-dimethoxyflavone (Fig. 4) show good inhibitory activity on both weeds and pathogens. Sakuranetin as a phytoalexin was identified from rice plants and inhibited spore germination of *Pyricularia oryzae* at very low concentration 10-20 μg/mL (Nakazato *et al.*, 2000). However, it demonstrated much less inhibitory activity on weeds, scarcely inhibiting weed growth at up to 100 μg/mL (Table 1). 5,7,4'-trihydroxy-3',5'-dimethoxyflavone was originally isolated from rice husk and later found in rice plant and root exudates (Kong *et al.*, 2004a, b). It significantly inhibited the weeds tested (Table 1).

Diterpenoids isolated from rice mainly involved oryzalexins A-F and S (Akatsuka *et al.*, 1985; Kodama *et al.*, 1992; Kato *et al.*, 1993, 1994), phytocassanes A-E (Koga *et al.*, 1995, 1997; Nakazato *et al.*, 2000), and momilactones A and B (Kato *et al.*, 1977). They are anti-microbial secondary metabolites accumulated in rice plants for disease resistance.

Oryzalexins (Fig. 5) inhibited the growth of *E. crus-galli*, *C. difformis* and *L. chinensis*, but their inhibition levels were different. At 100 μg/mL oryzalexin A and E significantly inhibited the growth of the tested weeds, but oryzalexin S showed little inhibitory activity (Table 1).

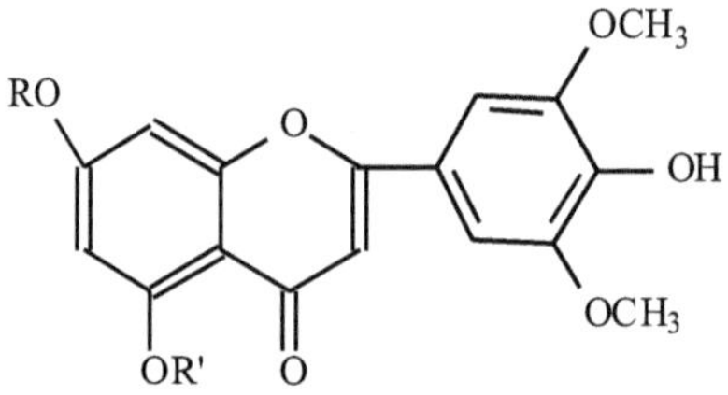

R: Glu, H, Rha. R: H, Glu(α), Glu(ß)
Flavonoid-glucosides

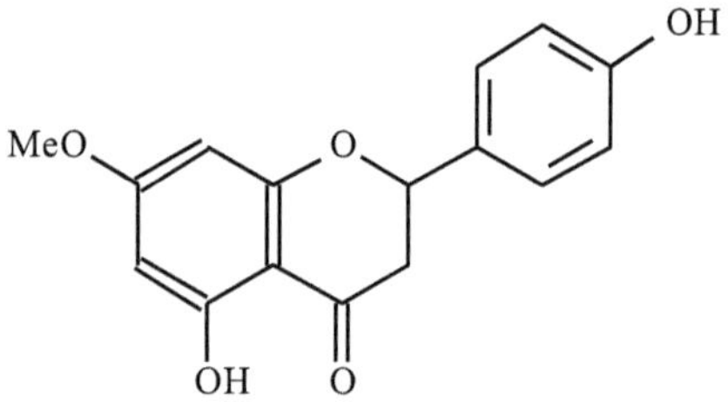

Sakuranetin

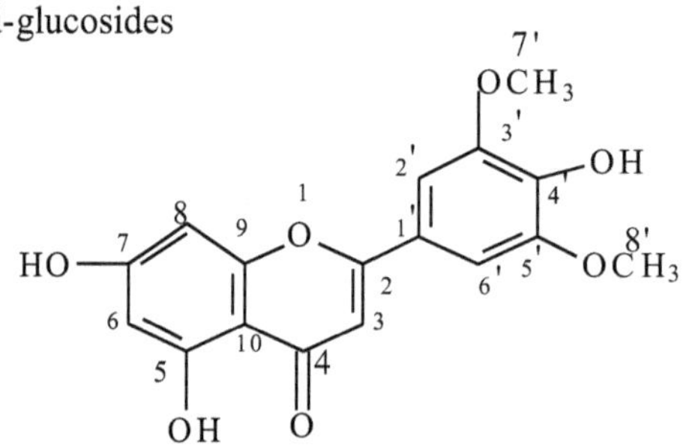

5,7,4'-Trihydroxy-3',5'-dimethoxyflavone

Fig. 4 Flavonoids and flavonoid glucosides from PI312777 rice

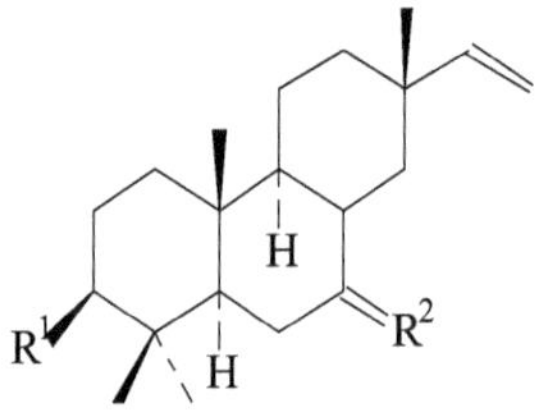

Oryzalexin A: $R^1 = OH, R^2 = CH_3$
Oryzalexin B: $R^1 = O, R^2 = OH$
Oryzalexin C: $R^1 = O, R^2 = O$
Oryzalexin D: $R^1 = OH, R^2 = O$

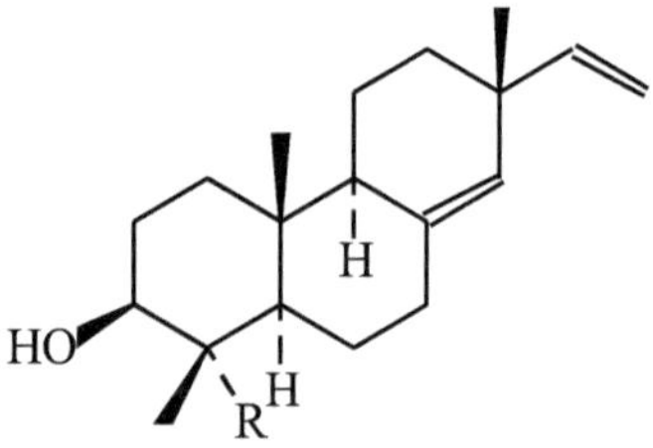

Oryzalexin E: $R = CH_3$
Oryzalexin F: $R = CH_2OH$

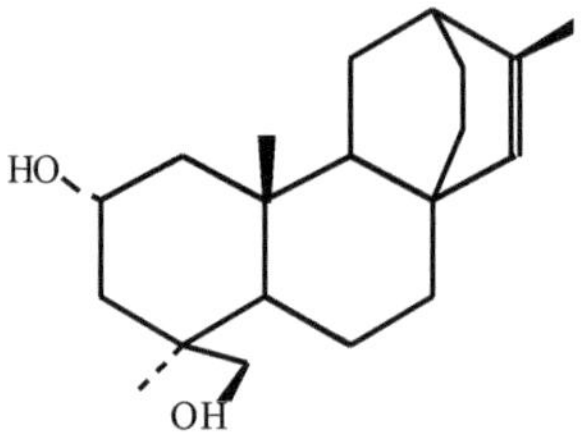

Oryzalexin S

Fig. 5 Oryzalexin diterpenoids from rice plants

Phytocassanes A-E (Fig. 6) are the diterpenes with a cassane skeleton (Koga *et al.*, 1995, 1997; Nakazato *et al.*, 2000). They were produced in rice leaves and stems, particularly those infected with *Magnaporthe grisea* and *Rhizoctonia solani*. Phytocassane A at the concentration tested, demonstrated high inhibitory activity against weed growth (Table 1).

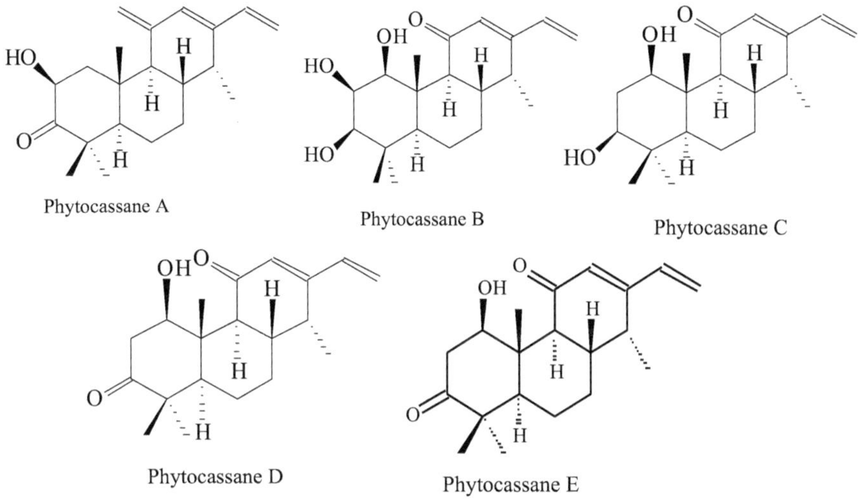

Fig. 6 Diterpenoidal phytocassanes from rice leaves and stems

Momilactones A and B (Fig. 7) are the first phytoalexins characterized from any members of the Gramineae family. They were isolated initially as plant inhibitors from rice husk (Kato *et al.*, 1977), and were subsequently found in rice responding to either infection by blast disease caused by the fungus *Pyricularia oryzae* or irradiation with UV light (Cartwright *et al.*, 1977, 1981). More recently, momilactones A and B have been found in rice root exudates participating in rice defense against weeds (Lee *et al.*, 1999; Kato-Noguchi *et al.*, 2002, 2003; Kong *et al.*, 2004a). They significantly inhibited the growth of weeds tested (Table 1).

Several triterpenes were isolated from the allelopathic rice seedlings. The structure of one triterpene glucoside (Fig. 8) was identified, but it had little inhibitory activity on weed growth (Table 1).

Rice allelopathy occurs if the compounds are not only biosynthesized in seedlings but also released by the living plants into their surroundings at ecologically relevant concentrations with inhibitory activity on associated weeds. Accordingly, the compounds mentioned above were produced in the allelopathic rice seedlings, but the presence of these compounds in rice seedlings does not necessarily mean that they can be released into the

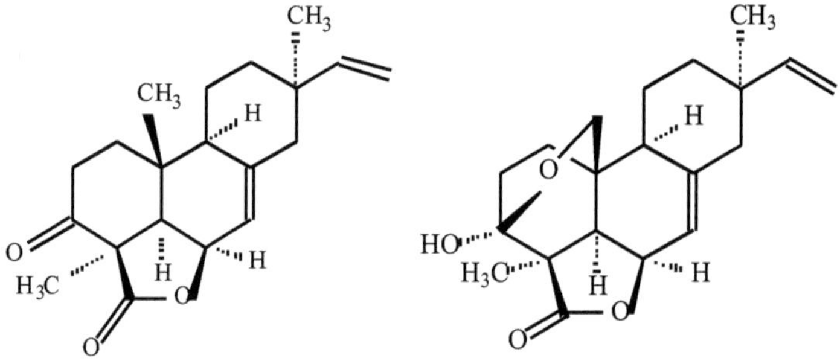

Momilactone A   Momilactone B

Fig. 7 The first phytoalexins described from the family Gramineae

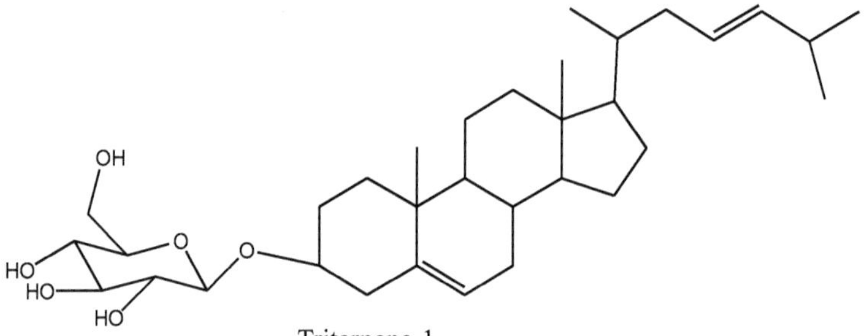

Triterpene-1

Fig. 8 One of the triterpene glucosides isolated from PI312777 rice

environment to demonstrate their allelopathic effects under natural conditions. It is not appropriate to call them allelochemicals until they have been shown to (1) be present in the soil (2) be active at ecologically relevant concentrations under natural conditions.

LC/MS determination and analysis demonstrated that phenolic acids, flavonoids, diterpenoids and cyclohexenone and urea derivatives were found in soil-grown allelopathic rice seedlings. Particularly, 3-isopropyl-5-acetoxycyclohexene-2-one-1, momilactone B and 5,7,4'-trihydroxy-3',5'-dimethoxyflavone occurred in the soil at day 15 after rice seedlings emergence, and reached the concentration initiated for inhibition on associated weeds at d 30 after rice emergence (Fig. 9) while other compounds were not detected or they were present as trace only in the soil. The results indicated that 3-isopropyl-5-acetoxycyclohexene-2-one-1, momilactone B and 5,7,4'-trihydroxy-3',5'-dimethoxyflavone were allelochemicals involved in rice allelopathy. It must be noted that several flavonoid-glucosides were identified in rice plants, but only 5,7,4'-trihydroxy-3',5'-dimethoxyflavone was found in soil-grown allelopathic

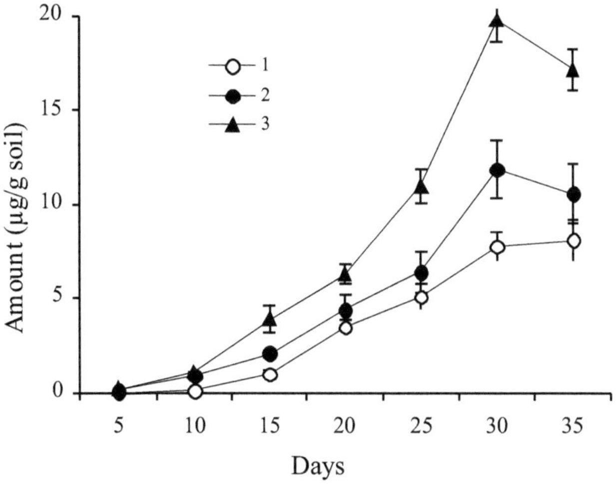

Fig. 9 The amounts of 3-isopropyl-5-acetoxycyclohexene-2-one-1 (1), momilactone B (2) and 5,7,4'-trihydroxy-3',5'-dimethoxyflavone (3) released into the soil growing PI312777 seedlings at different time intervals. Means ±SE from three independent experiments with 5 rice seedlings growing in 100 g soil for each determination are shown

rice seedlings. This suggests that all flavonoid glucosides may be changed into one aglycone, i.e. 5,7,4'-trihydroxy-3',5'-dimethoxyflavone, under natural conditions.

CONCLUSION

Rice allelopathy has been studied worldwide in recent years. However, the allelochemicals actually involved in rice allelopathy still remain a controversial issue. Our studies showed that allelopathic rice seedlings could produce many kinds of compounds. However, only those compounds with ecologically relevant concentrations, such as 3-isopropyl-5-acetoxycyclohexene-2-one-1, momilactone B and 5,7,4' -trihydroxy-3',5'-dimethoxyflavone, released from seedlings into the environment were allelochemicals involved in rice allelopathy. Generally, allelopathic rice plants produce and release allelochemicals at their early growth stages and the allelochemicals in living rice seedlings can usually be released via their root tissues (Olofsdotter *et al.*, 2002a; Hu *et al.*, 2004). Therefore, most studies were focused on searching and evaluating putative allelochemicals in rice root exudates. Although several classes of secondary metabolites including phenolic acids (Rimando *et al.*, 2001; Kim and Kim, 2002; Seal *et al.*, 2004a, b), cytokinins (Yoshida *et al.*,1971; Murofushi *et al.*, 1983), alkyl resorcinols (Bouillant *et al.*, 1994), momilactone B (Lee *et al.*, 1999; Kato-

Noguchi *et al.*, 2002, 2003; Kong *et al.*, 2004a) and flavonoids (Kong *et al.*, 2002b, 2004a, b), together with carbohydrates and amino acids (Bacilio-Jimenez *et al.*, 2003), have been determined from rice root exudates, it still remains obscure whether these compounds possess allelopathic function. In addition, the collection of active allelochemicals in rice root exudates was often performed under arbitrary conditions, rather than collecting from intact and living rice seedlings (Tang and Young, 1982). It remains a problem as to how to trap and identify the active allelochemicals in rice exudates. The design of our study offers a more realistic allelopathy model, closer to the actual field situation of a rice/soil system than does the examination of laboratory rice exudates. After all, rice allelopathy does occur in soil under natural conditions. Even if some compounds had high concentration in laboratory root exudates, it does not imply that the compounds would be available to exert an allelopathic effect in the soil.

It has been known for a long time that rice allelopathy occurs through the actions of phenolic acids produced and released in the root tissues (Rimando *et al.*, 2001; Seal *et al.*, 2004a, b). It has also been established that phenolic acids play certain important roles in rice allelopathy. However, most plant species usually release phenolic acids via their root systems. These phenolic acids in the exudates possess multiple functions in the chemical interactions with various organisms in the environment and allelopathy is only one of the multiple functions. The emphasis of our studies is that other types of allelochemicals in rice are predominantly involved in rice allelopathy. Besides the allelochemicals identified in our studies, there may be still other allelochemicals in rice plants awaiting detection and identification.

Allelochemicals in rice exudates or seedlings usually are determined by means of GC/MS or GC/MS/MS (Seal *et al.*, 2004a, b). GC/MS is a valuable instrument to determine the chemical constituents of mixture, in particular for volatile chemicals (Pelt *et al.*, 1998; Kong *et al.*, 2002a). However, most compounds produced and released from allelopathic rice seedlings, particularly in root exudates are not volatile. Moreover, analytical samples for GC/MS or GC/MS/MS have to be conducted at high temperatures. Therefore, GC/MS or GC/MS/MS is not good enough to determine fragile allelochemicals involved in rice allelopathy. The rice allelochemical phenomenon resulted from the complexity of the analytical samples and poor methodology. In our studies, the compounds were systematically isolated from allelopathic rice seedlings, identified one by one with NMR and MS, and then used to determine which compounds occurred in soil-grown allelopathic rice seedlings by applying gentle LC/MS at room temperature.

Rice can establish its own weed defense mechanisms through allelochemical production and release. However, to date, there has been limited success in finding allelochemicals that can really explain the allelopathic action of field rice (Olofsdotter *et al.*, 2002a). Further clarification of the chemical basis for rice allelopathy is warranted.

Acknowledgments

The author thanks Professor Chaoxian Zhang (Institute of Plant Protection, Chinese Academy of Agricultural Sciences) for correcting the earlier English manuscript. This work was supported by the National Natural Science Foundation of China (NSFC No. 30430460), Hundreds-talent Program of Chinese Academy of Sciences (BR04-135) and State Key Technologies R&D Programs of China (2004BA509B07).

References

Akatsuka, T., Kodama, Q., Sekido, H., Kono, Y., Takeuchi, S. (1985) Novel phytoalexins (Oryzalexins A, B and C) isolated from rice blast leaves infected with *Pyricularia oryzae*. Part I: Isolation, characterization and biological activities of oryzalexins. Agric. Biol. Chem. 49:1689-1694.

Bacilio-Jimenez, M., Aguilar-Flores, S., Ventura-Zapata, E., Perez-Campos, E., Bouquelet, S., Zenteno, E. (2003) Chemical characterization of root exudates from rice (*Oryza sativa*) and their effects on the chemotactic response of endophytic bacteria. Plant Soil. 249: 271-277.

Bouilant, M.L., Jacoud, C. and Zanella, I., Bally, R. (1994) Identification of 5-(12-heptadecenyl)-resorcinol in rice exudates. Phytochemistry. 35: 769-771.

Cartwright, D., Langcake, P., Pryce, R.J., Leworthy, D.P. (1977) Chemical activation of host defence mechanisms as a basis for crop protection. Nature. 267: 511-523.

Cartwright, D., Langcake, P., Pryce, R.J., Leworthy, D.P., Ride, J.P. (1981) Isolation and characterization of two phytoalexins from rice as momilactones A and B. Phytochemistry. 20: 535-537.

Chung, I.M., Ahn, J.K., Yun, S.J. (2001) Identification of allelopathic compounds from rice (*Oryza sativa* L.) straw and their biological activity. Can. J. Plant Sci. 81: 815-819.

Dilday R.H. (1994) Identification of allelopathy in USDA-ARS rice germplasm collection. Aust. J. Exp. Agric. 34: 907-910.

Hu, F., Kong, C.H., Xu, X.H., Zhang, C.X., Chen, X.H. (2004) Weed-suppressive effect and mechanism of allelopathic rice accession. Agric. Sci. China 3: 590-597.

Kato, T., Tsunakawa, M., Sasaki, N. (1977) Growth and germination inhibitors in rice husks. Phytochemistry. 16: 45-4

Kato, H., Kodama, O., Akatsuka, T. (1993) Oryzalexin E, a diterpene phytoalexin from UV-irradiated rice leaves. Phytochemistry. 33: 79-81.

Kato, H., Kodama, O., Akatsuka, T. (1994) Oryzalexin F, a diterpene phytoalexin from UV-irradiated rice leaves. Phytochemistry. 36: 299-301.

Kato-Noguchi, H., Ino, T., Sata, N., Yamamura, S. (2002) Isolation and identification of a potent allelopathic substance in rice root exudates. Physiol. Planta. 115: 401-405.

Kato-Noguchi, H., Ino, K. (2003) Rice seedlings release momilactone B into the environment. Phytochemistry. 63: 551-554.

Kim, K.W., Kim, K.U. (2000) Searching for rice allelochemicals. *In:* Rice Allelopathy K.U. Kim and D.H. Shin (eds). Kyungpook National University,Taegu, Korea. pp. 83-96.

Kim, K.U., Shin, D.H. (2000) Rice Allelopathy, Kyungpook National University, Taegu, Korea. pp. 1-2.

Kim, J.T., Kim, S.H. (2002) Screening of allelochemicals on barnyard grass (*Echinochloa crus-galli*) and identification of potentially allelopathic compounds from rice (*Oryza sativa*) variety hull extracts. Crop Prot. 21: 913-920.

Kodama, O., Li, W. X., Tamogami, S., Akatsuka, T. (1992) Oryzalexin S, a novel stemarane-type diterpene rice phytoalexin. Biosci. Biotech. Biochem. 56: 1002-1003.

Koga, J., Shimura, M., Oshima, K., Ogawa, N., Yamauchi, T., Ogasawara, N. (1995) Phytocassanes A, B, C and D, novel diterpene phytoalexins from rice, *Oryza sativa* L. Tetrahedron. 51: 7907-7918.

Koga, J., Ogawa, N., Yamauchi, T., Kikuchi, M., Ogasawara, N., Shimura, M. (1997) Functional moiety for the antifungal activity of phytocassane E, a diterpene phytoalexin from rice. Phytochemistry. 44: 249-253.

Kong, C.H., Hu, F., Xu, X.H. (2002a) Allelopathic potential and chemical constituents of volatile from *Ageratum conyzoides* under stress. J. Chem. Ecol. 28: 1173-1182.

Kong, C.H., Xu, X.H., Hu, F., Cheng, X.H., Lin, B., Tan, Z.W. (2002b) Using specific secondary metabolite as marker to evaluate allelopathic potential of rice variety and individual plant. Chin. Sci. Bull. 47: 839-843.

Kong, C.H., Liang, W.J., Xu, X.H., Hu, F., Wang, P., Jiang, Y. (2004a) Release and activity of allelochemicals from allelopathic rice seedlings. J. Agric. Food Chem. 52: 2861-2865.

Kong, C.H., CXu, X.H., Zhou, B., Hu, F., Zhang, C.X., Zhang, M.X. (2004b) Two compounds from allelopathic rice accession and their inhibitory effects on weeds and fungal pathogens. Phytochemistry. 65: 1123-1128.

Lee, C.W., Yoneyama, K., Takeuchi, Y., Konnai, M., Tamogami, S., Kodama, O. (1999) Momilactones A and B in rice straw harvested at different growth stages. Biosci Biotech Biochem. 63: 1318-1320.

Murofushi, N., Inoue, A., Watanabe, N., Ota, Y., Takahashi, N. (1983) Identification of cytokinins in root exudates of the rice plant. Plant Cell Physiol. 24: 87-92.

Nakazato, Y., Tamogami, S., Kawai, H., Hasegawa, M., Kodama, O. (2000) Methionine-induced phytoalexin production in rice leaves. Biosci. Biotech. Biochem. 64: 577-583.

Rice, E.L. (1984) Allelopathy, 2nd ed, Academic Press, Orlando, USA.

Rimando, A.M., Olofsdotter, M., Dayan, F.E., Duke, S.O. (2001) Searching for rice allelochemicals: An example of bioassay-guided isolation. Agron. J. 93: 16-20.

Romeo, J.T. (2000) Raising the beam: moving beyond phytotoxicity. J. Chem. Ecol. 26: 2011-2014.

Olofsdotter, M., Jensen, L.B., Courtois, B. (2002a) Improving crop competitive ability using allelopathy-an example from rice. Plant Breed. 121: 1-9.

Olofsdotter, M., Rebulanan, M., Madrid, A., Wand, D.L., Navarez, D., Olk, D.C. (2002b) Why phenolic acids are unlikely primary allelochemicals in rice. J. Chem. Ecol. 28: 229-241.

Seal, A.N., Pratley, J.E., Haig, T., An, M. (2004a) Identification and quantitation of compounds in a series of allelopathic and non-allelopathic rice root exudates. J. Chem. Ecol. 30: 1647-1662.

Seal, A.N., Haig, T., Pratley, J.E. (2004b) Evaluation of putative allelochemicals in rice root exudates for their role in the suppression of arrowhead root growth. J. Chem. Ecol. 30: 1663-1678.

Suzuki, Y., Esumi, Y., Hyakutake, H., Kono, Y., Sakurai, A. (1996) Isolation of 5-(8'Z-heptadecenyl)-resorcinol from etiolated rice seedlings as an antifungal agent. Phytochemistry. 41: 1485-1489.

Suzuki, Y., Esumi, Y., Saito, T., Kishimoto, Y., Morita, T. (1998) Identification of 5-*n*-(2'-oxo) alkylresorcinols from etiolated rice seedlings. Phytochemistry. 47: 1247-1252.

Tang, C.S., Young, C.C. (1982) Collection and identification of allelopathic compounds from the undisturbed root system of Bigalta Limpograss (*Hemarthria altissima*). Plant Physiol. 69: 155-160.

Van Pelt, C.D., Haggarty, H., Brenna, J.T. (1998) Quantitative subfemtomole analysis of α-tocopherol and deuterated isotopomers in plasma using tabletop GC/MS/MS. Anal. Chem. 70: 4369-4375.

Weidenhamer, J.D., Hartnett, D.C., Romeo, J.T. (1989) Density-dependent phytotoxicity: distinguishing resource competition and allelopathic interference in plants. J. Appl. Ecol. 26:613-624.

Yoshida, R., Oritant, T., Nishi, A. (1971) Kinetin-like factors in the root exudates of rice plants. Plant Cell Physiol. 12: 89-94.

SECTION 6

New Approach in Tree Allelopathy

18

Variation in Allelopathic Influence among Wide Range of Tree Species

Kanji Ito[1] and Misako Ito[2]

[1]*Field Science Education and Research Center, Kyoto University, Kitashirakawa, Sakyoku, Kyoto 606-8502, Japan*

[2]*Graduate School of Agriculture, Kyoto University, Kitashirakawa, Sakyoku, Kyoto 606-8502, Japan*

Hundreds of trees grown in an arboretum for 70 years under identical environment and management practices were found to exhibit an extreme variation in vegetational composition around their trunks. Comparison of vegetation and soil properties under trees of 200 different woody species in the arboretum was conducted, and the results were evaluated from an allelopathic viewpoint. Differences in existing vegetation surrounding tree trunks were similar to that in emergence/growth from soil seed banks of the same areas, suggesting species-depending allelopathic influences of the trees. Most coniferous species were categorized to the most inhibitory group, as compared with broadleaf species. The most promotive group only consisted of broadleaf species. Growth of test plant (lettuce) differed greatly, ranging from 0 mg to 358 mg/pot in dry weight, depending on the tree species/soil association. Soil pH and soil water EC also varied widely between 3.7 and 7.7, and between 0.06 mS and 2.02 mS/cm, respectively. The tendencies found for the arboretum trees coincided with those for tree species tested in a forest. On an average, growth of the test plants and soil pH were lower in conifer soils than in broadleaf soils. A high positive correlation was shown between growth of lettuce plants and soil pH for coniferous species. These results suggest that substances consecutively egressing into stem flow contribute to large variations in vegetation, that soil properties are specific to taxonomic groups, and at least in coniferous trees, substances which can lower soil pH appear to be the cause of growth inhibitory activities.

Keywords: *allelopathic influence, broadleaf trees, conifers, soil seed bank, soil pH, stem flow*

INTRODUCTION

Spatial arrangements in vegetation such as bare areas under trees are referred to as patterning or zoning. Rice (1984) pointed out the probable involvement of allelopathy in such phenomena and also that the vegetational patterns associated with trees are generally more striking than those associated with herbaceous species. This may be because trees must provide long term consecutive chemical influences to the soil under trees by means of stem flow, throughfall and/or litters. Thus, any established trees are assumed to preserve its own chemical specificity, whether it is extreme or not, in surrounding soils, providing diverse effects on the understory, although only prominent patterns caused by restricted species such as *Juglans nigra*, *Eucalyptus* spp. and *Podocarpus* spp. have been definitively attributed to allelopathy (Rice, 1984; Lovett, 1986).

In 2001, we discovered an evidence likely to support the above concept, in an arboretum of Kyoto University Forest, Japan. Hundreds of trees planted 70 years ago and grown unfertilized under the same environment and management practices exhibited extreme variation in vegetational patterns around their trunks, suggesting differential specific influences of continuous egression from trees. Therefore, comparison of vegetation and soil properties under trees of 200 different woody species in the arboretum was conducted, and results were evaluated from an allelopathic viewpoint. Vegetation and seed banks surrounding the tree trunk and physiological and chemical properties of the soils from the same areas were investigated. The relationships among tree taxa and promotive effects were also studied. This kind of comparative evaluation of allelopathic potential of a wide range of woody species may be helpful for ecological interpretation of vegetation and successional patterns, and perhaps also for practical selection of trees in landscaping uses.

DIFFERENCE IN VEGETATION UNDER TREES AMONG 120 SPECIES

The surveyed arboretum has been unfertilized and uniformly managed since planting of saplings 70 years ago. Therefore, all trees have grown in the same environment for a long period. In spite of this, extreme variation in vegetational appearance under trees was found among the species, particularly in the areas surrounding tree trunks. Based on visual observation, differences among tree taxa were hypothesized both in the amount of vegetative cover and in floristic composition. Therefore, coverage and plant height of all weed species appearing in the areas within 30 cm from tree bases were determined from randomly selected 120 tree species. The top 15 cm soil were collected from the same areas, and emergence and growth of weeds from the seed banks also determined. Vegetation differed from none to that more vigorous. Forty-eight weed species were recorded, including 22 perennials, mostly herbaceous, and *Galium sprium, Oxalis corynbosa, Stellaria media* and *Poa acroleuca* were most common. Emergence and growth of weeds from seed banks also varied both in amount and grass/broadleaf ratio between tree species (Fig. 1).

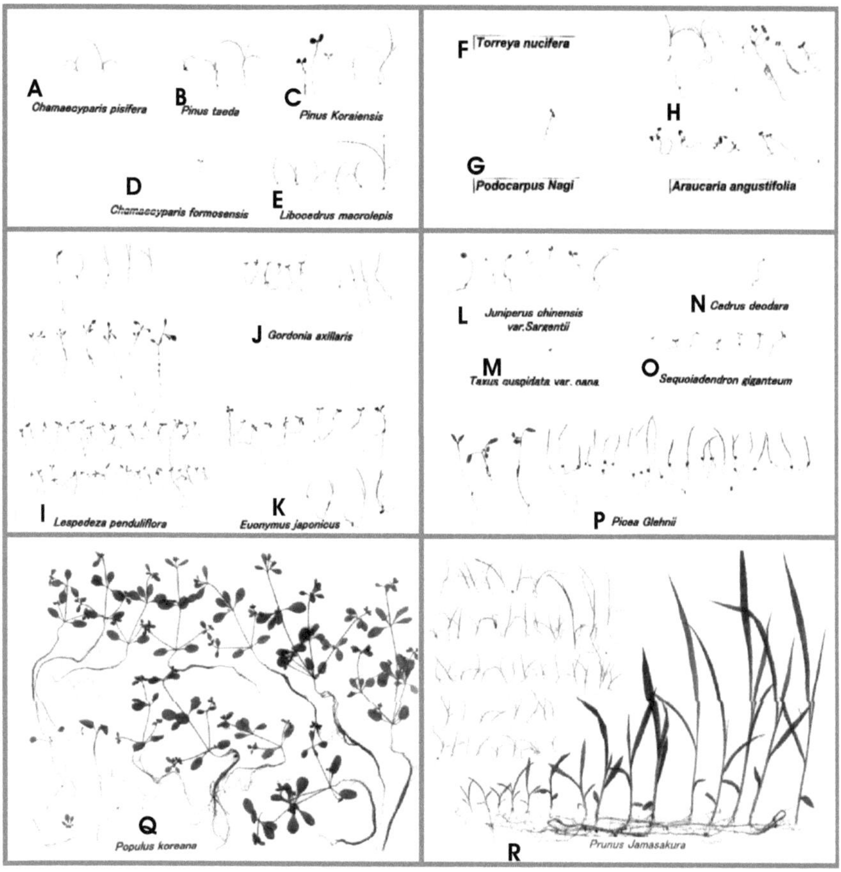

Fig. 1 Example of variations in emergence and growth of weeds from soil seed banks under different species of trees. Soil samples collected from 0-15 cm depth were potted in plastic cups and the pots were placed in a greenhouse and watered for 4 wks. A: *Chamaecyparis pisifera*, B: *Pinus taeda*, C: *P. koraiensis*, D: *C. formosensis*, E: *Libocedrus macrolepis*, F: *Torreya nucifera*, G: *Podocarpus Nagi*, H: *Araucaria angustifolia*, I: *Lespedeza penduliflora*, J: *Gordonia axillaris*, K: *Euonymus japonicus*, L: *Juniperus chinensis* var. *sargentii*, M: *Taxus cuspidate* var. *nara*, N: *Cedrus deodata*, O: *Sequoiadendron giganteum*, P: *Picea glehnii*, Q: *Populus koreana*, R: *Prunus jamasakura*.

Differences among the tree species were summarized according to the values of existing and latent vegetation under trees (Table 1). A trend observed for seed banks was similar to that for existing vegetation as a whole. Most of the soil from the areas without any vegetation expressed no emergence or inhibited growth from soil seed banks. In contrast, growth promotion of weeds from seed banks mostly appeared in soil collected under vigorous vegetation. A large proportion of trees surveyed exhibited relatively poor vegetation around them, varying from no emergence to species-depending growth inhibition. Coniferous species were regarded as more inhibitory than broadleaf species as a whole. Estimating from both existing and latent vegetation, the most promotive group consisted exclusively of broadleaf species. It is concluded that these vegetational differences among trees are taxonomically specific phenomena, as they were not explicable in terms of special difference between trees.

DIFFERENCE IN SOIL PROPERTIES UNDER TREES AMONG 175 SPECIES

In order to clarify the various influences of tree species on soil that were suggested above, growth regulating activity, acidity and soil water EC of the top 15 cm soil around tree trunk in the arboretum were compared among 175 species belonging to 55 families. In a bioassay of soil, growth of lettuce and Italian ryegrass expressed in dry weight ranged from 0 mg to 358 mg/pot and from 24 mg to 512 mg/pot, respectively, depending on the tree species/soil association. Soil pH varied ranging from 3.7 to 7.7, and soil water EC also widely varied from 0.06 mS to 2.02 mS/cm. On an average, test plant growth and soil pH were lower in conifer soil than in broadleaf soil. A positive correlation was found only between growth of lettuce plants and soil pH, among all combinations of the three parameters through species tested. The correlation was higher for coniferous species (0.815) than for broadleaf species (0.513) (Table 2). Among conifers, it was as high as 0.945 for Taxodiaceae and 0.898 for Cupressaceae but 0.687 for Pinaceae soil. In broadleaf soil, the relation between lettuce growth regulating ability and pH of the soil varied depending on the taxa, but the trend in growth was not explicable by pH alone. Soil water EC exhibited some positive correlations with test plant growth for broadleaf species, but no correlation for coniferous species.

The same determination was conducted for surface soil collected from under the trees of nine different species in a cool temperate forest in Kyoto, Japan. The tendency of lettuce growth in the bioassay across the species coincided with that found in the arboretum (Fig. 2), as did soil pH (data not

Table 1 Classification of 120 tree species by degree of abundance of existing and latent vegetation under tree

Existing vegetation	*Taxonomic group*[1)]	*Latent vegetation (Seed bank)*		
		No emergence	*Inhibited growth*	*Promoted growth*
None	CONIF	*Torreya nucifera*	*Araucaria angustifiolia*	*Larix Kaempferi*
		Podocarpus Nagi	*Pinus Koraiensis*	
		Pinus taeda	*Pinus Armandii*	
		Cedurs deodara	*Abies holophylla*	
		Keteleeria Davidiana	*Pseudotsuga japonica*	
		Cunninghamia lanceolata	*Sequoia sempervirens*	
		Chamaecyparis obutusa	*Sequsoiadendron giganteum*	
		Chamaecyparis pisifera	*Thuja Standishii*	
		Chamaecyparis formosensis	*Thujopsis dolabrata*	
		Libocedrus sempervirens		
		Libocedrus decurrens		
	BRODL	*Qurecus gilba*	*Pasania glabra*	*Phelledendron amurense*
		Fagus japonica	*Tilia japonica*	
		Magnolia grandiflora	*Fatsia japonica*	
		Illicium anistum	*Nerium indicum*	
		Spiraea Thunbegii		
		Koelreuteria paniculata		
		Fatsia japonica		
Poor	CONIF	*Taxus cuspidata*	*Taxus baccata*	*Metasequoia glyptostroboides*
		Pinus densiflora	*Podocarpus macro phllus*	*Curessus funebris*
		Pinus Koraiensis	*Cephalotaxus Harringtonia*	*Juniperus virginiana*
		Pinus Armandii	*Pinus thunbergii*	
		Pinus strobus	*Pinus Bungeana*	
		Pseudolarix amabilis	*Pinus palustris*	
		Cryptomeria japonica	*Pinus Elliotti*	
		Taiwania cryptomeria	*Pinus echinata*	
			Taxodium mucromatum	
			Sciadopitys verticillata	
			Juniperus chinensis	

(Table 1 Contd.)

Table 1 Contd.

	BRODL	*Quercus crispula*	*Corylus heterophylla*	*Petrocarya stenoptetera*
		Fagus crenata	*Quercus salicina*	*Castanea crenata*
		Magnolia obovata	*Cudrania tricusoidata*	*Pasania edulis*
		Ternstroemia gymnanthera	*Cleyera monadelpha*	*Castanopis cuspidata*
		Gymnocadus chinensis	*Gordonia axillars*	*Quercus myrsinaefolia*
		Daphnipplyllum macropodum	*Schima Wallichii*	*Quercus phillyraeoides*
		Koelreuteria paniculate	*Prunus spinulosa*	*Quercus serrata*
		Meliosma rigida	*Chaenomeles sinensis*	*Broussnetia papyrifera*
		Sabiaceae oldhamii	*Cladrastis Platycarpa*	*Clinnamomum japnicum*
		Ilex Buergeri	*Cladrastis sikokiana*	*Camellia japonica*
		Hedera herlix	*Acer palmatum*	*Distyium racemosum*
		Osmanthus insularies	*Acer pycnanthu*	*Prunus Grayana*
		Callicarpa japonica	*Ilex rotunda*	*Prunus Ssiori*
		Viburnum suspensum	*Euonymus alatus*	*Sophora japonica*
			Rhododendron Metternichii	*Melia Azedarach*
			Rhododendron macrosepalum	*Elaeorapaceae sylvestris*
			Enkianthus perulatus	*Camptheca acumminate*
			Pieris japonica	
			Leucothus perulatus	
			Gardenia jasminoides	
			Serissa japonica	
Vigrous	CONIF		*Taxodium disticum*	
	BRODL		*Juglans mandshurica*	*Populus koreana*
			Juglans nigra	*Carya ovata*
			Platycarya stobilacea	*Aphananthe aspera*
				Celtis sinensis
				Ulmus Davidiana
				Mallotus japonicus
				Firmiana simplex
				Ehretia ovalifolia
				Paulownia Kawakamii

1) CONIF : Coniferous species, BRODL : Broadleaf species

Table 2 Corelations between the growth of test plants [1)] and pH and EC of soils collected from under trees of different species

Test plant	*Correlation coeficient*			
	Conifer (n = 47)		*Broadleaf (n = 128)*	
	pH	*EC*	*pH*	*EC*
Lettuce	0.815*	0.042	0.513*	0.487*
Italian ryegrass	0.118	–0.089	0.034	0.332*

[1)]Dry weight at 4 wks after planting
*Significant at the 1% level

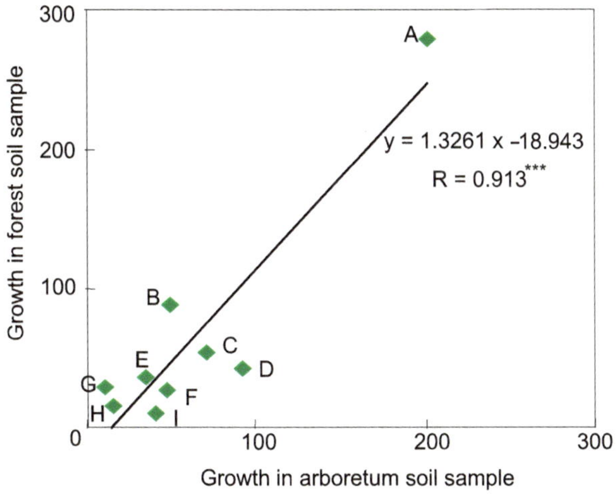

Fig. 2 Correlation in the growth of test plants (lettuce) between the soil of the same species grown in forest and in arboretum. Growth was presented by relative dry weight of plants after 4 wks of growing. A: *Pterocarya rhoifolia*, B: *Aesculus turbinata*, C: *Betula grossa*, D: *Acer mono.*, E: *Quercus crispula*, F: *Carpinus tschonoskii*, G: *Q. salicina*, H: *Cryptomeria japonica*, I: *Fagus japonica*. (Ito & Hirayama, unpublished)

shown). Soil water EC varied within the range found in the same species in the arboretum (data not shown).

DISCUSSION

Every tree manages the environment under its canopy by physical and chemical means. Chemical influences on the environment may occur by depositing substances to soil in three ways, litters, throughfall and stem flow caused by rainfall (Eaton *et al.*, 1973; Falkengren-Greup and Bjork, 1991). Naturally, these influences appear stronger in tree species than in herbaceous species because of greater biomass in trees and longer consecutive inflows from trees compared to herbs. Rainfall brings down substances, both common and specific, not only released from the tree organism but also from many life forms dwelling on the tree, such as microorganisms, insects, lichens and mosses.

In our studies, prominent differences in vegetation and soil properties were found under trees among a wide range of taxa. We assume that these occurred mainly due to stem flow, although there has been no previous report indicating the role of stem flow in allelopathy. This is supported by the fact that extreme vegetational changes were mostly limited to the area immediately surrounding the tree trunk. In stem flow, water that washes out all parts of the tree canopy congregates and flows into a restricted area of soil. Stem flow release not only leacheable substances of leaves and twigs but also some tree components and lives existing on the inner and outer barks. Therefore, stem flow is generally assumed to contain more concentrated and diverse compounds than throughfall. Extreme differences in soil pH and soil water EC demonstrated in this paper may support this view. Probable ways of chemical influence of trees on soil presumed in the above discussion are illustrated in Fig. 3.

Various evidence confirms the stable and rather pervasive influences caused by stem flow in mature trees. A high negative correlation recognized between breast height diameter of *Cryptomeria japonica* tree and the degree of inhibitory effect of soil beneath the tree indicates the increased influences as the tree grows larger (Ito and Hirayama, unpublished). Inokura and Nakano (2000) reported that pH of stem flow was lower in coniferous species but higher in broadleaf species than that of rainfall and throughfall. In *C. japonica,* the low stem flow pH was common in all forests, and the pH of stem flow was found to be uniquely modified by this species when passing on or through barks (Inokura and Nakano, 2000; Sato *et al.*, 2000; Tomizawa *et al.*, 2000). *Fagus crenata* stem flow pH changed little between seasons and habitats (Tsukahara, 1997). Moreover, stem flow pH tended to converge in ranges specific to tree species and cationic components of stem flow were characterized by the morphological differences of bark (Takenaka *et al.*, 2000).

The influence of substances deposited by trees in the soil on surrounding vegetation under trees is manifested both as direct allelopathic effects of compounds and/or indirectly through environmental changes in the soil. Direct effects often occur through complex interactions with existing soil properties. del Moral *et al.* (1978) concluded that allelopathic suppression by *Eucalyptus baxteri* of heath beneath the canopy was associated with phenolic acids and that large quantities of phenolic glycosides, which are known allelochemicals,

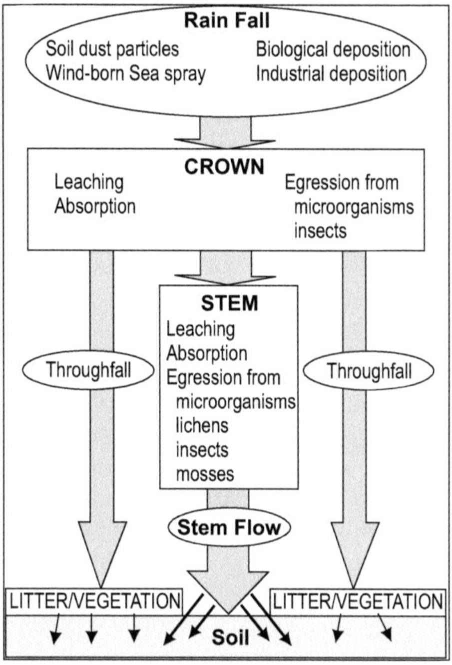

Fig. 3 Schematic diagram showing the probable view of chemical influences of a tree on soil

leached from leaf canopy. The soil was rich in acidic resins and, possibly, terpenoids but free phenolic acids were not obtained. This suggests that chemical change occurred when phenolics in leaf leachates entered the soil, but that soil processes did not remove their toxicity. Many reports describe the toxic effects of phenolics on plants. However, in comparison of phytotoxic activity of 21 naturally occurring phenolics by Kobayashi and Ito (1998), test plant growth varied from inhibitory to promotive depending on the compound. Activity also varied depending on the seed size of receiver species, in which the smaller seed species were more sensitive, suggesting that seed banks differences could affect allelochemical vegetation change.

In the coniferous species that we investigated, inhibition of vegetation and test plant growth were largely associated with soil acidity. Schles and Nowick (1998) showed that acidifying effects in a pine forest occurred via leaching of organic acids from decomposed pine litter. Stronger inhibition by the soil with lower pH is thought to be attributed to the direct activities of some acidic allelochemicals and/or indirect effects of acidic soil: increase of toxic aluminum, decrease of availability of Ca^{++} and Mg^{++} and solubilization of manganese occurring in the strongly acidic soils.

CONCLUSION

It is concluded that while vegetation surrounding trees is influenced by a variety of complex and interacting processes, the evidence shown here strongly suggests that species-specific chemical cocktails found in tree stem flow interact with the soil ecosystem to influence unique patterns of vegetation surrounding trees. Further research should focus on the confirmation of the direct effects of stem flow on plants among different tree taxa and also to clarify the behavior of acidic compounds occurring in stem flows from conifers that may change the soil to inhibitory.

Acknowledgement

We gratefully thank Osamu Rinoie, Tatsuichiro Kaku, Satoru Tanaka and staffs of Kitashirakawa Arboretum, Kyoto University, Japan for their technical assistance. We would also like to thank Dr. John Breen and Dr. Yoshiharu Fujii for their comments on the manuscript. This work was partially supported by the Grant-in-Aid of the Ministry of Education and Science of Japanese Government.

References

del Moral, R., Willis, R.J., Ashton, D.H. (1978) Suppression of coastal heath vegetation by *Eucalyptus baxteri.* Aust. J. Bot. 26: 203-219.

Eaton, J.S., Likens, G.E., Bormann, F.H. (1973) Throughfall and stemflow chemistry in northern hardwood forest. J. Ecol. 61: 495-508.

Falkengren-Greup, U., Bjork, L. (1991) Reversibility of stemflow-induced acidification in Swedish beech forest. Environ. Pollu. 74: 31-37.

Inokura, Y., Nakano, T. (2000) Throughfall and stem flow in *Cryptomeria japonica. In* Research Report, Grant-in-aid for the Scientific Research, No. 0856001, S. Kawanabe (ed.). pp. 515-533 (In Japanese).

Lovett, J.V. (1986) Allelopathy: the Australian Experience. *In* The Science of Allelopathy, A. R. Putnam, C.S. Tang (eds.), Wiley Interscience, NY. USA. pp. 75-100.

Kobayashi, Y., Ito, M. (1998) Valuation of phytotoxic activity of naturally occurring phenolic compounds. J. Weed Sci. Tech. 43: 341-348.

Rice, E.L. (1984) Allelopathy 2nd Ed., Acad. Press, NY. USA. pp. 130-188.

Sato, Y., Inokura, Y., Osaki, S., Sugihara, S., Yoshimura, K., Ogawa, S. (2000) Water chemistry in throughfall and stem flow between two species with different crown structure. *In* Research Report, Grant-in-aid for the Scientific Research, No. 0856001, S. Kawanabe (ed.) pp. 431-442 (In Japanese with English summary).

Schles, M.C., Nowick, T.E. (1998) Effect of pines on soil properties and processes. *In* Ecology and Geography of *Pinus.* E.D. Richardson (ed.), Cambridge Univ. Press, Cambridge. UK, pp. 341-353.

Takenaka, C., Suzuki, M., Yamaguchi, N., Imaizumi, Y., Sibata, E. (2000) Chemical characteristics of stem flow collected from deciduous trees. *In* Research Report, Grant-in-aid for the Scientific Research, No. 0856001, S. Kawanabe (ed.). pp. 172-177 (In Japanese with English summary).

Tomizawa, H., Okikawa, M., Nisio, F., Haraguchi, A. (2000) Chemical modification of rain water by *Alnus japonica* wetland forest in Kiritapp Mire, Eastern Hokkaido, Japan. *In* Research Report, Grant-in-aid for the Scientific Research, No. 0856001, S. Kawanabe (ed.). pp. 212-215.

Tsukahara, H. (1997) Seasonal and inter-individual changes of stem flow pH and EC in *Fagus crenata. In* Research Report, Grant-in-aid for the Scientific Research, No. 05506001, K. Fujiwara (ed.). pp. 102-107 (In Japanese).

19

Monitoring Allelopathic Expression and Functional Performance of Tamarind (*Tamarindus indica* L.): A Case Study

Mohammad Masud Parvez[1], Syeda Shahnaz Parvez[1,2,3], Hiroshi Gemma[2] and Yoshiharu Fujii[1]

[1]Biodiversity Division, National Institute for Agro-Environmental Sciences, 3-1-3 Kannondai, Tsukuba, Ibaraki 305-8604, Japan

[2]Laboratory of Pomology, Institute of Agriculture and Forestry, University of Tsukuba, 1-1-1 Tennondai, Tsukuba, Ibaraki 305-8752, Japan

[3]Present address: Post-harvest Science and Technology Division, Japan International Research Center for Agricultural Sciences, 1-1 Ohwashi, Tsukuba, Ibaraki 305-8686, Japan

Detailed monitoring of the allelopathic expression and functional performance of leaves, roots and seeds of the tamarind tree was carried out, based on bioassay-guided investigation, physiological experimentation and pot culture observations using several common weed and edible crop species. Tamarind leaves induced strong growth inhibition of tested species as demonstrated by the application of aqueous extraction, physiological and pot culture experimentation. Most importantly, the tested species regenerated their growth upon removal of the blended tamarind leaves from the growth medium, suggested that tamarind leaves function as biologically active material. The roots of the tamarind tree caused severe growth suppression of the tested species in both laboratory and greenhouse conditions as demonstrated by the "plant-box method", "soil-agar sandwich method" and physiological experiments. Interestingly, the magnitude of growth suppression was increased with the aging of tamarind seedlings, thus displaying strong competence of tamarind roots that exert function as a true growth regulator for plant growth regulation in the environment. Further studies on the seed and seed-coat of tamarind fruits to evaluate their functional capability as plant growth regulators revealed that these two parts possess species-specific expression through inhibitory and stimulatory effects. It is evident that the leaves and roots of the tamarind tree, as well as the seed and seed-coat of ripened tamarind fruits contain strong bio-active chemical(s) that function as true growth regulator(s) and are involved in plant-specific expression for growth regulation, particularly in weed species, thus, contributing significant importance in the environment. Differential allelopathic effects (inhibitory and stimulatory) of seed and seed-coat of ripened tamarind fruits are biologically important and could be utilized either as a bio-herbicide for weed control as well as bio-fertilizer for growth in agricultural production. The isolated bio-active molecules when chemically identified and characterized may yield excellent templates for use as plant growth regulators and/or bio-herbicide.

Keywords: *allelopathy, plant growth regulation, tamarind*

INTRODUCTION

Allelopathy is a natural phenomenon by which any plant or tree provides itself with a competitive advantage by releasing phytotoxins into the surrounding environment. Allelopathic plants or trees release chemical compounds into the environment through a number of ways including root exudation, leaching by dew and rain, and volatilization or decaying plant tissues which are considered the most significant processes (Rice, 1984). Allelopathy plays a major role in agro-ecosystems leading to a wide range of influences and interactions in biotic communities. It is evident that the majority of allelopathic effects so far reported are identified as inhibitory, thus, the allelopathy phenomenon has been well recognized as a biological means for integrated weed management with several advantages (Altieri and Doll, 1978; Leather, 1983; Putnam *et al.*, 1983; Rice, 1984; Weston, 1996; An *et al.*, 1998; Wu *et al.*, 2000).

Allelopathic competence of numerous plants and trees has been investigated by several researchers, yielding identification of many novel chemical compounds having potential use as natural products for herbicide formulation. However, a good number of potential trees are unexploited. An example of this species is the ever-green legume tree *Tamarindus indica* L. (syns. *Tamarindus occidentalis* Gaertn.; *Tamarindus officinalis* Hook.) popularly known as the "tamarind". Of all tropical and subtropical fruit trees, none is more widely distributed nor significantly recognized as an ornamental and medicinal tree than the tamarind

(Morton, 1987). It is surprisingly evident that the tamarind tree is poorly exploited commercially, although the fruit is highly priced, common in all local and international markets, and despite the fact that the fruit and seed have high medicinal and industrial values. In many developing tropical and sub-tropical countries, particularly in Asia and Africa, a vast majority of people especially the rural population rely on this species for a significant part of their food and energy requirements.

Many people particularly in India and Africa, have experienced a tent getting holes in the fabric within short span of time when pitched under a tamarind tree. This, coupled with the fact that almost no, or very few plants and weeds can grow or survive under any tamarind tree, suggests the corrosive effects of leaching of allelochemicals from the fallen tamarind leaves as well as the influence of the root exudates into the soil around the trunk. These facts clearly indicate that both leaves and roots of the tamarind tree have a definite mode of action on the growth regulation of other plants, thus implying a significant role in the environment. However, to date no evidence in the literature is available either on the elucidation of the process of such an impact, nor on the involvement of a definite chemical compound that acts as a true growth regulator for keeping a weed-free surrounding around the tamarind tree itself.

In this chapter, we discuss the careful monitoring of leaves and roots of the tamarind tree, and the seed and seed-coat of tamarind fruits, paying special attention on their allelopathic activity and functional performance through bioassay-guided studies, using several weeds and edible crops species under laboratory and greenhouse conditions. Detailed studies were employed – (i) to examine the physiological and biochemical events related to plant growth regulation in order to obtain new insights for further elucidation of the roles of tamarind leaves in allelopathy under laboratory conditions; (ii) to evaluate the allelopathic competence of the tamarind root under both laboratory and greenhouse conditions, and to elucidate the role of the tamarind roots in plant growth suppression; and (iii) to study the seed and seed-coat of tamarind fruits to elucidate whether any allelopathic influence existed in these two organs under laboratory conditions.

MATERIALS AND METHODS

Allelopathic Potentiality of Leaf

Plant materials

We collected fresh leaves of *Tamarindus indica* L. from a 10-y old tamarind tree grown in a greenhouse of the Tsukuba Botanical Garden, National

Science Museum, Tsukuba, Japan. The harvested leaves were put in a paper bag followed by oven drying at 70°C for approximately 3 d. The dried leaves were used for further analysis.

We carried out bioassay studies, germination tests and physiological experiments using seeds of 14 plant species under different families belonging to seven weed (viz. barnyard grass - *Echinochloa crus-galli* Mill., chinese milk vetch - *Astragalus sinicus* L., perennial ryegrass - *Lolium perenne* L., phacelia - *Phacelia* spp., timothy grass - *Phleum pratense* L., white clover - *Trifolium repens* L. and wild ginger - *Asarum canadense* L.) and edible crop (viz. asparagus - *Asparagus officinalis* L., cucumber - *Cucumber sativus* L., lettuce - *Lactuca sativus* L., radish - *Raphanus sativus* L., sesame - *Sesamum indicum* L., tomato - *Lycopersicum esculentum* L., and Welsh onion - *Allium fistulosum* L.) species.

Bioassay

Using a sandwich method developed by Fujii *et al.* (2003), bioassay studies for elucidating the allelopathic activity of the tamarind leaf were carried out in a triplicate randomized complete block design.

Crude water-soluble extraction of tamarind leaf

Appropriate amounts of oven dried tamarind leaves corresponding to 1%, 5% and 10% (w/v) were used to obtain crude water-soluble extracts using sterilized distilled water under dark conditions with continuous shaking at 70 rpm for 2 d at 25°C in an orbital shaker. The filtered extracted solution was either used freshly as a crude water-soluble extract or stored at 5°C in a refrigerator until further use.

Germination test

The germination test of weed seeds and edible crop species (as described earlier) were carried out using 1 mL of 3-different filtered crude water-soluble extracts (1%, 5% and 10%) (w/v) separately. Seeds of each species were placed in separate Petridishes (9 cm diameter and 1.7 cm deep) fitted with 9 cm filter paper in a triplicate randomized complete block design. The incubation of the seeds were carried by keeping covered Petridishes in a non-illuminated incubator at 25°C for 3 d followed by recording of the percentage of germination, physiological observation of the root and shoot, and the length of both organs.

Physiological experiment

Appropriate amount corresponding to 1%, 5% and 10% (w/w) of dried powered tamarind leaves were mixed with 25 g vermiculite in vinyl pots (7.5 cm diameter and 6.5 cm deep) for conducting the physiological experiment using 4 plant species belonging to two weed (barnyard grass and white clover) and two edible crop (lettuce and radish) species in a growth chamber under continuous fluorescent light (approximately 3,000 lux) at 80% relative humidity and 25°C for 2 wk. After 2 wk, the appropriate numbers of seedlings from each species were harvested and the root and shoot length were measured.

In order to determine whether the growth of the two weeds (barnyard grass and white clover) and two edible crops (lettuce and radish) species may recover if the tamarind leaves were removed from the growth medium, another set of the experiment was established: seedlings of each species were grown for 1 wk in vinyl pots containing blended tamarind leaves at three-different concentrations (as before) in vermiculite in a growth chamber with a randomized complete block design, as described earlier. Afterwards, the vermiculite including the tamarind leaves was removed and the plants were thoroughly washed with distilled water, followed by re-planting each species separately into a new vinyl pot filled with vermiculite only. Seedlings were further grown in a growth chamber for one more week, then the appropriate numbers of seedlings of each species were collected and the lengths of roots and shoots were measured.

Allelopathic Potentiality of Root

Plant materials

The tamarind (*Tamarindus indica* L., 10-year old) tree that was previously used was also included in this experiment. As to the control trees, a red sandalwood/coral tree (*Adenanthera pavonina* L., 10-y old) and a mango tree (*Mangifera indica* L. nine-year old), grown in the same greenhouse (as the tamarind tree) were used, since it had not been reported that these trees had any allelopathic characteristics. From the *Tamarindus indica* L. tree, the control trees were located approximately 2.2 m and 3.0 m apart (in the opposite direction).

From our previous study, it has been revealed that 2 weeds (barnyard grass – *Echinochloa crus-galli* Mill. and white clover – *Trifolium repens* L.) and 2 edible crops (lettuce – *Lactuca sativus* L. and radish – *Raphanus sativus* L.) species were most sensitive to the effects of the tamarind leaf, and were used as test plants using the "plant-box method" and "soil-agar sandwich method" under laboratory and in soil experiments in greenhouse conditions.

Plant-box method experiment

The plant-box method experiment was conducted according to the method of Fujii (1992) and Fujii *et al.* (1992). Briefly, the tamarind seedlings were grown in sandy soil in vinyl pots (5.5 cm diameter and 6.0 cm deep) in a greenhouse. To evaluate the allelopathic activity of root exudates, tamarind seedlings aging 7, 14 and 21 d after germination (DAG) were used for bioassay in the plant-box, as shown in Fig. 1A. A total of 33 seeds (at 21-different distances from the tamarind roots in the plant-box) of each 4 species (2 weed and 2 edible crop species) were sown vertically into the agar medium (Fig. 1B). The plant-boxes were kept at 25°C/20°C under 200 $\mu mol\ m^{-2}\ s^{-1}$ light intensity with a 12 h/12 h photoperiod for 5 d in an incubator. A control was established without putting a tamarind seedling into the plant-box. The length of root and shoot of all the tested species were measured, after 3-d incubation.

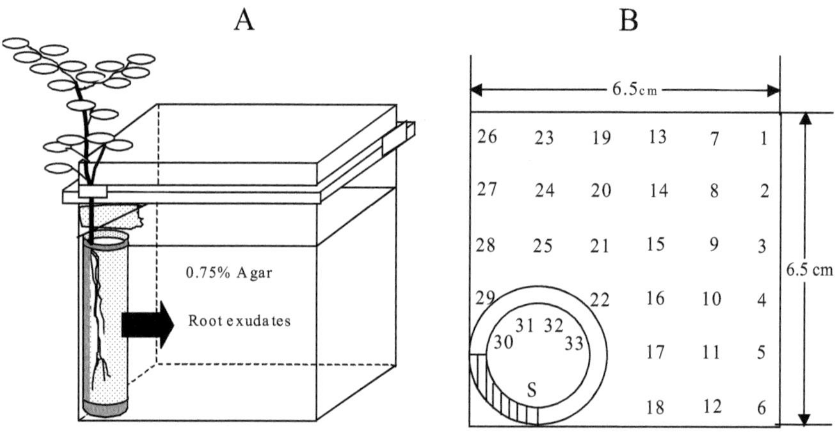

Fig. 1 Diagrammatic model of the plant-box method. In the plant-box (A) a tamarind seedling, and (B) sowing position (numbers 1 to 33) of seeds of different weed and edible crop species was tested at different distances from the tamarind seedling S represents position of the tamarind seeding.

The roots of the tamarind seedlings used (7- DAG, 14-DAG and 21-DAG) were carefully removed from the plant-box and gently washed with water after conducting the plant-box experiments. Roots and shoots were then cut-off from each seedling and placed in separate paper bags followed by drying at 70°C for approximately 3 d in an oven, and, finally, the dried roots and shoots were weighed separately.

Soil preparation and greenhouse experiment

A number of different weeds were found growing under both control trees (*Adenanthera pavonina* L. and *Mangifera indica* L.) while no weeds were visible in the soil under *Tamarindus indica* L. All weeds growing in the soil around the control trees were removed and a general cleaning of the soil around the roots of the three trees was carried out in order to prepare a suitable bed for sowing the seeds of the species tested. An appropriate numbers of seeds of each plant species were sown in rows at 2 cm apart in the soil bed under the three trees and, finally, the seedlings were grown for 2 wk providing adequate water without fertilizer during the experimental period. The experiment was established in triplicate randomized complete block design. A transparent net was set approximately 2 m above ground level to prevent any leaves from falling on the soil.

Soil sampling

Soil sampling was carried out from a 3-y old tamarind tree grown in a big-pot, using commercially available soil in the greenhouse. The soil from the big-pot where the tamarind tree was grown was carefully removed with the least possible disturbance of the roots. For conducting a bioassay of the test plant species mentioned above, fresh soil was used in the "soil-agar sandwich method" experiment. In addition, surface soil (approximately 7 cm deep, 20 cm from the main trunk of the trees) under the *Tamarindus indica* L. and two other control trees (*Adenanthera pavonina* L. and *Mangifera indica* L.) was collected for conducting another set of experiments. The collected soil from each tree was separately sieved through a 1-mm mesh then kept separately in glass beakers, and finally autoclaved at 115°C for 20 min. The autoclaved soil samples were used separately for bioassay of weed and edible crop species using the "soil-agar sandwich method", as described below.

Soil-agar sandwich method experiment

To evaluate the allelopathic activity of tamarind root exudates through bioassay-guided studies, a new method termed – "soil-agar sandwich method" was developed. Briefly, a portion of either fresh or autoclaved soil was placed between two layers of 0.75% agarose (total 8.2 ml) in each well of a six well multi-dish plastic plate, as described previously for the "sandwich method" (Fujii *et al.*, 2003). Seeds of two weeds and two edible crops species (as mentioned above) were used for the bioassay for incubation at 25°C in the dark for 3 d. After incubation, the lengths of both roots and shoots of the species tested were measured. For control treatment, fresh commercial soil, which was identical to the soil in the big-pot where the tamarind tree was grown, was used. The entire experiment was conducted in a randomized complete block design with three replications.

Allelopathic Potentiality of seed and seed-coat

Preparation of seed and seed-coat

To obtain the seed exudates, an appropriate number of seeds of ripened tamarind fruits were soaked in sterilized distilled water overnight at 25°C in complete darkness in a non-illuminated growth chamber. Using a sharp knife, the brown seed-coat of the tamarind seeds were separated from the kernels and the seed-coats were blended to a powder for water extraction at 3-different concentrations (1%, 5% and 10%) (w/v), as described earlier.

Germination test

An appropriate numbers of seeds of tested species were placed separately in Petridishes (9 cm diameter and 1.7 cm deep) fitted with a filter paper, as described earlier. The germination test was conducted in a randomized complete block design using 1 mL of filtered seed exudates or crude water-soluble extracts of seed-coat at three-different concentrations (1%, 5% and 10%) (w/v) in a non-illuminated incubator at 25°C for 3 d. Percentage of germination, physiological observation of the roots and shoots, and the length of both organs were then measured.

Pot culture experiment

For the pot culture experiments, a total of six plant species belonging to two weed (barnyard grass and white clover) and four edible crop (lettuce, radish, tomato and cucumber) species were selected based on the results obtained in the germination test using crude-water soluble extracts of seed-coat. The pot culture experiment was conducted using vermiculite in vinyl pots, as described earlier (see *1.5 Pot culture experiment*). The seedlings of each tested species were grown separately for 7 d, and then the experimental schedule was established in a triplicate randomized complete block design, as follows: (i) at 1 d-interval, crude water-soluble extracts of seed-coat at three-different concentrations (1%, 5% and 10%) (w/v) were sprayed for 3 d – experiment 1; (ii) seedlings were washed with distilled water and transplanted to new pots filled with vermiculite, and grown for another 4 d – experiment 2; and finally (iii) at 1 d-interval, crude-water soluble extracts of seed-coat at three-different extracts (1%, 5% and 10%) (w/v) were sprayed for an additional 3 d-experiment 3. An appropriate numbers of seedlings of each species were randomly harvested in each independent experiment and the lengths of roots and shoots were measured.

RESULTS AND DISCUSSION

The occurrence of allelopathy (Gressel and Holm, 1964; Rice, 1984) and its role in ecosystems and crop husbandry has received increased attention in the past few decades due to its potential for weed control and as an effective tool for increasing agricultural productivity. It is also believed to be a method for plants to influence the growth and development of competing plants (Bell and Koeppe, 1972; Muller and Moral, 1966).

Various mechanisms have been reported by which this phenomena manifest it, however, three possible routes are widely accepted: leaching from leaves and stems, exudation from roots, and volatilization from leaves and decaying tissues (Rice, 1984). In this chapter, we report on the allelopathic expression and functional performance of leaves and roots of the tamarind tree, and seed and seed-coat of tamarind fruits under both laboratory and greenhouse conditions.

Compared to the corresponding controls, the tamarind leaf caused significant growth inhibition in all the 14 tested species with the highest value in barnyard grass (root growth 79% and shoot 75%) and the lowest in Welsh onion (root growth 36% and shoot growth 33%), as demonstrated by the sandwich method (Fig. 2). Edible crop species were less affected than the weed species. A similar trend was also observed when the crude water-soluble extracts of tamarind leaves at 3-different concentrations (w/v) were applied on the tested species (Table 1). Numerous works have reported the reduction of growth and germination of different plant species when aqueous extracts of other plants were applied (Riffle *et al.*, 1988, 1990, Sahid and Sugau 1993, Russo *et al.*, 1997).

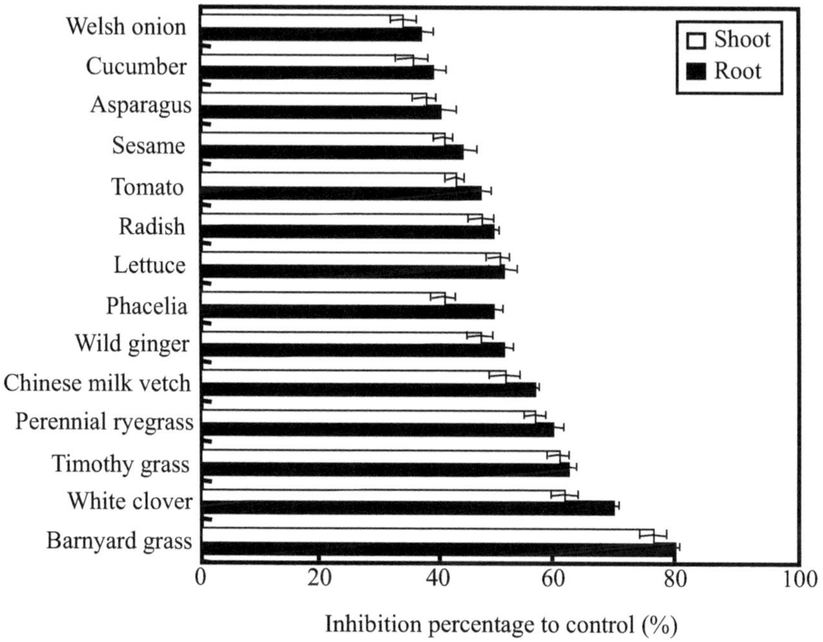

Fig. 2 Effect of the tamarind leaf on the inhibition of root and shoot growth in several weed and edible crop species as assayed by a sandwich method. The inhibition is expressed as a percentage to the control. Bars indicate standard errors.

Table 1 Effect of the tamarind leaf crude water-soluble extracts at three-different concentration (1%, 5% and 10%) (w/v) on the inhibition percentage to control (%) of the root and shoot growth in several weed and edible crop species ± indicates standard.

Family	*Scientific name*	*Common name*	*Inhibition percentage to control at different concentrations (w/v) of tamarind leaf extracts*					
			1%		*5%*		*10%*	
			Root	*Shoot*	*Root*	*Shoot*	*Root*	*Shoot*
Gramineae	*Echinochloa crus-galli* Mill.	Barnyard grass	40.1±1.5	38.2±2.5	65.6±2.2	62.3±1.9	80.3±4.1	76.5±3.2
	Lolium perenne L	Perennial ryegrass	33.1±1.5	31.4±2.1	52.8±1.8	49.5±3.8	62.3±3.0	59.9±2.9
	Pheleum pratense L	Timothy grass	35.5±3.2	33.0±1.8	58.4±2.6	56.1±2.0	66.5±3.4	62.4±3.1
Leguminosae	*Trifolium repense* L.	White clover	37.2±2.2	35.1±2.9	60.1±2.6	59.0±1.5	74.2±1.5	70.1±2.2
	Astragalus sinicus L	Chinese milk vetch	31.2±2.4	29.8±2.7	48.7±1.3	42.3±1.9	60.5±3.4	55.3±4.1
Aristo chiaceae	*Asarum siebolidii* Miq.	Wild ginger	29.8±2.5	26.1±2.5	42.7±3.5	37.2±4.1	56.5±4.1	50.3±2.4
Hydrophyllaceae	*Phacelia spp.* L	Phaceslia	26.5±2.4	24.2±1.8	39.3±1.7	31.5±2.6	51.3±2.8	40.1±3.3
Compositae	*Lacututuca sativa* L	Lettuce	34.2±2.8	30.1±1.9	46.4±1.7	39.1±2.8	59.5±2.4	54.3±3.0
Brasscaese	*Raphanus ssativus* L	Sesame	32.6±4.2	28.4±1.5	43.7±4.2	35.4±2.7	56.1±2.0	52.0±2.8
Solanaceae	*Lycopersicum esculentum* L	Tomato	30.5±1.6	26.8±3.5	40.3±2.6	32.3±3.3	52.3±4.1	50.1±2.7
Pedaliaceae	*Sesamum indicum* L	Sesame	28.6±4.3	25.0±2.8	37.3±1.8	30.0±1.5	50.1±2.0	46.3±3.2
Cucurbitaceae	*Cucumis sativus* L	Cucumber	24.8±1.8	22.4±2.8	33.4±2.9	26.1±3.0	45.6±3.2	38.0±1.9
Liliaceae	*Aspagus officinalis* L	Asparagus	26.3±3.8	23.5±2.5	35.1±2.9	28.2±1.8	48.3±2.5	40.0±2.1
	Allumfirstulosum L	Welsh onion	22.5±1.9	21.9±2.7	31.3±2.4	2.4.5±2.4	41.9±2.5	36.3±2.7

We observed that weed species were more susceptible than edible crop species to the allelopathic activity of tamarind leaves. Therefore, in our further analysis, we selected four plant species based on the severity of growth inhibition to explore the effect of tamarind leaves on the physiological growth of both weeds and edible crops species. In the pot culture experimentation with blended tamarind leaves at three-different concentrations (1%, 5% and 10%) (w/w), the growth of both roots and shoots were inhibited and the inhibitory magnitude increased proportionately with an increase in the concentration (w/w) of tamarind leaves (Fig. 3A). In addition, a strong inhibition of seedling growth occurred (20-60% of the control) when all species were initially grown for 1 wk with incorporated tamarind leaves (data not shown). All seedlings recovered and grew well when blended tamarind leaves were removed from the growth medium. The growth recovery was 76-97% of the control in roots and shoots (Fig. 3B) coinciding with a similar growth recovery in several crop and weed species in recent findings when leaves of the red pine tree were removed from the growth medium (Matsumoto *et al.*, 1999). Further confirmation of this idea was carried out by growing all the tested species in vermiculite containing 80% MeOH extracted oven-dried blended tamarind leaves at different concentrations. The growth of treated seedlings was identical to the corresponding controls (data not shown). These results clearly suggest that tamarind leaves contain biologically active allelopathic compound(s) and are true growth regulators. Thus,

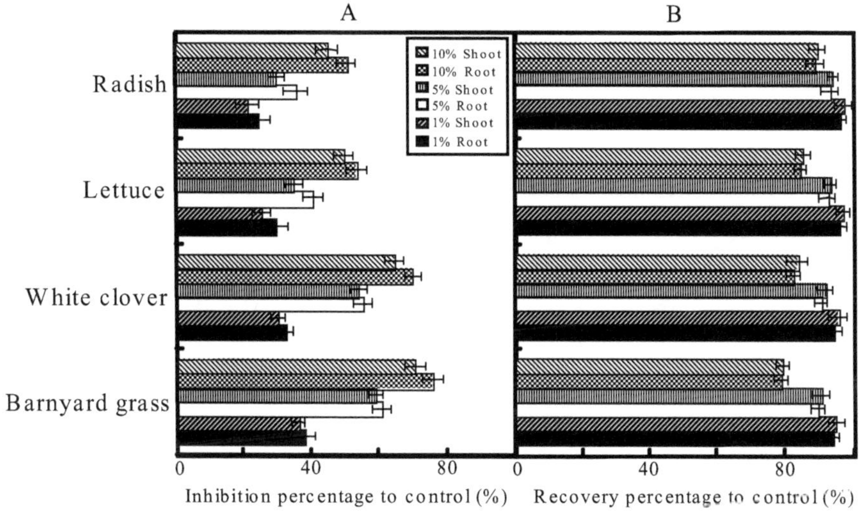

Fig. 3 Effects of addition (A) and removal (B) of blended tamarind leaves in pot culture experiments on the growth of root and shoot growth of two weed (barnyard grass and white clover) and two edible crop (lettuce and radish) species. Bars indicate standard errors.

species tested can recover, at least in the seedling stage, upon removal of incorporated leaves or elimination of existing organic chemical(s) in the leaves from the growth medium.

As demonstrated in the plant-box method experiments, the growth of tested plant species was inhibited by root exudates from tamarind seedlings and the magnitude of growth inhibition was inversely proportional to the distance of seedlings from the tamarind roots (Fig. 4). The accelerated growth inhibition of species tested occurred with an increase in the age of the tamarind seedlings used (data not shown) coinciding with the fact that inhibitory substances exuded from the tamarind roots at a very early stage (5-DAG) of tamarind seedling growth and continued for a long duration during plant growth as reported by Mandal (2002) and Mandal and Tapaswi (1997). This may be due to increment root weights and the amount of root exudation from aging tamarind seedlings as shown by the high correlation coefficient (> 0.92, $P = < 0.01$). Some reports claim that exudation tends to increase with plant growth (Rovira, 1959, 1969; Vancura, 1967; Rice, 1984; Yu *et al.*, 2000).

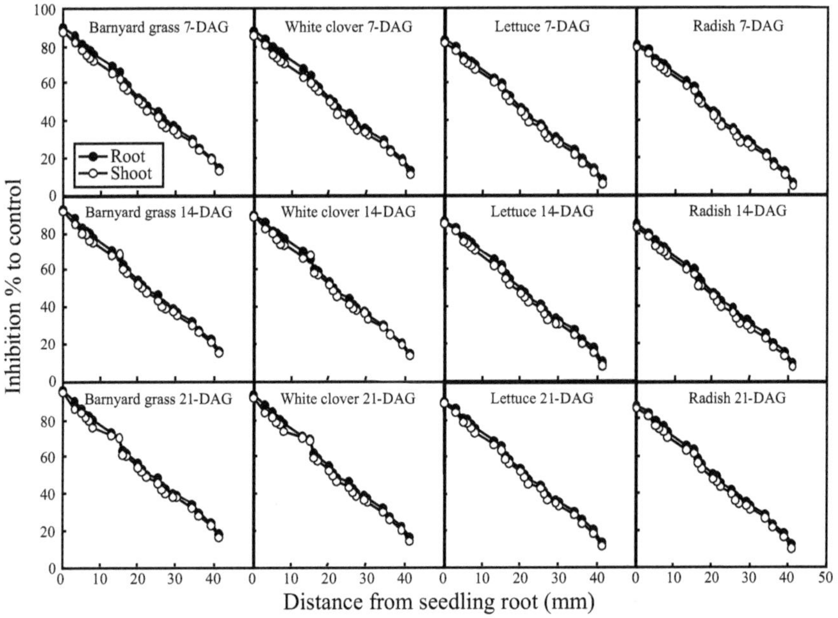

Fig. 4 Effect of tamarind seedlings of different ages on the growth inhibition of the root and shoot of weed and edible crop species in the plant-box experiment. DAG – days after germination.

Significant growth inhibition of the tested species occurred when seedlings were grown in the soil under the tamarind tree but not under the control trees (Fig. 5A). The percentage of growth reduction was higher in weed species, particularly in barnyard grass, than in the edible crops. Numerous reports showed that it is difficult to identify allelopathic events under field conditions (Bonner, 1950; Kommedahl *et al.*, 1959; Rice, 1965; Horsley, 1977; Putnam and Duke, 1978; Drost and Doll, 1980; Stachon and Zimdhal, 1980; Sahid and Sugau, 1993). However, we were successful in elucidating the inhibitory effects of tamarind roots on the growth of 4 species in the greenhouse tests. Furthermore, it was observed that the growth of plant species in the soil where there was no vegetation in the same greenhouse was parallel to those of the control treatments (data not shown). It is widely believed that root exudates and root residues are two main sources of allelopathic compounds present in soil (Yu *et al.*, 2000). Research efforts on the elucidation of the possible role for soil microbes and the effect of soil texture are sparse, even though they could be considered to be associated with the allelopathic phenomenon under field conditions. Therefore, to eliminate the possible involvement of microbes and the influence of soil texture, we carried out bioassay tests of all the plant species by the "soil-agar sandwich method" using autoclaved soils obtained from under the *Tamarindus indica* L., *Adenanthera pavonina* L. and *Mangifera indica* L. trees in a greenhouse. *Adenanthera* and *Mangifera* trees were considered as control trees and the growth of tested species in the soil of *Tamarindus* tree was identical to that of control trees (Fig. 5B). Root and shoot growth was significantly inhibited (66.9-76.1% to controls) in the tested species in the soil from the *Tamarindus* tree (Fig. 5B), and the magnitude of growth inhibition was lowered by 18.4-22.0% of the natural soil in the greenhouse (for comparison, see Figs. 5A and 5B). In addition, using both commercially available and greenhouse (where there was no vegetation) autoclaved soil, results were obtained which were identical (data not shown) to those using autoclaved control soil, as found in Fig. 5B. Soil microbes and soil texture are responsible for increasing or decreasing allelopathic effects under field conditions (Bonner, 1950; Kommedahl *et al.*, 1959; Rice, 1965; Stachon and Zimdhal, 1980). Our results clearly showed that the effects of soil microbes and soil texture were associated with an approximate 20% increase in the allelopathic value of tamarind root exudates in natural soil conditions. These results clearly suggested that tamarind root exudates are potent allelochemical(s) and are involved in a weed-free surrounding around the tamarind tree trunk.

The effects of tamarind seed exudates on the growth of several weed and edible crop species are presented in Fig. 6. Interestingly, the growth inhibition in both root and shoot of 11 species occurred while three species (lettuce, radish and sesame) showed stimulatory expression due to the

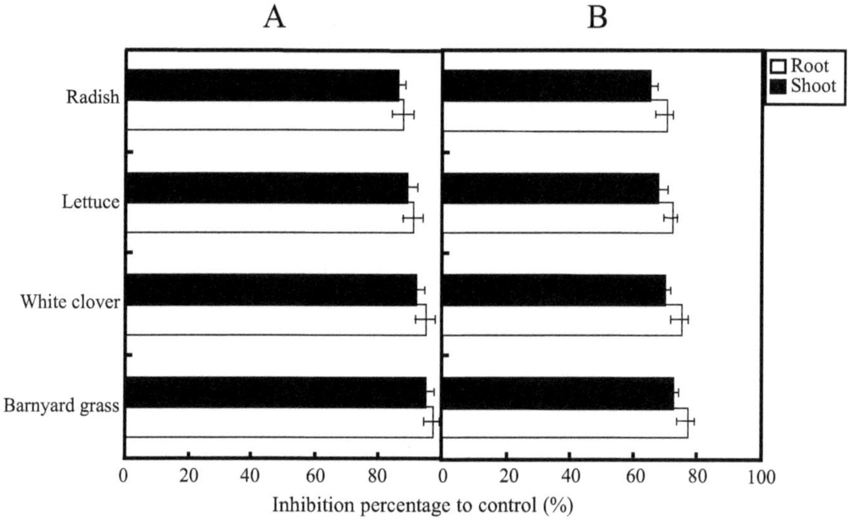

Fig. 5 Effect of the tamarind root on the growth inhibition of the root and shoot in weed and edible crop species using soils under the tamarind tree and control trees; (A) under the natural soils in greenhouse, and (B) autoclaved soils in the laboratory using the soil-agar sandwich method (see materials and methods). The growth inhibition is expressed as a percentage to the control. Bars indicate standard errors.

influence of the applied seed exudates. These results clearly indicated that these exudates have plant-specific influence on weeds (only inhibitory) and edible crops (both inhibitory and stimulatory) species coinciding with similar plant-specific inhibitory and stimulatory results exhibited by the seed exudates of watermelon (Kushima *et al.*, 1998).

Based on the findings in Fig. 6, we have been prompted to elucidate the biological activity of the crude water-soluble extracts of tamarind seed-coat on some selective plant species in the pot culture experiments. Fig. 7A-F represents the effects of spraying crude water-soluble extracts of tamarind seed-coat at three-different concentrations (1%, 5% and 10%) (w/v) on the growth of six tested species (lettuce, radish, tomato, cucumber, barnyard grass and white clover) in the pot culture experiments.

The growth of two weeds (barnyard grass and white clover) and two edible crops (tomato and cucumber) species was severely inhibited (29.7-61.4%) (Fig. 7A) while two other edible crops species (lettuce and radish) showed strong growth promotion (32.5-62.5%) (Fig. 7B).

After turning off the spraying of seed-coat crude extracts and grew up the transplanted seedlings for four more days in new pot filled with vermiculite, both the weed (barnyard grass and white clover) and edible

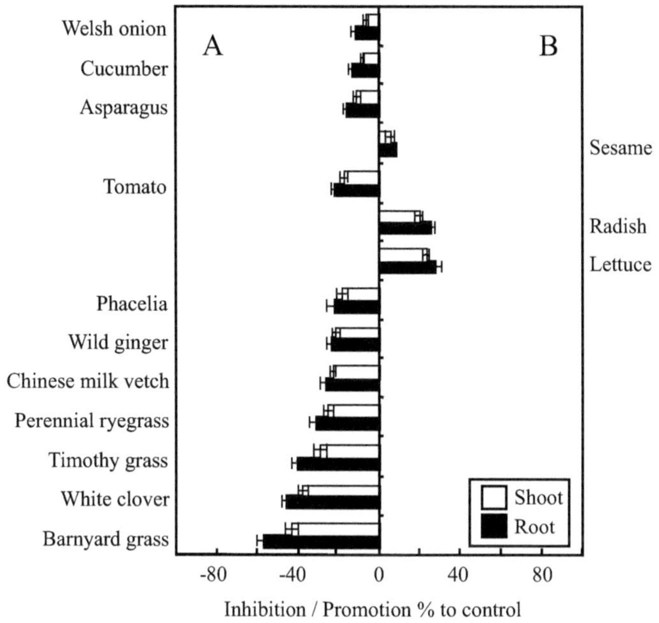

Fig. 6 Growth inhibition (A) and promotion (B) percentage to control (%) of the root and shoot in several weed and edible crop species as affected by tamarind seed exudates. Bars indicate standard errors.

crops (tomato and cucumber) species recovered well (78.1-82.6%) (Fig. 7C); although, the growth of other two edible crops species (lettuce and radish) continued to be stimulated by approximately 4.1-7.9% (Fig. 7D).

From these results, it is evident that the crude water-soluble seed-coat extracts of tamarind seeds possess certain biologically active true growth regulator(s), thus, either by turning off the spraying schedule and transplanting of affected seedlings into a new growth medium could make them recover well (Fig. 7C), as observed in the case of tomato, cucumber, barnyard grass and white clover; or once the stimulatory activity begins within the plant body as a result of the effect of existing chemical compound(s) in the extracts it probably induces further stimulation/ biosynthesis for continued enhanced growth as was observed in lettuce and radish (Fig. 7D). At this point, we continued our experimentatto understand how the seedling reacts to further application of seed-coat extracts spraying: the two weeds and four edible crops species (as before) exhibited reduction in growth (Fig. 7E) at the on-set of spraying of seed-coat extracts again while the other two edible crops species (lettuce and radish) displayed stimulatory expression (Fig. 7F). As described earlier, a similar recovery in growth upon removal of the incorporated tamarind

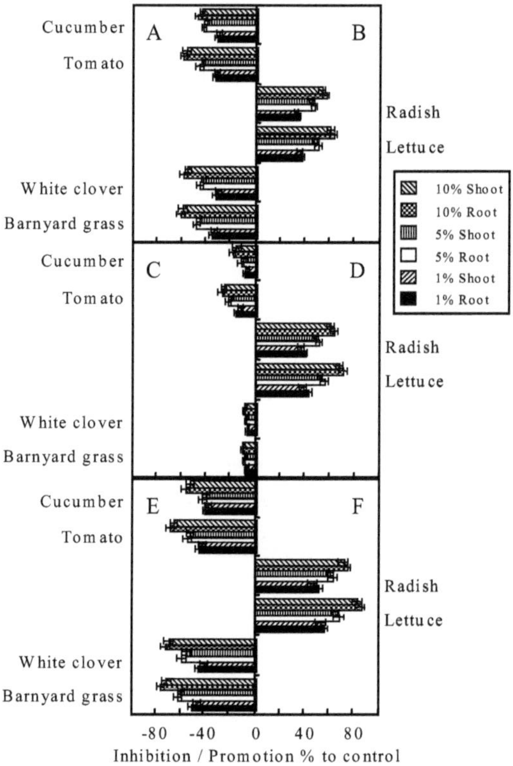

Fig. 7 Effects of crude-water soluble extracts of seed-coat of tamarind at three-different concentrations (1%, 5% and 10%) (w/v) on the growth of several agronomic crop (lettuce, radish, tomato and cucumber) and weed (barnyard grass and white clover) species in the pot culture experiments. Growth inhibition (A, C and E) and promotion (B, D and F) percentage to the control (%) of the root and shoot as affected by the onset of crude-water soluble extracts spraying (A-B), turning off the spraying (C-D) and again onset of spraying (E-F). Bars indicate standard errors.

leaf from the growth medium was found. The findings of our research is very interesting, particularly the generation of further inhibitory or stimulatory growth in the tested species due to the spraying onset once after turning off spraying and no reports of this phenomenon in any plant species has been previously reported elsewhere.

CONCLUSION

Recent research efforts with improved analytical methods have shown that the allelopathy phenomenon is operative in plant ecosystems (Rice, 1964; Muller and Moral, 1966; Rice and Parenti, 1967; Whittaker, 1971;

Tinnian and Muller, 1972; Stachon and Zimdhal, 1980). Researchers and crop growers rely on the biochemical activities and events of released allelochemicals through allelopathy because they provide an alternative to synthetic herbicides, insecticides and nematocides in agro-ecosystems. Additionally, in order to reduce environmental pollution and increase agricultural production for sustainable agriculture, allelopathy research in various aspects has been applied to the fields of agriculture and forestry (Qasem and Foy, 2001).

In this chapter, we demonstrated that both the leaves and roots of the tamarind tree, as well as seed and seed-coat of tamarind fruits contain biologically active allelopathic compound(s) and are true growth regulators. Therefore, all these findings could be exploited for weed control as bio-herbicides or for growth stimulators in crop plants as a bio-fertilizer during crop husbandry. Tamarind and its various parts are natural products that are easily bio-degraded and the use of its various parts (such as, the leaf, seed, etc.) are sure to play important roles for an eco-friendly environment, coupled with a decreased reliance on synthetic products (herbicide, fertilizer, etc.) for crop husbandry, thus, reducing human health hazards. Identification and characterization of the chemical compound(s) involved in the allelopathic expression and functional performance of various parts of the tamarind tree would be of interest for work in the future.

Acknowledgment

A part of the data presented here was previously published in Plant Growth Regulation, and we thank Springer (www.springeronline.com) for their generous permission of reproducing it here.

References

Altieri, M.A., Doll, J.D. (1978) The potential of Allelopathy as a tool for weed management in field crops. Pest Artic. News Summ. (PANS). 24: 495-502.

An, M., Pratley, J.E., Haig, T. (1998) Allelopathy: from concept to reality. *In* the Proceedings of the 9th Australian Agronomy Conference, D.L. Michalk., J.E. Pratley (eds). Australian Agronomy Society, Wagga Wagga, Australia, pp. 563-566.

Bell, D.T., Koeppe, D.E. (1972) Noncompetitive effects of giant foxtail on the growth of corn. Agron. J. 64: 321-325.

Bonner, J. (1950) The role of toxic substances in the interaction of higher plants. Bot. Rev. 16: 51-65.

Drost, D.C., Doll, J.D. (1980) The allelopathic effect of yellow nutsedge (*Cyperus esculentus*) on corn (*Zea mays*) and soybeans (*Glycine max*). Weed Sci. 28: 229-233.

Fujii, Y. (1992) The potential biological control of paddy weeds with Allelopathy (allelopathic effects of some rice varieties). *In* Proceedings. Int. Symp. Biological Control and Integrated Management of Paddy and Aaquatic Weeds in Asia, Natl. Agric. Res.

Cent., Tsukuba, Japan. Food and Fert. Tech. Cent. for the Asian and Pacific region, China. pp. 305-320.

Fujii, Y., Shibuya, T., Yasuda, T. (1992) Allelopathy of velvetbean: Its discrimination and identification of L-DOPA as a candidate of allelopathic substances. JARQ. 25: 238-247.

Fujii, Y., Shibuya, T., Nakatani, K., Itani, T., Hiradate, S., Parvez, M.M. (2003) Assessment method for allelopathic effect from leaf litter leachates. Weed Biol. Manage. 4: 19-23.

Gressel, J.B., Holm, L.G. (1964) Chemical inhibition of crop germination by weed seed and the nature of the innhibition by *Abunlon theophrasii*. Weed Res. 4: 44-53.

Horsley, S.B. (1977) Allelopathic interference among plants. II. Physiological Modes of Action. *In* Proceedings of the Fourth North American Forestry Biology Workshop, Syracuse University Press, Syracuse, NY, USA, pp. 39-136.

Kommedahl, T., Kotheimer, J.B., Bernardine J.V. (1959) The effects of quackgrass on germination and seedling development of certain crop plants. Weeds. 7: 1-12.

Kushima, M., Kakuta, H., Kosemura, S., Yamamura, S., Yamada, K., Yokotani-Tomita, K., Hasegawa, K. (1998) An allelopathic substance exuded from germinating watermelon seeds. Plant Growth Regul. 25: 1-4.

Leather, G.R. (1983) Weed control using allelopathic crop plants. J. Chem. Ecol. 9: 983-990.

Mandal, S., Tapaswi, P.K. (1997) Allelopathic agents in *Tamarindus indica*. Indian Biol. 29: 31-35.

Mandal, S. (2002) Role of *Tamarindus* seed coat extract as a biofertilizer. *In* Prococeedings Of the 3rd World Conference on Allelopathy. pp. 137.

Matsumoto, N., Gemma, H., Nakatani, K., Fujii Y. (1999) Allelopathy of leaves from red pine tree and application for weed control in orchard garden. J. of Weed Sci. Tech. 44: 184-185.

Morton, J.F. (1987) Tamarind. *In* Fruits of the Warm Climates C.F. Dowling (ed). Morton J.F., 20534 SW 92 Ct. Miami, FL, USA. pp. 115-121.

Muller, C.H. and Moral, R.D. 1966. Soil toxicity induced by terpenes from *Salvia leucophylla*. Bull. Torrey Bot. Club. 93: 130-137.

Putnam, A.R., Duke, W.B. (1974) Biological suppression of weeds: Evidence for allelopathy in accessions of cucumber. Science. 185: 370-372.

Putnam, A.R., Duke W.B. (1978) Biological suppression of weeds: Evidence for allelopathy in accessions of cucumber. Science. 185: 370-372.

Putnam, A.R., Defrank, J., Barnes, J.P. (1983) Exploitation of Allelopathy for weed control in annual and perennial cropping systems. J. Chem. Ecol. 9: 1001-1011.

Qasem J.R., Foy, C.L. 2001. Chemical weed control and response to competition in majoram (*Origanum syriacum*). *In* the Proceedings of the Southern Weed Science Society. 54: 221.

Rice, E.L. (1964) Inhibition of nitrogen-fixing and nitrifying bacteria by seed plants. I. Ecology. 45: 824-837.

Rice, E.L. (1965) Inhibition of nitrogen-fixing and nitrifying bacteria by seed plants. II. Characterization and identification of inhibitors. Physiol. Plant. 18: 255-268.

Rice, E.L., Parenti, R.L. (1967) Inhibition of nitrogen-fixing bacteria by seed plants. V. Inhibitors produced by *Bromus japonicus* Thumb. Southwest Natur. 12: 97-103.

Rice, E.L. (1984) Allelopathy, 2nd Ed., Academic Press, NY, USA.

Riffle, M.S., Thilsted, W.E., Murray, D.S., Ahring, R.M., Waller, G.R. (1988) Germination and seed production of unicorn-plant (*Proboscidea louisianica*). Weed Sci. 36: 787-791.

Riffle, M.S., Waller, G.R., Murray, D.S., Sgaramello, R.P. (1990) Devil's-claw (*Proboscidea louisianica*), essential oil and its components : Potential allelochemical agents on cotton and wheat. J. Chem. Ecol. 16: 1927-1940.

Rovira, A.D. (1959) Root excretions in relation to the rhizosphere effect. IV. Influence of plant species, age of the plant, light, temperature and calcium nutrition on exudation. Plant Soil. 1: 53-64.

Rovira, A.D. (1969) Plant root exudates. Bot. Rev. 35: 35-57.

Russo, V.M., Webber, C.L., Myers, D.L. (1997) Kenaf extract affects germination and post-germination development of weed, grass and vegetable seeds. Indus. Crops Produc. 6: 59-69.

Sahid, I.B, Sugau, J.B. (1993) Allelopathic effect of lantana (*Lantana camara*) and siam weed (*Chromolaena odorata*) on selected crops. Weed Sci. 41: 303-308.

Stachon, W.J., Zimdhal, R.L. (1980) Allelopathic activity of Canada thistle (*Cirsium arvense*). Weed Sci. 28: 83-86.

Tinnian, R.O., Muller, C.H. (1972) The allelopahtic influences of *Avena fatuca* : The allelopathic mechanism. Bull. Torrey Bot. Club. 99: 287-292.

Tukey, H.B.J. (1969) Implications of Allelopathy in agricultural plant science, Bot. Rev. 35: 1-16.

Vancura, V. (1967) Root exudates of plants. III. Effect of temperature and 'clod shock' on the exudation of various compounds from seeds and seedlings of maize and cucumber. Plant Soil. 27: 319-327.

Weston, L.A. (1996) Utilization of Allelopathy for weed management in agroecosystems. Agron. J. 88: 860-866.

Whittaker, R.H. (1971) The chemistry of communities. *In* Biochemical Interactions Among Plants. Environmental Physiology. Subcomm. US National Comm., IBP, Nat. Acad. Sci., Washington, DC, USA, pp. 10-18.

Wu, H., Pratley, J., Lemerle, D., Haig, T. (2000) Evaluation of seedling allelopathy in 453 wheat (*Triticum aestivum*) accessions by Equal-Compartment-Agar-Method. Aust. J. Agric. Res. 51: 937-944.

Yu, J.Q., Shou, S.Y., Qian, Y.R., Zhu, Z.J., Hu, W.H. (2000) Autotoxic potential of cucurbit crops. Plant Soil. 223: 147-151.

20

Influence of Water Extract from *Uncaria tomentosa* Bark on Ultrastructure of *Capsicum*

Teresa Tykarska[1], Alicja Zobel[3], Julita Nowakowska[1], Krzysztof Gulewicz[2] and Mieczysław Kuvaś[1]

[1]*Department of Ecotoxicology, Warsaw University, Miecznikowa str. l, 02-096 Warsaw, Poland*

[2]*Laboratory of Phytochemistry, Institute of Bioorganic Chemistry PAN, Z. Noskowskiego str.12/14, 61-704 Poznañ, Poland*

[3]*Trent University, Peterborough, Ontario, K9J 7B8 Canada*

Potted pepper plants were treated with *Uncaria tomentosa* bark extract applied to the soil @ 0.4 mg/ml or 1.6 mg/ml. The plants were grown under this condition for next 3 wk and then examined with electron microscopy methods to analyze the reaction of the first small leaf at the top of the pepper plant, then the one in the center of the plant and the lowest leaf at the bottom. During ontogenesis of control plants, black deposits in leaf vacuoles increased in number forming larger spheres. Nuclear chromatin gradually became slightly condensed. Chloroplasts contained more plastoglobuli but the number of starch grains decreased a little. Plastoglobuli in older leaves were extruded from chloroplasts. Large spherical deposits, dark droplets and opalescent crystal in the middle layer of mesophyll, which were visible in semithin preparations of control plants, disappeared after the treatment. Nuclei showed very strong condensation of chromatin. Chloroplasts showed degradation of grana and dilatation of thylakoids appearance as early as in youngest leaves. Plastoglobuli in plastids were enlarged. They were less homogenous and did not get expelled from the chloroplasts. Starch grains declined. Vesiculation of the ground cytoplasm was much stronger than in the control. Much more multivesicular bodies appeared, invigilating into the central vacuole. These observations suggested that *Uncaria tomentosa* extract enhanced natural ontogenesis until senescence of *Capsicum* leaf and additionally caused changes in cell structure and physiology.

Key words: *allelochemicals, Uncaria tomentosa, Capsicum leaves, ultrastructure*

INTRODUCTION

Uncaria tomentosa is a liana growing in the jungles of South and Central America. Owing to the content of numerous biologically-active substances such as alkaloids (Laus, 1997) pvoanthocyanidins (Motegro de Matta, 1976), many sterols (Senatore, 1989), quinovic acid glycosides and numerous triterpenes (Aquino, 1997). *Uncaria tomentosa* seems to offer a wide range of applications in the treatment of various diseases, including cancer (Keplinger, 1982). Thus it might also influence small plants growing in its vicinity.

Capsicum annuum (pepper) is an agricultural crop important both in countries of warm climatic zones and those of cold climates, where it is grown in greenhouses. Numerous alkaloids, carotene and flavonoid dyes, sugars, vitamins, ethereal oils and other compounds were found in plants of this species (Somos, 1984). It is very often used as a model system for analysis of influence of different environmental conditions and changes made by humans by genetical, cytological, ultrastructural evaluations.

Allelochemicals are compounds synthesized by a plant affecting another plant in its neighborhood. These compounds can be released directly by roots, rinsed off by rain from the surface of leaves, or can come from decaying leaves which have fallen down from the plant. Also whole plants can be dug in the ground by animals, so that leaves of different sizes and age could release the compounds into the soil. If taken up by roots, such compounds can rapidly immigrate to the shoot apex, possibly affecting all the leaves and changing their physiology and structure, leading to retardation or stimulation of growth and even to death. We wanted to investigate whether extract from leaves of *U. tomentosa* after a single application into the soil affects ultrastructure of cells in *Capsicum* leaves located at different levels above the ground.

MATERIAL AND METHODS

The experiment was carried out using pepper plants (*Capsicum annuum* L.) var. KoŸlak. The plants were cultivated in pots with garden soil kept in a greenhouse, under natural light and temperature conditions. After 3 month of culture, just before blossoming, a group of six plants was treated with 200 ml of *Uncaria tomentosa* bark extract at the concentration of 0.4 mg/ml; another group was treated with the same amount of the extract at the concentration of 1.6 mg/ml, and the third one – the control group –

was treated with 200 ml of distilled water. After 3 wk of the treatment, in each group of plants, leaves from three levels of plants were picked: from the top (a young leaf), middle (a mature leaf) and bottom of the plant (the oldest leaf). Leaf samples of 1 mm^2 cut from the middle part of the leaf blade were fixed in 2.5% glutaraldehyde at pH 7.2 (0.1 M cacodylate buffer) with 0,1% caffeine (added to precipitate some phenolics, according to Mueller and Greenwood, 1978) for 2 h, and then post-fixed in 1% Osmium tevoxide (OsO_4) for 12 h. The material was dehydrated in ethanol and embedded, through propylene oxide, in Epon/Spurr mixture. Semi-thin sections (about 0.5 μm thick) were cut with LKB ultramicrotome, stained with 1% toluidine blue in 1% borax, observed in a NU ZEISS light microscope and photographed by the standard method with a photographic camera. Ultra-thin sections (60-90 nm thick) were contrasted with uranyl acetate and lead citrate according to Reynolds (1963). Observations and electronograms were made in a JEOL JEM 100C transmission electron microscope. The object of the observations was ontogenesis of the leaf cell structure in control plants and in plants after a single treatment, and a comparison between them.

RESULTS

The pepper plants were retardant in growth after the treatment. Flowering and fruiting were enhanced and young fruits fell faster. The leaves were a bit smaller (Fig. lB) in comparison to the control (Fig. lA), and yellow edges appeared in treated leaves, whereas in the control only small parts at the very edge of the oldest leaf were affected.

The thickness of leaves was similar in the control and treated plants (Fig. lC, D). In semi-thin sections, the main difference was related to dark deposits. They occurred in control leaves mesophyll (palisade parenchyma - Fig. 1C, white arrowhead; and spongy parenchyma - Fig. 1C, black arrowhead) and both upper epidermis (white double arrowhead - Fig. 1C) and lower epidermis (black double arrowhead - Fig. 1C). After the treatment, these deposits were not observed (Fig. 1D). The second visible difference was the disappearance of crystal deposits which were observed in the spongy parenchyma layer directly adjacent to the palisade layer of control plant leaves (Fig. lC, white and black arrows, and Fig. 1D - lack of crystals). The mesophyll cells were slightly deformed after the treatment (Fig. 1D, arrows).

In electron microscope observations, the main changes in ultrastructure after the treatment with *U. tomentosa* extract were: disappearance of large, spherical, dark deposits in vacuoles (Figs. 2A and 3A), condensation of euchromatin to heterochromatin, enlargement of plastoglobuli,

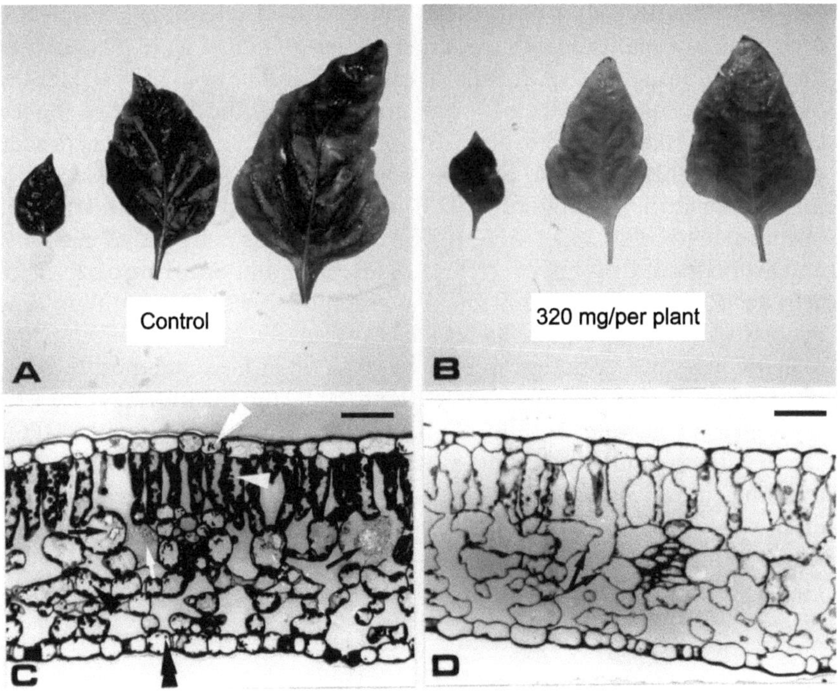

Fig. 1 Morphology of leaves of a control plant (A) and a plant treated with the extract at the concentration of 1.6 mg/ml (B), at three developmental stages: 1 – a young leaf (the third one from the top of the plant), 2 – a mature leaf (the middle part of the plant), 3 – an old leaf (the oldest one at the plant). Leaves of a treated plant are smaller; the mature and old leaves are decolorized, especially at the borders and top of the leaf blade. Trans-sections of an old control leaf (C) and a leaf treated with *U. tomentosa* extract at the concentration of 1.6 mg/ml (D). Semi-thin sections, stained with 1% toluidine blue, observed with a light microscope. Bars = 20 mm.

condensation of mitochondrial matrix and progressing development of the system of endomembranes, forming numerous multivesicular bodies.

In the control, the process of leaf ontogenesis (as observed in ultrastructure) from maturation to senescence was distinct when leaves were analyzed, starting from the top to the bottom of each plant: the highest (Fig. 2B), central (Fig. 2C, D) and bottom leaf (Fig. 3E, F).The number of dark deposits in vacuoles increased from the top to the bottom.

The degree of condensation of nuclear chromatin changed progressively from its dispersion to small concentrations (Fig. 3E - black arrows).

The structure of chloroplasts in developing leaves changed as well, in respect of grana, thylakoids, starch grains and plastoglobuli. In the top leaves, many starch grains, distinguished thylakoids and grana (Fig. 2B), as well as small plastoglobuli (Fig. 2B, white arrowheads) were observed. The centrally located leaf contained larger plastoglobuli (Fig. 1D – arrows). Plastoglobuli in the central leaf often were pushed to the edge of the chloroplast deforming its membrane (**pg** – Fig. 2C). Some of them were extruded from chloroplasts. Many chloroplasts additionally contained long paracrystalline structures (**pc** – Fig. 2C), similar to the structure of a chromoplast, which drastically changed the shape of the chloroplasts (Fig. 2C, white arrow). Such structures were only very sporadically observed in the top and bottom leaves. In the oldest leaves, only slight dilatation of thylakoids were visible (Fig. 2E, thin black arrow), some starch grains disappeared (Fig. 2E, black arrowheads), and plastoglobuli were expelled out of the chloroplasts more often than in middle leaves (Fig. 2F, black arrows–**pg**). Lipid-like homogeneous matter (probably coming from plastoglobuli), often containing electron-lucent small "vacuoles", was usually found within the cytoplasm in the vicinity of chloroplasts (**ll** – Fig. 2E).

At all levels, mitochondria had numerous cristae and rather dense matrix (**M** – Figs. 2C, D). In the oldest leaves, cristae were dilated (Fig. 2E, white arrow) and mitochondrial matrix was condensed. At all developmental stages of the control leaves, many peroxisomes containing an electron-dense paracrystalline core were observed (**Pe** – Figs. 2B, D). The endomembrane system ontogenesis, their differentiation and degradation could be described here as changes from small vesicles (**v**) and concentric heterostructural multivesicular bodies (**mv**) and membrane structures (**ms**), visible in the top and middle leaves (Fig. 2B, C, D), to agglomeration of many separated membranes which looked like having undergone lysis of their interior. Conglomerations of several vesicles and membranes of different size and shape were pushed into the central vacuole (**mv**–Fig. 2D). In the bottom leaves, conglomerations of larger structures were observed at the edge of the central vacuole, resembling a system of many neighboring vacuoles (**v**) which had not fused (Fig. 2F). Sometimes electron-dense deposits were visible among them (**d** – Fig. 2C, **sd** – Fig. 2F). The treatment enhanced this process of ontogenesis and added extra changes. In treated cells, the nuclei contained large centers of very condensed chromatin (Fig. 3E, arrows), which in the case of treatment with both concentrations was observed as early as in the top leaves.

Chloroplasts showed structural changes suggesting acceleration of their ontogenesis. Their thylakoids, grana, starch and plastoglobuli structure indicated degradation. Thylakoids underwent gradual dilatation starting from the earliest stages of development (Fig. 3B, C, E – white arrowheads). In the oldest leaves, grana showed very strong changes or sometimes were

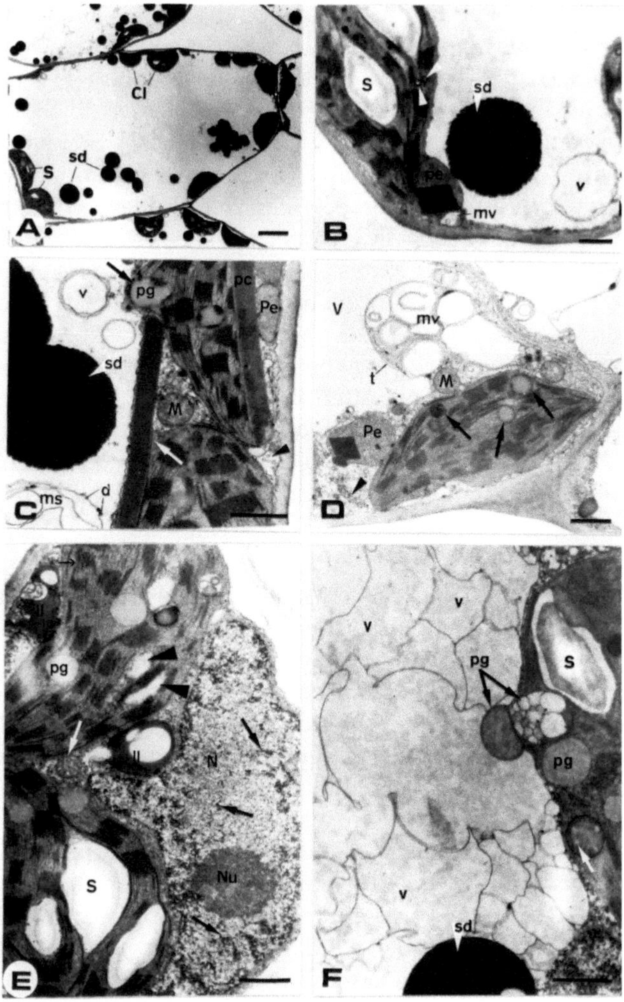

Fig. 2 Ultrastructure of a control plant leaf cells. Ultrathin sections, contrasted according to Reynolds. Palisade parenchyma cells of an old leaf (A). The huge central vacuole contains spherical electron-dense deposits (**sd**); chloroplasts, located adjacent to cell walls (**Ch**) contain single small starch grains (**S**). Bar = 5 μm. A portion of a spongy parenchyma cell of a young leaf (B). The chloroplast contains numerous, high grana, intergranal thylakoids, starch grains (**S**) and single, small plastoglobuli (arrowhead). Close to the chloroplast, a peroxisome (**Pe**) is visible, having a typical electron-dense paracrystalline core and a forming multivesicular body (**mv**). The central vacuole contains a spherical electron-dense deposit (**sd**) and an electron-lucent vesicle (**v**), probably excreted here by endocytosis (the black arrowhead points at a connection with the cytoplasm). Bar = 1 μm. A portion of spongy parenchyma of a mature leaf: chloroplasts containing oblong electron-dense paracrystalline bodies (**pc**), deforming them, and large plastoglobuli with dark borders (**pg**) (C). One of them

(Contd.)

not visible at all (Fig. 3E), and some starch grains shrunk or disappeared. Plastoglobuli (**pg**) were extremely enlarged (Fig. 3E). The process of "pushing" plastoglobuli out of chloroplasts or their presence in the central vacuole was not observed.

The mitochondria (**M**) in cells of treated plants showed extremely dense matrix and very distinguished cristae (Fig. 3B, C). The amount of typical peroxisomes with paracrystalline core decreased.

The endomembrane system changed throughout the ontogenesis. At the beginning, in the top leaves, it took the form of multivesicular bodies with numerous small vesicles in the ground cytoplasm (Fig. 3B – thin black arrows indicate the membrane surrounding them). Then, such vesicles, additionally enveloped with tonoplast (**t**), were "pushed into" the center of vacuole as typical multivesicular bodies (**mv** - Fig. 3D), often accompanied by ER scrolls (Fig. 2F – **ms**). Apart of them, also transparent vesicles (**v**) of different sizes were formed and gathered under the tonoplast (Fig. 3D). They were also observed in the central vacuole (**v** – Fig. 3F).

Similar but usually less pronounced changes were also observed after treatment with the lower concentration of *U. tomentosa* extract (0.4 mg/ml).

CONCLUSION

Influence of one plant on another by releasing mixtures of different compounds into soil is known as allelopathy. The second plant is usually

---

(Contd.)

presses the chloroplast envelope (thick arrow); moreover, 2 microbodies (**pe**), a mitochondrion (**M**) with numerous cristae, and a forming multivesicular body (black arrowhead) are visible within the cytoplasm; the central vacuole contains electron-empty vesicles (**v**), or membranous structures (**ms**) with small osmophilic deposits (**d**) and large spherical electron-dense deposits (**sd**). Bar = 1 μm. A portion of a spongy parenchyma cell of a mature leaf (D). A chloroplast with plastoglobuli of various contents, mitochondria (**M**), peroxisome (**Pe**) with a paracrystalline core, rER (black arrowhead) and a multivesicular structure (**mv**) covered with tonoplast (**t**), with numerous membranes, invaginating into the central vacuole (**V**). Bar = 1 μm. A portion of spongy parenchyma of an old leaf: the nucleus (**N**) with small portions of condensed chromatin (arrows) and nucleolus (**Nu**) of fine-grained structure; chloroplasts showing symptoms of thylakoid dilatation (thin arrows) and starch grain degradation (arrowheads), and plastoglobuli (**pg**) with light spots at the borders, and lipid-like bodies (**ll**), containing electron-lucent small vacuoles, outside the chloroplasts; mitochondria (**M**) with dense matrix and numerous dilated cristae (E). Bar = 1 μm. Palisade parenchyma of an old leaf: at the border of the cytoplasm, merged vesicles (**v**) with osmophilic spherical deposits stuck among them (**sd**) (F). Below, a chloroplast with numerous plastoglobuli (**pg**) of different density and size. One of them, at the border of the chloroplast, releases its gray contents, leaving light vesicles (**pg** – forked arrow). Another one contains light spots (white arrow). Bar = 1 μm.

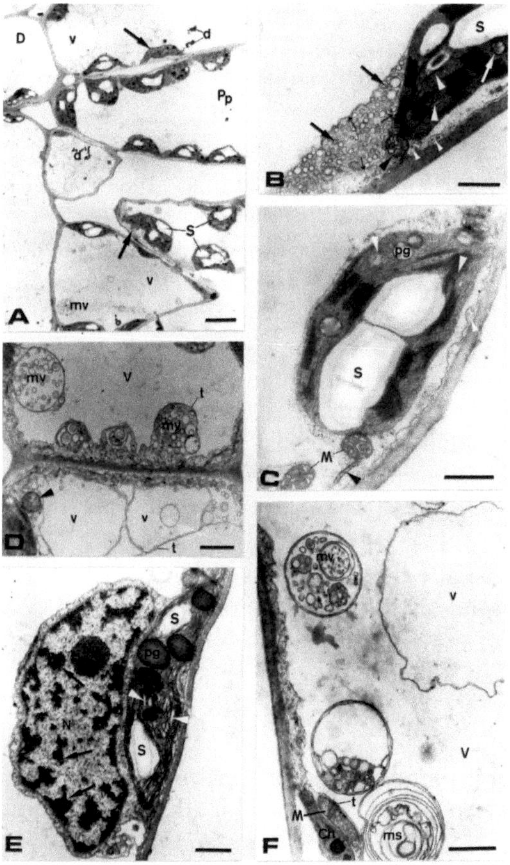

Fig. 3 Ultrastructure of pepper leaves after treatment with the extract at the concentration of 0.4 mg/ml (A-B) and 1.6 mg/ml (C – F). Mature leaf, a portion of the upper epidermis (**D**) and palisade parenchyma (**Pp**): The cytoplasm in cells forms a very thin layer adjacent to the cell wall. Chloroplasts are often flattened and deformed (arrows) (A). They contain numerous starch grains (**S**). In vacuoles, few small, irregular, electron-dense deposits (**d**) and numerous vesicular structures (**v**, **mv**) are found. Bar = 1 μm. A young leaf, a portion of spongy parenchyma (B). The cytoplasm is filled with small vesicles (arrows). Most of these are located within enveloped bodies (thin arrows indicate the membrane surrounding them). Note dilatations of intergranal thylakoids (white arrowhead). Plastoglobuli with vesicular light spots (white arrow). Mitochondrion with very dense matrix and dilated cristae (black arrowhead). Under the plasmalemma, aggregations of very small vesicles are seen (small white arrowheads). Bar = 1 μm. A young leaf, a portion of palisade parenchyma (C). Ultrastructure of mitochondria is similar as in case of lower concentration (B), chloroplasts show stronger dilatation of thylakoids (arrowheads), larger vesicles than in the previous case (Fig. B) are excreted under the plasmalemma (arrowhead). rER – black arrowhead. Bar = 1 μm. A mature leaf, a portion of palisade parenchyma: in the cytoplasm of the upper cell, at the border of the central vacuole, numerous multivesicular structures are formed (**mv**), pressing at the tonoplast (**t**); in the lower cell, right under the tonoplast (**t**), large electron-

(Contd.)

not passive and defends itself by adaptative changes in its structure and cell physiology, and by producing phytoalexins in its cells, which in turn may be released into the soil.

Growth of pepper plants was retarded and their fruits fell off too early as a result of single watering of the potted plant with a dose of 80 or 320 mg of the *Uncaria tomentosa* bark extract per plant. Continuos feeding might bring even more dramatic results. Speeding up of the leaf maturation and yellowing leaf edges pointed to enhanced plant ontogenesis. Over a 3-wk period, a part of the normal process, additional changes, caused specifically by the treatment, were visible. The most obvious one was the disappearance of dark deposits, most likely of phenolic compounds, precipitated with caffeine for better contrast (Mueller and Greenwood, 1978).

Vesicles and vacuoles were now electron-empty, containing no deposits precipitated with caffeine. Such change suggests diversification of treated plants defense. This was accompanied by a marked decrease in the amount of vesicles under the tonoplast and inside the central vacuole, which could contain various compounds typical of *Capsicum*. These phenomena may result from the observed changes in condensation of chromatin.

In the plants treated with the extract, the endomembrane system was changed, resulting in formation of many additional vesicles from endoplasmic reticulum, further forming numerous membranous complexes, taking form of multivesicular bodies, accumulating at the borders of the central vacuole. This suggests degradation or a change in the mechanism of biochemical synthesis. Also, the number of paracrystalline structures in peroxisomes decreased, suggesting changes in cell physiology. Enzyme production should be investigated in future by analyzing ATPases and hydroxylases responsible for processing energy status and possible lysis continuation. Schwitzgguebel and Siegenthaler (1984) explained that the observed increased number of microbodies, plastids and mitochondria with active ultrastructure is a result of photorespiration. Although the appearance of mitochondria in treated

---

(Contd.)

lucent vesicles (**v**) are located; mitochondrion (arrowhead) (D). Bar = 1 μm. An old leaf, a portion of palisade parenchyma: cell nucleus (**N**) with nucleolus (**Nu**) of granular structure, with extremely condensed chromatin (arrows) and light karyolymph (E); below – a chloroplast undergoing destruction. It contains large gray plastoglobuli (**pg**), degraded grana and dilated thylakoids (arrows) and rather large starch grains (**S**). Bar = 1 μm. An old leaf, a portion of spongy parenchyma (F). The central vacuole containing numerous complex multivesicular (**mv**) and membranous (**ms**) structures, and single vessels (**v**); a mitochondrion (**M**) with large amount of cristae, and a portion of a chloroplast (**Cl**) are visible under the tonoplast (**t**). Bar = 1 μm.

plants, even in the youngest leaves, resembled their ultrastructure in the oldest control leaves; destruction of plastids was observed very early as well. The above facts, together with high condensation of nuclear euchromatin, similar as in aging cells (Simeonowa *et al.*, 2000), lead to the conclusion that the *Uncaria tomentosa* extracts accelerate processes of aging. However, lack of black spherical deposits in the vacuoles and excessive production of vesicular structures following the treatment may indicate a possibility of variation in quality and quantity of natural products.

Connection of cytophysiological changes to *U. tomentosa* extract treatment and ultrastructural modifications will be further investigated.

References

Aquino, R., de Tommasi, N., de Simone, F., Pizza C. (1997) Triterpenes and quinovic acids glycosides from *Uncaria tomentosa*. Phytoc. Hem. 45: 1035-1040.

Keplinger, K. (1982) Cytostatic, contraceptive and antiinflammatory agents from *Uncaria tomentosa*. PTC Int. Appl., WO 8201. pp.130.

Laus, G., Brossner, D., Keplinger, K. (1997) Alkaloids of Peruvian *Uncaria tomentosa*. Phytochemistry. 45: 855-860.

Montenegro de Matta, S., Delle Monache, F., Ferrari, F., Marini-Bettolo, G.B. (1976) Alkaloids and procyanidins of an *Uncaria* sp. from Peru. Farmaco Sci. 31: 527-535.

Mueller, W.C., Greenwood, A.D. (1978) The ultrastructure of phenolic storing cells fixed with caffeine. J. Exp. Bot. 29: 757-64.

Reynolds, E.S. (1963) The use of lead citrate at high pH as electron-opaque stain for electron microscopy. J. Cell. Biol. 17: 208-213.

Schwitzgguebel, J-P., Siegenthaler, P-A. (1984) Purification of peroxisomes and mitochondria from spinach leaf by gradient centrifugation. Plant Physiol. 75: 670-674.

Senatore, A., Cataldo, A., Iaccarino, F.P., Elberti, M.G. (1989) Ricerche fitochimische e biologische sull' *Uncaria tomentosa*. Boll. Soc. Ital. Biol. Sper. 65: 517-520.

Simeonova E., Sikora A., Charzyñska, M., Mostowska, A. (2000) Aspects of programmed cell death during leaf senescence of mono- and dicotyledonous plant. Protoplasma. 214: 93-101.

Somos, A. (1984) De paprika. Akademiai Kiado, Budapest.

SECTION 7

New Field in Allelopathy (Aquatic Plants, Mushrooms, Insects, Animals)

21

Production of Allelochemicals by an Aquatic Plant, *Myriophyllum spicatum* L.

Satoshi Nakai
Graduate School of Engineering, Hiroshima University 1-4-1, Kagamiyama, Higashi-Hiroshima, Hiroshima 739-8527, Japan

Myriophyllum spicatum is highly competitive and exhibits potent growth and dispersal strategies. Therefore, *M. spicatum* has been studied to understand the mechanisms underlying its replacement of other aquatic plant species in aquatic ecosystems, and its allelopathic effects have been intensively investigated. This chapter reviews the literature on the allelochemicals released from *M. spicatum* and their modes of action.

Keywords: *cyanobacteria, fatty acids, modes of action, Myriophyllum spicatum, polyphenols*

INTRODUCTION

Aquatic plants of the genus *Myriophyllum* are distributed worldwide. *Myriophyllum spicatum* L. is one representative of this genus that is highly competitive and exhibits potent growth and dispersal strategies. This enables the plant to rapidly dominate aquatic ecosystems, and *M. spicatum* sometimes forms monospecific communities (Grace and Wetzel, 1978; Smith and Barko, 1990). For this reason, *M. spicatum* has been studied carefully to understand the mechanisms underlying its replacement by other plant species in aquatic ecosystems.

Shoots of *M. spicatum* grow fast and form canopies at the water surface, thus monopolizing photosynthesis and interfering with the use of light by other aquatic organisms. For example, Madsen *et al.* (1991) reported that *M. spicatum* exhibited the highest maximum photosynthetic rate in a comparison with six other submerged aquatic plants. Therefore, coverage by epiphytes and phytoplankton development are both frequently low in ecosystems dominated by *M. spicatum*. Many papers have reported the highly competitive morphological and physiological strategies adopted by *M. spicatum*, such as the development of the above mentioned canopy (Adams *et al.*, 1974), removal of nutrients from open water, heavy vegetative spread by means of autofragmentation resulting in the dissemination of vegetative propagules (Smith *et al.*, 2002), and low light and CO_2 compensation points (Grace and Wetzel, 1978).

Allelopathy is another powerful growth strategy that can eventually eliminate other aquatic plants and microorganisms such as algae and cyanobacteria. *Myriophyllum spicatum* is one of the aquatic plants whose allelopathic effects have been intensively investigated. Several papers have reported an allelopathic potential against other aquatic plants (Agami and Waisel, 1985; Elakovich and Wooten, 1989; Jones, 1993) and phytoplankton (Planas *et al.*, 1981; Nakai *et al.*, 1996). For example, Agami and Waisel (1985) confirmed a reduction in the biomass of *Najas marina* by *M. spicatum* using containers in which *N. marina* and *M. spicatum* were planted or in which *M. spicatum* had been grown. The allelopathic effect of *M. spicatum* on cyanobacteria was subsequently demonstrated by assaying a solution in which *M. spicatum* had been cultured (Nakai *et al.*, 1999) and a medium to which an extract of an *M. spicatum* culture solution had been added (Gross and Sütfeld, 1994).

To understand the mechanism of allelopathy, the released allelochemicals and their growth inhibition modes must be elucidated. The present paper describes the kinds of compounds that have been reported to be allelochemicals released by *M. spicatum* and their proposed modes of action.

POLYPHENOLS

Planas *et al.* (1981) extracted phenolic compounds from *M. spicatum* using a methanol - 1% HCl solution. After analyzing the extract using thin-layer chromatography, 12 kinds of phenols and polyphenols and three organic acids were identified, and their relative amounts were determined (Table 1). For other species in the genus *Myriophyllum*, Saito *et al.* (1989) demonstrated the presence of two hydrolyzable tannins (eugeniin and 1-desgalloyl eugeniin) in *M. brasiliense* as well the presence of ellagic acid and gallic acid, which are both components of these tannins.

Table 1 Major compounds identified in *M. spicatum* and relative amounts (Planas *et al.*, 1981)

Relative amount (3 = highest)	*Major compounds identified*
3	ellagic acid, gallic acid, tannic acid, 3,5-dimethoxy-4- hydroxycinnamic acid
2	protocatechuic acid, shikimic acid
1	caffeic acid, cinnamic acid, coumaric acid, ferrulic acid, gentisic acid, pyrogallol, quinic acid, sinapic acid, syringic acid

In a study of the exudates of *M. spicatum*, Gross and Sütfeld (1994) found a galloyl ester and derivatives of ellagic acid by analyzing the solid-phase extract of an *M. spicatum* culture solution using a high-performance liquid chromatograph (HPLC) equipped with an ultraviolet (UV) detector. In subsequent research, eugeniin (tellimagrandin II), ellagic acid, and gallic acid were identified in the ethyl acetate fraction of a 50% acetone extract of *M. spicatum*; eugeniin and ellagic acid were also detected in a solid-phase extract of an *M. spicatum* culture solution (Gross *et al.*, 1996). In a more recent study, (+)-catechin and pyrogallic acid, in addition to ellagic and gallic acids, were found in both a *M. spicatum* culture solution and the plant itself using a HPLC equipped with a mass spectrometer (MS) (Nakai *et al.*, 2000). The release of gallic acid, ellagic acid, and pyrogallic acid were expected as based on previous results (Planas *et al.*, 1981), but the (+)-catechin and eugeniin represented new discoveries. The structures of the polyphenols released from *M. spicatum* are illustrated in Fig. 1.

Many studies have investigated the allelopathic effects of phenolic compounds, including tannins, on terrestrial plants (Rice, 1984). These compounds are believed to play an important role in the defense against pathogens and herbivores (Choi *et al.*, 2002; Smolders *et al.*, 2000; Walenciak *et al.*, 2002) and in competition with other plants (Quayyum *et al.*, 1999) and with organisms such as algae and cyanobacteria (Gross and Sütfeld., 1994; Pillinger *et al.*, 1994; Gross *et al.*, 1996; Nakai *et al.*, 2000). In some papers, plant extracts were used (Planas *et al.*, 1981; Ayoub and Yankov, 1985; Saito *et al.*, 1989; Gibson *et al.*, 1990; Gross and Sütfeld., 1994; Gross *et al.*, 1996; Nakai *et al.*, 1996). The anti-algal and anti-cyanobacterial activities of the polyphenols that have been found to be released by *M. spicatum* have been investigated in various species: eugeniin and gallic acid have been found to exhibit allelopathic effects on two cyanobacteria, *Microcystis aeruginosa* and *Anabaena flos-aquae* (Saito *et al.*, 1989); eugeniin, ellagic acid, and gallic acid using the cyanobacterium *Trichormus* var. P-9 (Gross *et al.*, 1996); and (+)-catechin, ellagic acid, gallic acid, and pyrogallic acid using the cyanobacteria *M. aeruginosa* (Nakai *et al.*, 2000), *Phormidium*

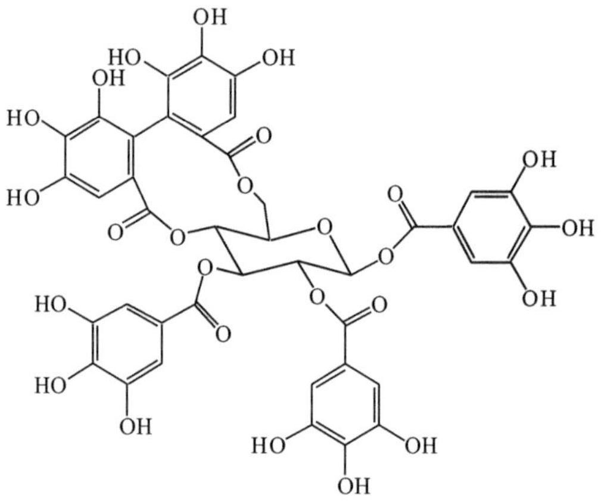

Eugeniin (tellimagrandin II)

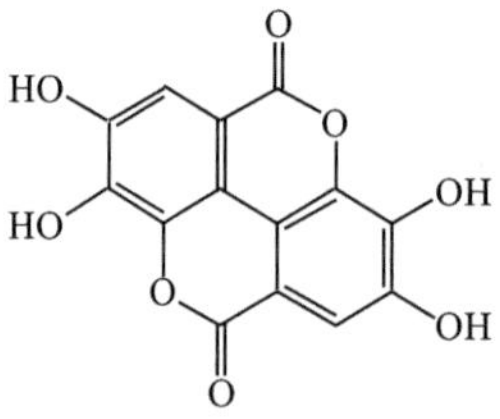

Ellagic acid

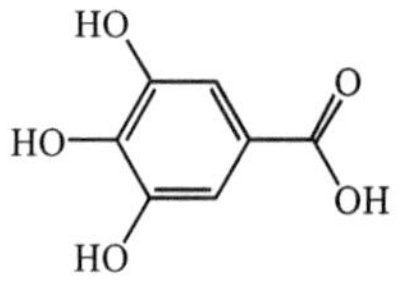

Gallic acid

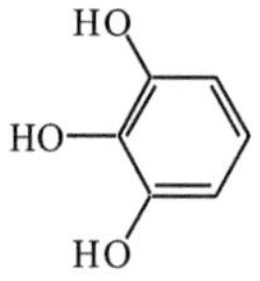

Pyrogallic acid

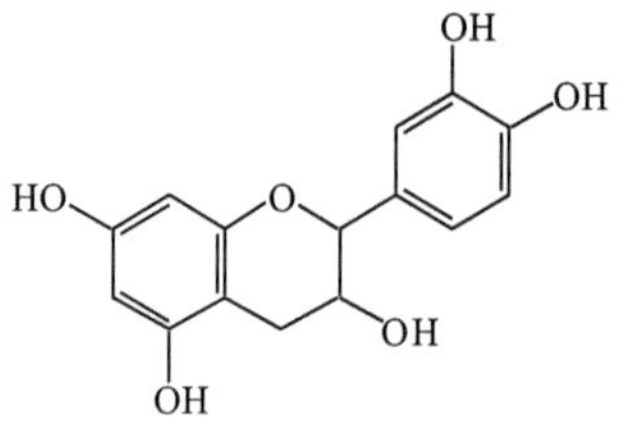

(+)-Catechin

Fig. 1 Polyphenols released from *M. spicatum*

tenue, the green algae *Selenastrum capricornutum* and *Scenedesmus quadricauda,* and the diatoms *Achnanthes minutissim* and *Nitzschia palea* (Nakai *et al.,* 2002a).

FATTY ACIDS

Fatty acids are also well known to cause allelopathic growth inhibition of algae (Rice, 1984). Although several papers have reported the release and allelopathic effects of polyphenols (as described in the previous section), only limited information is available for the fatty acids released by *M. spicatum*. Recently, our group (Nakai *et al.*, 2005) performed a series of analyses of solid extracts of *M. spicatum* culture solutions and methanol extracts of *M. spicatum* using a gas chromatograph equipped with a mass spectrometer (GC-MS) and showed that *M. spicatum* released the following fatty acids: nonanoic, tetradecanoic, hexadecanoic, octadecanoic, and octadecenoic acids. By comparing the observed mass spectrometry pattern and the retention time with that of authentic samples, we determined that the octadecenoic acids were present in *cis*-6- and/or *cis*-9- forms. An assay using *M. aeruginosa* demonstrated significant growth inhibition by nonanoic acid, *cis*-6-octadecenoic acid, and *cis*-9-octadecenoic acid; tetradecanoic acid, hexadecanoic acid, and octadecanoic acid showed no allelopathic effects. The structures of the three anti-cyanobacterial fatty acids are illustrated in Fig. 2.

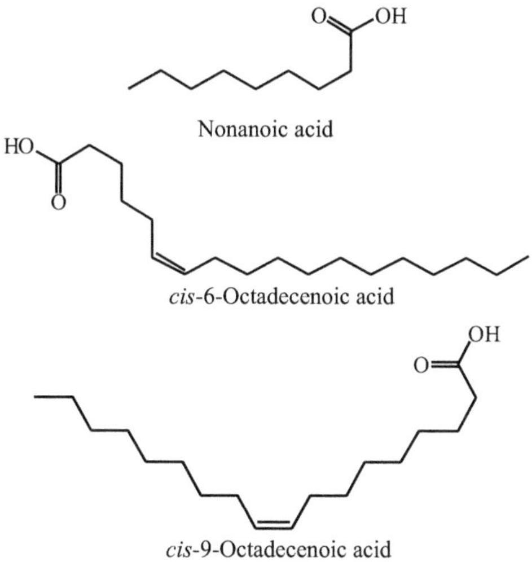

Fig. 2 Fatty acids released by *M. spicatum*

It is interesting that nonanoic acid was detected in both the culture solution and the methanol extract. Nonanoic acid was used in herbicides in Japan until 2003 (JT Agribusiness Division 1999). Although nonanoic acid sometimes has adverse effects on plants, some terrestrial plants are known to contain derivatives of nonanoic acid (Okuno *et al.*, 1993; Pelissier *et al.*, 2001). However, it is also well-known that plants generally produce fatty acids with an even number of carbon atoms. Since the *M. spicatum* used in this research was not grown in axenic culture, the possibility exists that microorganisms living in *M. spicatum* or its culture medium, rather than *M. spicatum* itself, produced the nonanoic acid.

CONTRIBUTION OF POLYPHENOLS AND FATTY ACIDS TO THE ALLELOPATHIC EFFECT ON CYANOBACTERIA

To determine which compounds cause allelopathic effects on cyanobacteria, it is essential to understand: (i) the quantitative effects of allelochemicals on the target organisms, (ii) the apparent concentrations of the compounds actually present in the aquatic ecosystem, (iii) the release rates of the compounds from *M. spicatum* and (iv) whether the compounds cause concurrent action. We chose to express the strength of the effect of each allelochemical as the 50% effective concentration (EC_{50}); this value represents the concentration at which 50% growth inhibition was observed. For example, we previously reported the EC_{50} values at which the polyphenols (+)-catechin, ellagic acid, gallic acid, and pyrogallic acid, as well as the three allelopathic fatty acids, inhibit normal growth of *M. aeruginosa* (Nakai *et al.*, 2000, 2005). Saito *et al.* (1989) also provided EC_{50} values for eugeniin, ellagic acid, and gallic acid for *M. aeruginosa*, as summarized in Table 2. Although differences certainly exist among the assay series, the data nonetheless allow us to predict and compare the inhibitory effects of each allelochemical. Note that the effects of these allelochemicals may also be species specific (Nakai *et al.*, 2002a).

Subsequently, a comparison of the EC_{50} values for allelochemicals with their apparent concentration in the aquatic system could be made to investigate the contribution of the identified allelochemicals. However, in all *M. spicatum* culture solutions investigated, the apparent concentrations of the allelochemicals were insufficient to cause the observed growth inhibition of *M. aeruginosa*. For example, our paper reported the following apparent concentrations of gallic acid, ellagic acid, pyrogallic acid, and (+)-catechin, respectively, in the *M. spicatum* culture solution at 100 g-wet/L for 3 d: about 63, 77, 5.2, and 17 μg/L (Nakai *et al.*, 2000). About 50 μg/L of nonanoic acid, the most inhibitory compound among the six allelochemicals (Table 2), was detected in another culture solution prepared in the same manner (Nakai *et al.*, 2005). Gross (2000) showed that the eugeniin content in *M. spicatum* changed seasonally, reaching a maximum of about 20 mg/g-dry wt; in contrast, Glomski *et al.* (2002) detected no eugeniin in an *M. spicatum* culture solution, although the extraction efficiency in later literature was not shown. Unfortunately, no study before that of Gross reported the presence of eugeniin in any *M. spicatum* culture solution. These results raised the question whether these compounds really contribute to the allelopathic effect of *M. spicatum* on *M. aeruginosa* in the field.

Table 2 EC_{50} values for the allelochemicals released by *M. spicatum* with respect to the growth of *M. aeruginosa* (mg/L)

Allelochemical	EC_{50}
Pyrogallic acid	0.7*
Gallic acid	1.0*
	3.2**
(+)-Catechin	5.5*
Nonanoic acid	0.5***
9-cis-Octadecenoic acid	1.6***
6-cis-Octadecenoic acid	3.2***

*Data reported by Nakai *et al.* (2000)
**Data reported by Saito *et al.* (1989)
***Data reported by Nakai *et al.* (2005)

Polyphenols are autoxidized in alkaline solutions to produce quinones (Inescu *et al.*, 1978; Doona and Kustin, 1993). Pillinger *et al.* (1994) focused on the production of quinones by autoxidation of phenols, including polyphenols, and confirmed the growth inhibition of *M. aeruginosa* by authentic quinones such as 1,2-naphthoquinone and 9,10-phenanthrenequinone. Furthermore, our previous study (Nakai *et al.*, 2000) confirmed the autoxidation of gallic acid, ellagic acid, and pyrogallic acid and of (+)-catechin after their release by *M. spicatum*, as well as the existence of anti-cyanobacterial autoxidation products, though the anti-cyanobacterial autoxidation products were not identified. Therefore, the effects of the autoxidation products must be estimated so that we can understand the role of these polyphenols in the allelopathic effects of *M. spicatum* on *M. aeruginosa*.

To accomplish this analysis, the release rates of the polyphenols are essential information. We estimated the release rates of four polyphenols (gallic acid, ellagic acid, pyrogallic acid, and (+)-catechin) by *M. spicatum* on the basis of a mass balance equation for each polyphenol in a *M. spicatum* culture solution, as expressed by Equation (1), on the assumption that microbial degradation of the four polyphenols did not occur. Note that the autoxidation rates of the four polyphenols are expressed as functions of pH (Nakai *et al.*, 2001).

$$d(CV)/dt = aW - kCV \quad (1)$$

where

C = apparent polyphenol concentration in the *M. spicatum* culture solution, μg/L

V = volume of the *M. spicatum* culture solution, L

W = fresh weight of *M. spicatum* in the culture solution, g *a* = release rate of the polyphenol from *M. spicatum*, μg/(g-fresh wt × h)

k = experimentally obtained autoxidation rate coefficient for the polyphenol, 1/h

Finally, we confirmed the contribution of these four polyphenols to the allelopathic effects of *M. spicatum* on *M. aeruginosa* by comparing the growth inhibition produced by a *M. spicatum* culture solution with those of simulated culture solutions prepared using these four polyphenols based on their estimated release rates. The results showed that together, the four polyphenols accounted for 10–100% of the allelopathic effect caused by *M. spicatum* (Nakai *et al.*, 2001). Unfortunately, the release rate of eugeniin could not be determined because no authentic sample was commercially available, though we believe that eugeniin must also contribute to the allelopathy. It should be noted that a mixture of gallic acid and pyrogallic acid with (+)-catechin caused synergistic inhibitory effects on *M. aeruginosa* (Nakai *et al.*, 2002a). This suggests that concurrent action may occur when *M. spicatum* releases a combination of (+)-catechin, eugeniin, ellagic acid, gallic acid, pyrogallic acid, nonanoic acid, *cis*-6-octadecenoic acid, and *cis*-9-octadecenoic acid.

MODES OF ACTION OF ALLELOCHEMICALS

It is well known that polyphenols can form complexes with proteins. Gross *et al.* (1996) focused on this fact in their investigation of the influence of eugeniin and gallic acid on the activity of an alkaline phosphatase. Their results confirmed inhibition of the exoenzyme, indicating that the released polyphenols might inhibit such extracellular enzymes and those bound to the outer cell membrane. Subsequently, her group confirmed the inhibition of photosystem II using lipophilic extracts of *M. spicatum* and reported that eugeniin was the active compound (Leu *et al.*, 2002). Unfortunately, no other papers have been published in which the target sites of the allelochemicals released by *M. spicatum* are described.

In regards to the behavior of the polyphenols, we have demonstrated the production of radicals at the beginning of the autoxidation of gallic acid and pyrogallic acid in a cyanobacterial medium (*M. aeruginosa*) using electron spin resonance (Nakai *et al.*, 2002b). Based on the observed lifetimes of the radicals produced by autoxidation of pyrogallic acid and gallic acid, we hypothesized that such radicals and other autoxidation products may be responsible for the observed growth inhibition of *M. aeruginosa*. As a result of this behavior of the polyphenols, the autoxidation products must be identified if we are to reveal how polyphenolic allelochemicals cause allelopathic effects.

CONCLUSION

Recent studies have revealed the allelopathic effects of chemical constituents of *M. spicatum* which inhibited the growth of *M. aeruginosa*. To further understand the effect of these, their apparent concentrations in the aquatic ecosystem and their release rates by *M. spicatum* is to be investigated. As the polyphenols are autoxidized after their release by *M. spicatum*, the influence of autoxidation on the inhibition of *M. aeruginosa* growth should also be studied. In particular, subsequent experiments should be designed to reveal the mechanisms of growth inhibition by the polyphenolic allelochemicals, and which target sites or processes they act upon.

In regards to the allelopathic effects of *M. spicatum* on other organisms such as aquatic plants, only limited information is available. To better define the role of allelopathy by *M. spicatum* in replacing other plants in the aquatic ecosystem, the influence of the species' exudates must be assayed using other aquatic plants. Analytical techniques used for the identification of anti-cyanobacterial allelochemicals (Planas *et al.*, 1981; Saito *et al.*, 1989; Gross *et al.*, 1996; Nakai *et al.*, 2000, Nakai *et al.*, 2005) may be useful in theidentification of allelochemicals.

References

Adams, M.S., Tistus, J., McCracken, M.D. (1974) Depth distribution of photosynthetic activity in a *Myriophyllum spicatum* community in Lake Wingra. Limnology and Oceanography. 19: 377-389.

Agami, M., Waisel, Y. (1985) Inter-relationships between *Najas marina* L. and three other species of aquatic macrophytes. Hydrobiologia. 126: 169-173.

Ayoub, S.M.H., Yankov, L.K. (1985) Algicidal properties of tannins. Fitoterapia. 56: 227-228.

Choi, C., Bareiss, C., Walenciak, O., Gross, E.M. (2002) Impact of polyphenols on growth of the aquatic herbivore *Acentria ephemerella*. J. Chem. Ecol. 28: 2245-2256.

Doona, C.J., Kustin, K. (1993) Kinetics and mechanism of pyrogallol autoxidation: Calibration of the dynamic response of an oxygen electrode. Int. J. Chem. Kinetics. 25: 239-247.

Elakovich, S.D., Wooten, J.W. (1989) J. Aquatic Plant Manage. 27: 78-84.

Gibson, M.T., Walch, I.M., Barrett, P.R.F., Ridge, I. (1990) Allelopathic potential of sixteen aquatic and wetland plants. J. Appl. Phycol. 2: 241-248.

Glomski, L.A., Wood, K.V., Nicholson, R.L., Lembi, C.A. (2002) The Search for Exudates from Eurasian Watermilfoil and Hydrilla. J. Aquatic Plant Manage. 40: 17-22.

Grace, J.B., Wetzel, R.G. (1978) The production biology of Eurasian watermilfoil (*Myriophyllum spicatum* L.): A review. J. Aquatic Plant Manage. 16: 1-11.

Gross, E.M. (2000) Seasonal and spatial dynamics of allelochemicals in the submersed macrophyte *Myriophyllum spicatum* L. Verhandlungen Internationale Vereinigung für Theoretische und Angewandte Limnologie. 27: 2116-2119.

Gross, E.M., Sütfeld, R. (1994) Polyphenols with algicidal activity in the submerged macrophyte *Myriophyllum spicatum* L. Acta Hort. 381: 710-716.

Gross, E.M., Meyer, H., Schilling, G. (1996) Release and ecological impact of algicidal hydrolysable polyphenols in *Myriophyllum spicatum*. Phytochemistry. 41: 133-138.

Inescu, G.H., Matei, F., Duca, A.L. (1978) Kinetic aspects of the oxidation reactions of gallic acid using an oxygen-selective electrode. Revue Roumaine de Chimie. 23(11/12): 1611-1617.

Jones, H.L. (1993) Allelopathic influence of various aquatic plant extracts on the growth of hydrilla (*Hydrilla verticillata*). M.S. thesis, Biological Science Dept., Univ. Southern Mississippi, USA .

JT Agribusiness Division (1999) Summary of toxicity studies on pelargonic acid. Pest. Sci. 24: 421-423. (in Japanese).

Leu, E., Krieger-Liszkai, A., Goussias, C., Gross, E.M. (2002) Polyphenolic allelochemicals from the aquatic angiosperm *Myriophyllum spicatum* L. inhibit photosystem II. Plant Physiol. 130: 2011-2018.

Madsen, J.D., Hartleb, C.F., Boylen, C.W. (1991) Photosynthetic characteristics of *Myriophyllum spicatum* and six submersed aquatic macrophyte species natice to Lake George, New York. Freshwater Biol. 26: 233-240.

Nakai, S., Hosomi, M., Okada, M., Murakami, A. (1996) Control of algal growth by macrophytes and macrophyte-extracted bioactive compounds. Water Sci. Tech. 34: 227-235.

Nakai, S., Inoue, Y., Hosomi, M., Murakami, A. (1999) Growth inhibition of blue-green algae by the allelopathic effects of macrophytes. Water Sci. Tech. 39(8): 47-53.

Nakai, S., Inoue, Y., Hosomi, M. (2000) Growth inhibition of blue-green algae (*Microcyctis aeruginosa*) by *Myriophyllum spicatum*-releasing four polyphenols. Water Res. 34: 3026-3032.

Nakai, S., Inoue, Y., Hosomi, M. (2001) Allelopathic effect of 4 polyphenols released by *Myriophyllum spicatum* on growth of the cyanobacterium *Microcystis aeruginosa*. Allelopathy J. 8: 201-210.

Nakai, S., Inoue, Y., Hosomi, M. (2002a) Algal growth inhibition effects of allelopathic polyphenols released by *Myriophyllum spicatum*. Allelopathy J. 10(2). 123-132.

Nakai, S., Inoue, Y., Lee, B.D., Hosomi, M. (2002b) Algal growth inhibition effects of plant-produced phenols. Japanese J. Limnology. 63: 201-208. (in Japanese)

Nakai, S., Yamada, S., Hosomi, M. (2005) Anti-cyanobacterial fatty acids released from *Myriophyllum spicatum*. *Hydrobiologia*, 543(1), 71-78.

Okuno, M., Kameoka, H., Yamashita, M., Miyazawa, M. (1993) Components of volatile oil from plants of Polypodiaceae. J. Japan Oil Chemists' Society. 42: 44–48. (In Japanese).

Pelissier, Y., Haddad, C., Marison, C., Miharu, M., Bessier, J.-M. (2001) Volatile constituents of fruit pulp of *Dialium guineense* wild. J. Essential Oil Res. 13: 103-104.

Pillinger, J.M., Cooper, J.A., Ridge, I. (1994) Role of phenolic compounds in the anti-algal activity of barley straw. J. Chem. Ecol. 20: 1557-1569.

Planas, D., Sarhan, F., Dube, L., Godmaire, H., Cadieux, C. (1981) Ecological significance of phenolic compounds of *Myriophyllum spicatum*. Verhandlungen Internationale Vereinigung für Theoretische und Angewandte Limnologie. 21: 1492-1496.

Quayyum, H.A., Mallick, A.U., Lee, P.F. (1999) Allelopathic potential of aquatic plants associated with rice (*Zizania palustris*) I. Bioassay with plant and lake sediment samples. J. Chem. Ecol. 25: 209-220

Rice, E.L. (1984) Allelopathy 2nd ed., Academic Press, NY, USA.

Saito, K., Matsumoto, M., Sekine, T., Murakoshi, I. (1989) Inhibitory substances from *Myriophyllum brasiliense* on growth of blue-green algae. J. Natural Products. 52: 1221–1226.

Smith, C.S., Barko, J.W. (1990) Ecology of Eurasian watermilfoil. J. Aquatic Plant Manage. 28: 55-63.

Smith, D.H., Madsen, J.D., Dickson, K.L., Beitinger, T.L. (2002) Nutrient effects on autofragmentation of *Myriophyllum spicatum*. Aquatic Bot. 74(1): 1-17.

Smolders, A.J.P., Vergeer, L.H.T., Velde, vander G. Roelofs, G.M. (2000) Phenolic contents of submerged, emergent and floating leaves of aquatic and semi-aquatic macrophyte species: why do they differ? OIKOS. 91: 307-310.

Walenciak, O., Zwisler, W., Gross, E.M. (2002) Influence of *Myriophyllum spicatum*-derived tannins on gut microbiota of its herbivore *Acentria ephemerella*. J. Chem. Ecol. 28: 2045-2056.

22

Fruiting Bodies of Mushrooms as Allelopathic Plants

Hiroshi Araya

Chemical Ecology Unit, Department of Biological Safety, National Institute for Agro-Environmental Sciences, 3-1-3 Kan-nondai, Tsukuba, Ibaraki 305-8604, Japan

**Department of Agriculture, Meiji University, 1-1-1 Higashimita, Tama-ku, Kawasaki 214-8571, Japan*

Mushrooms relate to a number of plant lives and influence numerous plants directly or indirectly. The influence of fruiting bodies of 83 mushrooms on the growth of lettuce (*Lactuca sativa* L.) seedling was examined by the "sandwich method" and this clarified that all the mushrooms showed inhibitory activity at 50 mg applications. On the other hand, intensity of inhibition varied at 10 mg application. Among the mushrooms tested, *Morchella esuculenta, Lycoperdon perlatum, Amanita sinensis, Tricholoma muscarium, Cortinarius violaceusm Agaricus subrutilescens, Coprinus comatu, Agrocybe cylindracea, Pholiota lubrica, Pleurocybella porrigens, Lyophyllum decastes, Lepista nuda, Ramaria Formosa, Cortinarius aureobrunneus, Xanthoconium affine, Pleurotus cornucopiae, Panellus serotinus, etc.* exhibited strong efficacy of inhibitory. The results showed that mushrooms produce plant growth regulating compounds which act as allelochemicals in nature. Furthermore, these compounds enable them to play a role as allelopathic plants in nature, since the amount of mycelia in the soil is very high as compared with fruiting bodies. A possibility of utilization of certain mushrooms and their products for weeds control will be discussed in this chapter.

Keywords: *allelopathy, fruiting body, herbicide, higher fungi, litter, mushroom, plant growth inhibitor*

**Author's Present Address*

INTRODUCTION

A mushroom is the fruiting body of a paricular plant in two kingdoms established by Carl von Linne (1735), the part of the fungus that typically appears above ground and contains its reproductive units, or spores. It seems that there are more than 8,000 mushrooms in the world. While being members of the plant kingdom, they differ from most plants in that they lack chlorophyll and must rely on organic material for nutrition (Lincoff, 1981). Their utilization methods for nutrients are classified in three ways; mycorrhizally, parasitically and saprophytically. It has been observed that many species of mushrooms influence plants directly or indirectly. It is well known that several mushrooms cause lawn sickness called "Fairy Ring disease", mushrooms develop and form large circles in lawns (Smiley, *et al.* 1992). In general, the circle expands year by year. As long as the fairy rings are small, there may be no effect on the grass, however as they expand, a zone of stimulation forms where the mushrooms come up. Inside the circle is an area of poor grass growth, or even dead grass. Another zone of stimulation may occur in the dead or dying zone. The mushroom produces plant growth regulators or gives some influence to the growing environment in nature. It is reported that *Calocybe georgii* and *Agaricus campester* release some plant growth regulators which cause growth stimulation of grass (Nada, 1975). The phenomenon is considered as that these mushrooms show "allelopathy" by releasing plant growth regulators.

The term "allelopathy" was coined by Professor Molisch, which refers to biochemical interactions between all kinds of plants including the microorganisms (Molisch, 1937). Chemical compounds of allelopathy are called allelochemicals or allelochemics. The phenomenon, where a plant species chemically interferes with the germination, growth, maturation of fruits or development of other plant species, has been known for over 2,000 years. Many researchers have studied allelopathy, and many allelochemicals (including candidates) have been isolated until now. However, most of target organisms are plants, and there are only a few reports regarding mushroom allelopathy in the literature. A lot of compounds have been isolated according to various bioassay guided fractionation from different mushrooms and are used for fungicides, anticancer drugs, antibiotics, etc. Some of compounds isolated from mushrooms have been used as anticancer drugs, enzyme inhibitors (biochemical tools for experiments). However, there are no commercial herbicides from mushrooms. There should be the possibility to find chemicals for weeds control by pursuing allelochemicals studies on mushrooms. In this chapter, studies on allelopathy of mushrooms will be reviewed, based on both the result of the conducted experiment and natural products isolated from mushrooms as plant growth inhibitors.

MUSHROOM ALLELOPATHY

So far only three studies of allelopathy of mushrooms have been reported. Molisch first reported the influence of fruiting bodies of *Agaricus campestris* on vetch (*Vicia sativa*) in 1937 (Molisch, 1937). He found that the volatiles released from the mushroom had a slight inhibitory effect (*ca.* 4.7%) on the 4-d growth in the length of the vetch seedlings. Fifty years later, Chaumont and Simeray reported that water extracts of 114 kinds of mushrooms showed allelopathic activity to red radish (*Raphanus sativus* var. radicicola) seedling growth (Chaumont and Simeray, 1985). Their results showed that the most inhibitory activities were provided by Boletales and Collybia, and many other extracts cause grave disturbance on radish shoots' growth. The allelopathic activity of litters of fruiting bodies to lettuce seedling growth, was recently reported, and proposed the possible release routes of allelochemicals from mushrooms (Araya, 2005). The following mushrooms exhibited strong inhibitory activity: *Agaricus subrutilescens, Amanita cokeri, Amanita sinensis, Boletus fraternus, Boletus reticulates, Coprinus comatus, Cortinarius violaceus, Gymnopilus spectabilis, Lanopila nipponica, Pleurocybella porrigens, Ramaria formosa, Lyophyllum decastes, Morchella esculenta, Panellus serotinus, Pleurotus ostreatus,* and *Hypsizigus marmoreus.*

In another study, Kitaya *et al.,* (1994) reported that the growth of lettuce plants (*Lactuca sativa* L. cv. Okayama) was promoted by carbon dioxide emission, which resulted from respiration of the mycelium of Shiitake mushroom (*Lentinus edodes*), in a CELSS (controlled ecological life support system). The result showed that carbon dioxide acts as an allelochemical in the closing system.

It is noteworthy that productions of plant hormones were also reported from mushrooms. An auxin, indole-3-acetic acid, was isolated from medium culture of *Schizophyllum commune* (Epstein and Miles, 1967), and cytokinins were first isolated from cultured liquid medium of *Rhizopogon roseolus* (Miller, 1967). Emission of ethylene gas was identified from both the vegetative (Tschierpe and Sinden, 1965; Richter, 1967) and reproductive phase (Lockard and Kneebone, 1962) of cultivated *Agaricus bisporus*. These compounds are considered as potential allelochemicals and further studies should be conducted.

On the other hand, several compounds possessing plant growth inhibitory activity have been isolated from fruiting bodies of mushrooms, *e.g.*, azetidine-2-carboxylic acid from *Clavaria miyabeana* S. Ito and several mushrooms were classified to the same *Clavariaceae* family, (the compound was identified by a TLC experiment from *C. fusiformis, C. inaequalis, C. vermiculata, C. purpurea, C. botrytis, C. Formosa, C. flara, C. pyxidata*) (Ikeda, *et al.*, 1977), neogrifolin, grifolin from *Polyporus confluens* (Ikeda, 1989), 2-

amino-3-cyclopropyl-butanoic acid (Yoshimura *et al.*, 1999; Wakabayashi i 2001; Morimoto *et al.*, 2002), 2-amino-5-chloro-4-pentenoic acid (Yoshimura, *et al.*, 1999; Wakabayashi, *et al.*, 2001) from *Amanita castanopsidis* Hongo, 2-geranylgeranyl-4-acetoxy resorcinol from *Suillus luteus* (Saito *et al.*, 1995), fasciculols from *Naematoloma fascicylare* (Ikeda *et al.*, 1978) (Fig. 1). These mushrooms may suppress surrounding plant growth by releasing these chemicals. L-DOPA (3,4-dihydroxy-phenylalanine, Fig. 1), identified as an allelochemical from *Mucuna pruriens* (Fujii, 1994), was found in many mushroom species (Oka *et al.*, 1981; Senatore *et al.*, 1987), especially *Strobilomyces floccopus* and *Hygrocybe conica* contain it at high concentrations [0.36% fresh fruiting bodies (Steglich and Esser, 1973), 3.2% dried fruiting bodies (Steglich and Preuss, 1975)] respectively. Thus, the compound might act as one of allelochemicals in these mushrooms and show allelopathic activity in their surroundings.

MATERIALS AND METHODS

Materials

All wild-fruiting bodies of mushrooms were collected from the mountain region of Ibaraki, Tochigi and Fukushima prefecture, Japan during 2001 to 2003. The cultivated mushrooms were provided generously by Dr. Sekiya, Forestry and Forest Product Research Institute, Japan. After removal of all stray material (soil, insects, leaves, etc.) the mushrooms were lyophilized. Voucher specimens of these mushroom fruiting bodies were kept at the Chemical Ecology Unit, National Institute for Agro-Environmental Sciences, Japan.

Bioassay

The estimation of allelopathic potential of mushrooms, the sandwich method (Fujii *et al.*, 2004), which was developed to assess the allelopathic activity of fallen plant materials, was applied. Lettuce (*L. sativa* L. var. Grate Lakes 366) was used as a test plant for the bioassay studies, because it is highly sensitive to bioactive substances. This method has an advantage that the whole plant (mushroom) materials can be used for the assay. Application of the sample amount was followed by the standard procedure for common plant samples; namely, 10 mg and 50 mg were subjected at three replications respectively. For the control treatment, mushrooms were not added. After 3 d incubation at 25°C under dark condition, the length of both radicle and hypocotyl was measured. The inhibition ratio of lettuce seedling growth was calculated by the following formula respect to control.

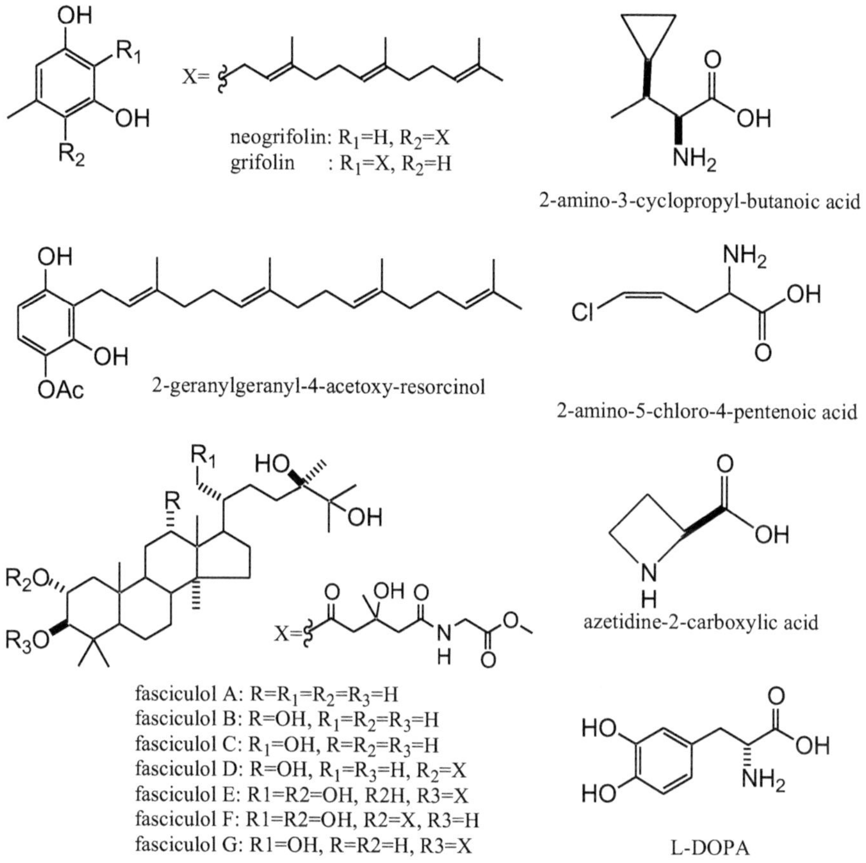

Fig. 1 Plant inhibitory compounds isolated from several mushrooms

Efficacy of Inhibition (%) = 100 – 100 × sample growth/control growth. 100% means completely killed, 0% means no effect, – 100% means two-fold growth.

RESULTS

The result of allelopathic activity of mushroom fruiting bodies using the sandwich method was shown in Fig. 2 for 50 mg application and Fig. 2 for 10 mg application respectively. Nearly all the mushrooms exhibited strong allelopathic activity. Application of 50 mg of mushroom inhibited lettuce radicle growth considerably (Fig. 2), and three-fourths of the mushrooms inhibited hypocotyl growth more than 50%. On the other hand, the inhibition was even spotted at application of 10 mg of fruiting body (Fig. 3). Four-fifths of the mushrooms showed more than 50% radicle growth

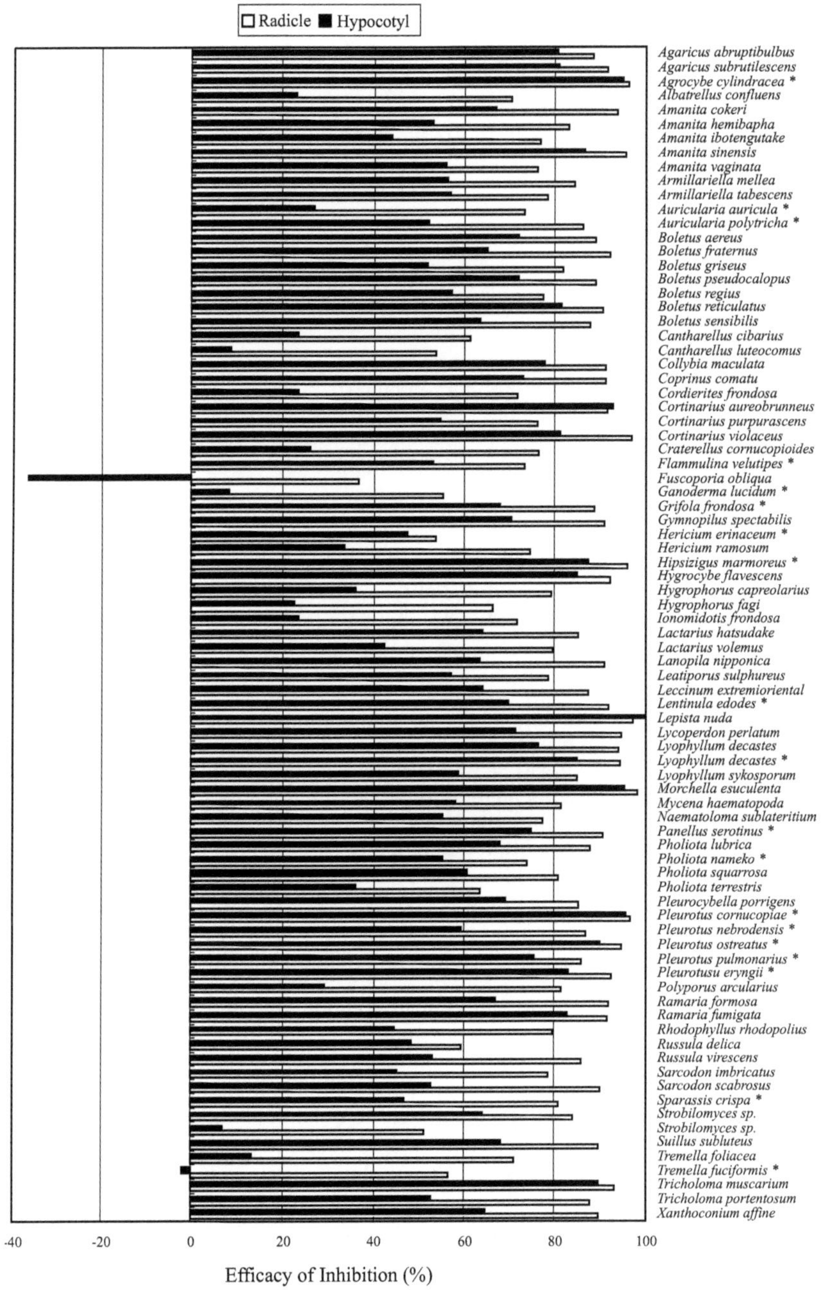

*Indicates cultivating mushrooms.

This figure was made from data of previous reports (Araya *et al.*, 2003, 2004; Araya, 2005)

Fig. 2 Efficacy of lettuce growth inhibition on application of 50 mg

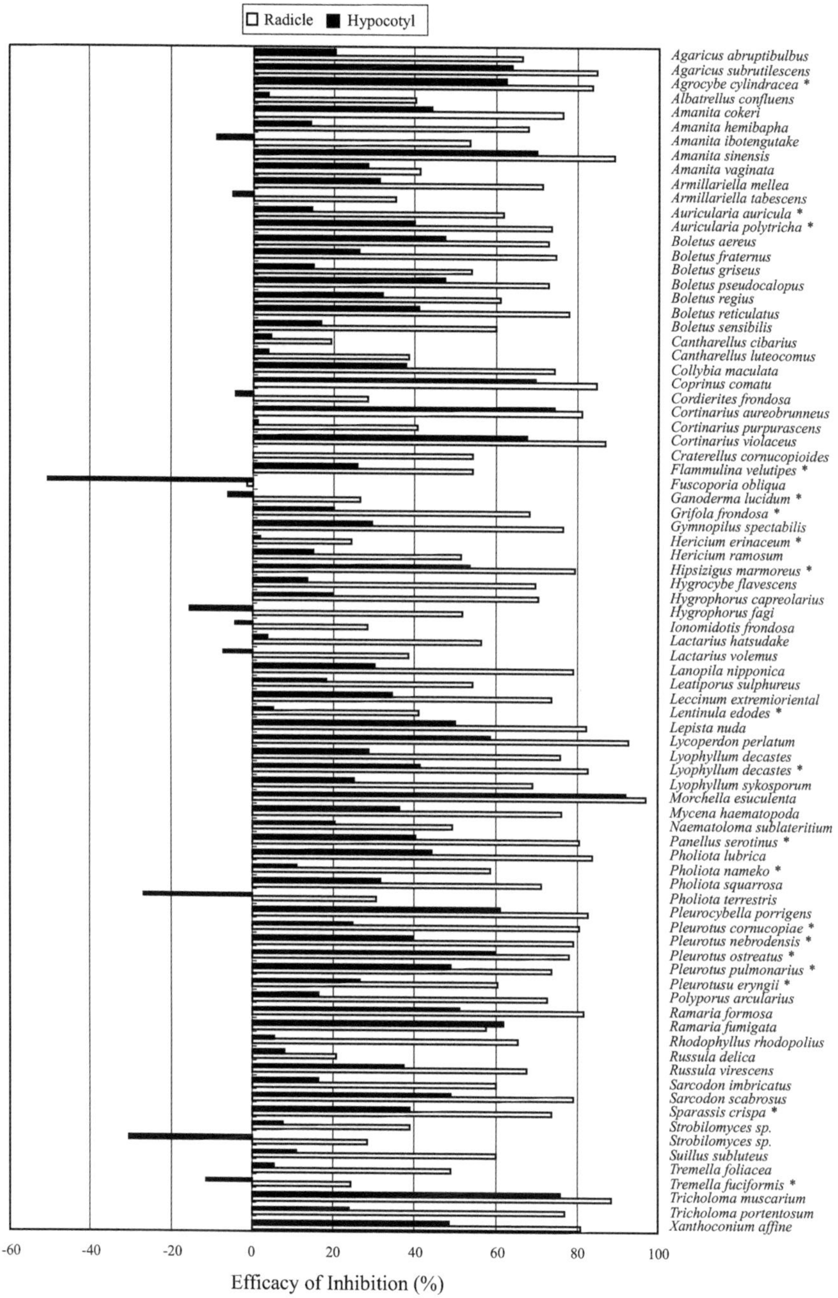

*indicates cultivating mushrooms

This figure was made from data of previous reports (Araya, *et al.*, 2003, 2004; Araya, 2005).

Fig. 3 Efficacy of lettuce growth inhibition on application of 10 mg

inhibition at 10 mg application. Hypocotyl growth inhibition was shown in almost all the mushrooms at various ranges.

The result shows that Morchella esuculenta, Lycoperdon perlatum, Amanita sinensis, Tricholoma muscarium, Cortinarius violaceusm, Agaricus subrutilescens, Coprinus comatu, Agrocybe cylindracea, Pholiota lubrica, Pleurocybella porrigens, Lyophyllum decastes, Lepista nuda, Ramaria Formosa, Cortinarius aureobrunneus, Xanthoconium affine, Pleurotus cornucopiae, Panellus serotinus, etc. exhibited strong efficacy of lettuce seedling growth. It was noteworthy that Fuscoporia oblique, Tremella foliacea showed growth stimulating effect to hypocotyls growth at both volume application, and in addition some of species showed growth stimulating effect to hypocotyls growth at lower concentrations. There is a possibility that these mushrooms might produce some kind of plant growth stimulating compounds such as plant hormones.

CONCLUSION

Overuse of the synthetic chemicals for pests control is posing a serious threat to the environment (Chou, 1995; Cutler and Cutler, 1999; Kohli, *et al.* 1999). It is well known that domestic plants producing compounds, being biodegradable, are friendly in various original ecosystems. In both managed and natural ecosystems, allelopathy plays an important role. In agroecosystems that are managed by artifical activities, the allelopathic interaction between crops and weeds gave reflects crops production yield. Many researchers have tried to utilize allelopathy by different strategies. They have applied various strategies to utilize allelopathy. For example rye (*Secale cereale* L.) produces allelochemicals and is used for weed management (Weston 1996). Many allelopathic researchers have stated that allelochemicals are ideal agrochemicals, as they are biodegradable and environmentally friendly (Rice, 1995; Cutler, 1999; Hoagland, 1999; Macías *et al.*, 1999; Duke *et al.*, 2000). However, this concept is still ambiguous as natural products are always environmentally safe. Furthermore, while almost all the allelochemicals have been isolated, it has been not confirmed whether these are safe for our health. Therefore, we must choose allelopathic plants (or microorganisms) by identification of the allelochemicals, and assess the safety of these natural products before application.

Various mushroom fruiting bodies having strong allelopathic activity was clarified by the "sandwich method" as mentioned above, and the result showed that mushroom fruiting body produce allelochemical(s).

Mushroom allelochemical(s) might act as plant growth inhibiting factors by leaching from fruiting bodies in the natural ecosystem. The main source of allelochemical(s) would be mycelia in the soil, because the life of fruiting bodies is generally short (Araya, 2005). Fruiting bodies would be suitable for searching for allelochemicals, as many wild mushroom fruiting bodies can be obtained in almost a pure form in several cases; it is difficult to isolate mycelia from the soil as a pure form; there are many reports that fermented fruiting bodies (mycelia) produce different secondary metabolites from wilds; many symbiotic mushroom could not grow in an artificial medium.

In literature, allelopathic-like phenomena of mushrooms have been reported, for example, soil sickness by Matsutake (*Tricholoma matsutake*) in their large habitat (Ogawa, 1978); stimulation of a species of bamboo growth by fungus in Ramariaceae (Hongo, 2003). The amount of plant biomass in the area *Naematoloma fascicylare* appeared for several years is smaller than around the area (Ikeda *et al.*, 1978). It was also noted that wheat (*Triticum aestivum* L.) was not healthy in fields fertilized by immature compost of bed log for lacquered bracket fungus (*Ganoderma lucidum*) (Araya *et al.*, 2003). This may result in their pathogenicity (Ariffin *et al.*, 2000).

The emission routes of allelochemicals from mushrooms were proposed recently by Araya (2005). Allelochemicals was released via several ways; volatiles from whole fungi, leaching from fruiting bodies, exudates from mycelium in soils, litters from fruiting bodies and its tissue (e.g., spores, warts, annuli, universal veil, etc.). The practical main source of allelochemicals would be mycelia in soil, because the life of fruiting bodies is generally short.

Studies on mushrooms metabolites concerning allelopathy are a newly developed area. Identification of mushroom allelochemicals is in process. However, there are a few problems. It is difficult to obtain sufficient amounts of mushrooms for research in many cases. Secondary metabolites between fermented mycelia and fruiting bodies is a well known fact. Furthermore, we must be careful about the condition of samples obtained. That is, differences in metabolites of fruiting bodies, that were the same species but different districts, were reported (Hatanaka and Terakawa, 1968; Wada, *et al.* 1995).

I look forward to a development of study of mushroom allelopathy and mushroom allelopathy in natural ecosystem could be proved strictly. Furthermore, The author expect the application of mushroom including their metabolites for weed management in the future.

Acknowledgments

I would like to express my thanks to Dr. Yoshiharu Fujii, Dr. Syuntaro Hiradate and the staff of Chemical Ecology Unit, NIAES for their helpful assistance and useful comments on this research. The author is also indebted to Dr. Atsushi Sekiya, Forestry and Forest Products Research Institute, Japan, for the generous gift of several cultivating mushrooms.

References

Araya, H., Fujii. Y., Sekiya, A. (2003) Unpublished data.

Araya, H., Hiradate, S., Fujii, Y. (2003) Allelopathic activity of fruiting bodies of higher fungi. J. Weed Sci. Tech. 48 (s): 198-199.

Araya, H., Sekiya, A., Hiradate, S., Fujii, Y. (2004) Allelopathy of fruiting bodies of higher fungi. J. Weed Sci. Tech., 49 (s): 176-177.

Araya, H. (2005) Allelopathic activities in litters of mushrooms. *In* New Discovery in Agrochemicals: J.M. Clark, H. Ohkawa (eds.). ACS Symposium Series 892, American Chemical Society, Washington, DC, USA. pp. 63-72.

Ariffin, D., Idris, A. S., Singh, G. (2000) Status of Ganoderma in oil palm. *In Ganoderma* Diseases of Perennial Crops, J. Flood, P.D. Bridge, M. Holderness (eds.). CABI Publishing, Oxon, UK. pp. 49-68.

Chaumont, J.P., Simeray, J. (1985) Propriétés allélopathiques de 114 extraits de carpophores de champignons sur la germination de semences de radis. Rev. Écol. Biol. Sol. 22: 331-339.

Chou, C.H. (1995) Allelopathy and sustainable agriculture. *In* Allelopathy: Organisms, Processes and Applications. Inderjit, K.M.M. Dakshini, F.A. Einhellig (eds.). ACS Symposium Series 582, American Chemical Society, Washington, DC, USA. pp. 211-223.

Cutler, H.G. (1988) Perspectives on discovery of microbial phytotoxins with herbicidal activity. Weed Tech. 2: 525-532.

Cutler, H.G. (1999) Potentially useful natural product herbicides from microorganisms. *In* Principles and Practices in Plant Ecology: Allelochemical Interactions, CRC Press, Boca Raton. Fl, USA. pp. 497-516.

Cutler, H.G., Cutler, S.J. (1999) Agrochemicals and pharmaceuticals: The connection. *In* Biologically Active Natural Products: Agrochemicals, H.G. Cutler, S.J. Cutler, (eds.). CRC Press, Boca Raton, London, NY, Washington DC. pp. 1-14.

Duke, S.O., Dayan, F.E., Romagni, J.G., Rimando, A.M. (2000) Natural products as sources of herbicides: current status and future trends. Weed Res. 40: 99-111.

Epstein, E., Miles, P.G. (1967) Identification of indole-3-acetic acid in the basidiomycete *Schizophyllum commune*. Plant Physiol. 42: 911-914.

Fujii, Y. (1994) Screening of allelopathic candidates by new specific discrimination, and assessment methods for allelopathy, and the identification of L-DOPA as the allelopathic substance from promising velvetbean (*Mucuna pruriens*). Bull Natl. Inst. Agr.-Env. Sci. 10: 115-218.

Fujii, Y., Shibuya, T., Nakatani, K., Itani, T., Hiradate, S., Parvez, M.M. (2004) Assessment method for allelopathic effect from leaf litter leachates. Weed Biol. Manage. 4: 19-23.

Hatanaka, S., Terakawa, H. (1968) Biochemical studies on nitrogen compounds of fungi I. Distribution of some nonprotein amino acid 1. Bot. Mag. Tokyo. 81: 259-266.

Hoagland, R.E. (1999). Biochemical Interactions of the microbial phytotoxin phosphinothricin and analogs with plants and aminmals. *In* Biologically Active Natural Products: Agrochemicals, H.G. Cutler, S.J. Cutler, (eds). CRC Press, Boca Raton, London, NY, Washington DC. pp.107-125.

Hongo, T. (2003) Narrow road to mushrooms (in Japanese): Tombow Publishing Co., Ltd., Tokyo, Japan. pp. 30-31.

Ikeda, M., Naganuma, Y., Ohta, K., Sassa, T., Miura, Y. (1977) Isolation and identification of a plant-growth inhibitor, azetidine-3-carboxylic acid, from *Clavaria miyabeana* S. Ito and its occurrence in the family *Clavariaceae*. Nippon Nôgeikagaku Kaishi. 51: 519-522.

Ikeda, M., Sato, Y., Sassa, T., Miura, Y. (1978) Structures of fasciculols, new plant growth inhibitors from *Naematoloma fasciculare*. Symposium Papers 21st Symposium on the Chemistry of Natural Products, Sapporo, Japan. pp. 584-591.

Ikeda, M., Kanou, H., Nukina, M. (1989) Plant growth inhibitors from fruit bodies of *Polyporus confluens* Bull. Yamagata Univ., Agr. Sci. 10: 849-852.

Kitaya, Y., Tani, A., Kiyota, M., Aiga, I. (1994) Plant growth and gas balance in a plant and mushroom cultivation system. Adv. Space Res. 14: 281-284.

Lincoff, G.H. (1981) National Audubon Society Field Guide to North American Mushrooms: Chanticleer Press Inc., NY, USA. pp. 11-13.

Lockard, J.D., Kneebone, O.R. (1962) Investigation of the metabolic gases produced by *Agaricus bisporus* (Lange) Sing. Mushroom Sci. 5: 281-299.

Macías, F.A., Molinillo, J.M.G., Galindo, J.C.G., Varela, R.M., Torres, A., Simonet, A.M. (1999) Terpenoids with potential use as natural herbicide templates. *In* Biologically Active Natural Products: Agrochemicals, H.G. Cutler, S.J. Cutler, (eds). CRC Press, Boca Raton, London, NY, Washington DC. pp. 15-31.

Miller, C.O. (1967) Zeatin and riboside from a Mycorrhizal fungus. Science. 157: 1055-1057.

Molisch, H. (1937) Der Einfluβ einer Pflanze auf die andere Allelopathy: Fischer, Jena, Germany. pp. 41

Morimoto, Y., Takaishi, M., Kinoshita, T., Sakaguchi, K., Shibata, K. (2002) Total synthesis and determination of the stereochemistry of 2-amino-3-cyclopropylbutanoic acid, a novel plant growth regulator isolated from the mushroom *Amanita castanopsidis* Hongo. Chem. Commun. pp. 42-43.

Nada, G. (1975) Hormonalni vpliv gob *Calocybe georgii* Fr. in *Agaricus campester* Fr. na rast in razvoj trav v èarovniškem risu. Biol. Vestn. (Ljubljana). 23: 89-96.

Ogawa, M. (1978) Biology of Matsutake (in Japanese). Tsukiji Shokan Publishing Co., Ltd., Tokyo, Japan. p. 326.

Oka, Y., Tsuji, H., Ogawa, T., Sasaoka, K. (1981) Quantitative determination of free amino acids and their derivatives in the common edible mushroom, *Agaricus bisporus*. J. Nutr. Sci Vitaminol. 27: 253-262.

Rice, E.L. (1995) Biological Control of Weeds and Plant Diseases - Advances in Applied Allelopathy. Univ. Oklahoma Press, Norman, USA. p. 439.

Richter, E. (1967) Über den einfluss des verschliessens von champignon-brut-flaschen mit polypropylenfolie auf den stoffwechsel des wachsenden mycels. Gartenbauwissenschaft. 32: 331-342.

Saito, A., Murayama, T., Nazarova, M., Ikeda, M. (1995) Studies on metabolites in Numeriiguchi (*Suillus luteus*, collected in Russia). Nippon Nôgeikagaku Kaishi Sapporo, Japan.

Senatore, F., Dini, A., Cerri R., Schettino, O. (1987) Chemical constituents of some Tricholomataceae. Biochem. System. Ecol. 15: 639-641.

Senatore, F., Dini, A., Marino, A., Schettino, O. (1988) Chemical constituents of some Basidiomycetes. J. Sci. Food Agric. 45: 337-345.

Smiley, R.W., Dernoeden, P.H., Clarke, B.B. (1992) Fairy rings. *In* Compendium of Turfgrass Diseases 2nd Ed., R.W. Smiley, P.H. Dernoeden, B.B. Clarke (eds.). APS Press, Minnesota, USA. pp. 61-63.

Steglich, W., Esser, F. (1973) L-3,4-Dihydroxy-phenylalanin aus *Strobilomyces floccopus*. Phytochemistry. 12: 1817.

Steglich, W., Preuss, R. (1975) L-3,4-Dihydroxyphenylalanine from carpophores of *Hygrocybe conica* and *H. ovina*. Phytochemistry. 14: 1119.

Tschierpe, H.J. Sinden, J.W. (1965) Über leicht flüchtige produkte des aeroben und anaeroben stoffwechsels des culturchampignons, *Agaricus campestis* var. bisporus (L.) Lge. Archiv. Für Mikrobiologie. 52: 231-241.

Wakabayashi, K., Soga, K., Hoson, T., Kamisaka, S., Yoshimura H., Shibata, K. (2001) Growth inhibition of lettuce (*Lactuca sativa* L.) roots by α-amino acids, 2-amino-3-cyclopropyl-butanoic acid and 2-amino-5-chloro-4-pentenoic acid, isolated from *Amanita castanopsidis* Hongo. Plant Growth Regul. 33: 169-173.

Wada, T., Kobata, K., Hayashi Y., Shibata, H. (1995) Two chemotypes of *Boletus cavipes*. Biosci. Boiotech. Biochem. 59: 1036-1039.

Weston, L.A. (1996) Utilization of allelopathy for weed management in agroecosystems. Agron. J. 88: 860-866.

Yoshimura, K., Takegami, K., Doe, M., Yamashita, T., Shibata, K., Wakabayashi, K., Soga, K., Kamisaka, S. (1999) α-Amino acids from a mushroom, *Amanita castanopsidis* Hongo, with growth-inhibiting activity. Phytochemistry. 52: 25-27.

23

Allelopathic Action of Triticale Allelochemicals Towards Grain Aphid

B. Leszczynski, A. Wójcicka, S. Golawska and H. Matok
Department of Biochemistry, University of Podlasie,
ul. B. Prusa 12, 08-110 Siedlce, Poland

Chemical interactions between cereals and the grain aphid, focusing on the role of the cereal allelochemicals have been studied. Cereal allelochemicals showed allelopathic action towards the grain aphid at the surface, within peripheral tissues (epidermis and mesophyll), and within the phloem sap. The grain aphid accepted cereal cultivars heavily covered with surface waxes considerably less. GLC-MS analysis of dichloromethane extracts from flag leaves and ears showed that such cultivars contained high level of β-diketones and lauric acid derivatives. High content of the cereal allelochemicals: phenolic compounds, hydroxamic acids and indole alkaloids within peripheral tissues as well as low concentration of nutrients and feeding stimulators in phloem sap strongly reduced the abundance and development of the grain aphid population on the studied cultivars. The relationship between the cereal allelochemicals and acceptance of the cereals during host-plant selection by the grain aphid is discussed.

Keywords: *aphid feeding behavior, cereal allelochemicals, EPG, epicuticular waxes, grain aphid, Sitobion avenae, surface compounds*

INTRODUCTION

Aphids are important phytophagous pests equipped with sucking-piercing mouthparts, which mostly feed on phloem sap. Host plant selection by the aphids consists of a series of the following events: 1) flight (migration); 2) landing on a selected plant; 3) preliminary probing/testing; 4) feeding; and 5) reproduction (Dixon, 1987; Dabrowski, 1988). At all the stages of the host-plant selection by the aphids, there are strong allopathic interactions between cereals and the cereal aphids.

The aphids abundance on cereals is dependent on the level and distribution of plant nutrients and allelochemicals within the plant tissues (Niraz *et al.*, 1996; Leszczynski, 1999). These chemical interactions are grouped into at least three different categories which vary in time and location. The first step of the host-plant selection by cereal aphids is located at the plant surface and involves volatile and non-volatile surface allelochemicals. The second step considers allelochemicals occurring mostly within vacuoles of peripheral tissues (epidermis and mesophyll) that affect the aphids during the preliminary probing of the potential new host plant. The third step is connected with cereal allelochemicals occurring usually in trace doses within cells of the vascular system. Each of the three stages of the host-plant selection by the aphids is crucial and may affect the final aphid decision, which is either to stay on the selected plant or leave it (Niemeyer, 1990).

There are several aphid species occurring on cereals in Europe, including the grain aphid (*Sitobion avenae* Fabricius), bird cherry-oat aphid (*Rhopalosiphum padi* L) and (*Metopolophium dirhodum* Walker). The grain aphid is a dominant species on wheat, rye and triticale in Poland (Leszczynski *et al.*, 1997). The present chapter reports the effect of the plant allelochemicals located at the plant surface, in the peripheral tissues, and in the vascular system on the grain aphids' biology.

MATERIAL AND METHODS

Plant Material

Selected cultivars of winter triticale accepted and moderately resistant to the grain aphid were used in the field experiments. The tested cultivars were obtained from the Institute of Plant Breeding and Acclimatization (IHAR), Radzików/Blonie. The experiments were performed at IHAR's experimental station and on small experimental plots of Podlasie University located in Aleksandria Park at Siedlce.

Field Experiments

The field observations were carried out on experimental fields and on small plots (0.5 m × 0.5 m), during 1999-2001. Abundance of *S. avenae* on the studied cultivars was evaluated according to Wratten *et al.* (1979) and Lykouressis (1984). The number of aphids was counted weekly on 10 plants selected at random and the experiment was conducted in three independent replicates. The field observations were carried out from the beginning of the grain aphid colonization period until triticale maturity (G.S. 47-80; in Tottman and Broad scale) (Tottman and Broad, 1987). The aphid performance on the studied triticale cultivars was expressed as cumulative aphid index/stem and average percentage of the infested plants.

Laboratory Tests

Seeds of the studied cultivars were germinated in a glasshouse kept at 20-25°C and under 16 h daylight. The seedlings were grown in a medium nutrient, fine structure compost with sand in 7 cm × 7 cm × 9 cm plastic pots, one per pot. The pots were regularly watered and no extra fertilizers were added. Adult apterous females of the grain aphid, used in the experiment came from a stock culture kept at the University of Podlasie in Siedlce. The females were caged individually on 5 days old seedlings of the tested cultivars and allowed to deposit nymphs. After 24 h, one nymph remained on a single plant and the other offsprings and the adult were removed. Development time and reproduction of the individuals were observed daily until its death. Pre-reproductive period (time from birth until maturity), total and daily fecundity, intrinsic rate of natural increase (r_m), net reproduction (R_o), multiplication of the population growth (λ) and an average generation development time (T) were determined (Leszczynski, 1987).

Aphid Feeding Recordings

Feeding behavior of the grain aphid on the studied winter triticale cultivars was monitored with help of EPG (electrical penetration graphs) method (Tjallingii, 1988). The 8 h EPG recordings were carried out, for each cultivar on 10 different plants and aphids. The duration and number of such aphid activities as: non-probing (Np), penetration of peripheral plant tissues (ABC), salivation into sieve elements (E1), phloem sap ingestion (E2) and xylem sap ingestion (G) were measured. Analysis of the EPGs were performed using the computer program Stylet (DOS PCs).

Extraction of Surface Waxes

The surface waxes of the triticale were extracted from flag leaves and ears at the florescence stage (G.S. 65). Dipping the studied triticale organs into cold dichloromethane for approximately 30 s carried out the extraction (Bianchi and Figin, 1986). Then the extracts were filtered and treated with (BSTFA-PIRYDYNE (2:3) reagent). The silylation mixture was composed of one part wax extracts and ten parts of BSTFA-PIRYDYNE reagent.

GC-MS Analysis

Chemicals occurred in the surface wax layer of the triticale were analysed by combined gas chromatography-mass spectrometry (SHIMADZU GC-MS, QP 5050A, equipped with Zebron ZB-5 column (30 m × 0.25 mm). The sample was introduced *via* a column injector and the GC separation was performed at the temperature - programmed from an initial 280°C to 340°C (a rise 9.8°C per min). Epicuticular waxes composition was identified by their mass spectra which were matched by computer search with the Mass Spectral Library (MS – Windows CLASS - 5000).

Cereal Allelochemicals and Nutrients Assays

While the aphid occurred on the studied cereals, the flag leaves were collected, freeze dried and used for analyses of the chemical constituents. The content of phenolic compounds within the flag leaf tissues of the studied cereals was measured spectrophotometrically after Leszczynski *et al.* (1989). Cyclic hydroxamic acids were determined with the help of HPLC method according to Leszczynski and Dixon (1990). Ion-exchange chromatography was used to determine amino acids level (Leszczynski *et al.*, 1999). The content of sucrose was estimated spectrophotometrically after Leszczynski (1996).

Statistics

The differences among the cultivars in the resistance parameters were tested with one-way ANOVA, followed by Duncan's test.

RESULTS AND DISCUSSION

Effect of Surface Allelochemicals on Host-plant Selection by the Grain Aphid

Field observations indicated that heavy waxy-covered triticale significantly affected the biology of the grain aphid *S. avenae* and were less accepted than wax-less ones. The grain aphid was less abundant and formed a smaller colony on the waxy cereals. As a result, significant differences in cumulative aphid index value and average percentage of infested plants of both types of the studied cultivars were found (Table 1).

EPG recordings showed that aphids spent about 78% of the total time with stylets inserted into tissues of the waxy covered cultivars, in comparison to 90% on wax-less ones. Significant differences between the cultivars in regard to the number of initial probes and the duration of the non-probing phase were observed (Table 1).

Table 1 Abundance and probing behavior of the grain aphid on waxy and wax-less of triticale during the initial period of tissues penetration

Studied parameters	*Cereals*	
	Waxy cultivar	*Wax-less cultivar*
Cumulative aphid index	4.90 b	11.90 a
Percentage of infested plants	16.10 b	32.80 a
Number of initial probes	10.30 a	6.80 ab
Duration of non-probing (min)	106.30 a	49.30 b

The values in the same rows followed by different letters are significantly different at 0.05 (Duncan's test)

Extracts of surface chemicals from the triticale cultivars heavily covered with waxes contained five major classes of the chemical constituents: alkanes, aldehydes, acids, esters and fatty acids derivatives. Within the surface extracts the following 19 compounds were identified: propanoic acid, nonanal, octanoic acid, tetradecanoic acid, hexadecanoic acid, octadecanoic acid, 2-methyl eicosane, eicosanoic acid, tricosane, pentacosane, docosanoic acid, hexacosane, tetracosanoic acid, octacosane, hexacosanoic acid, triacontane, hentriacontane, 14,16-hentriacontanedione and lauric acid 2-(hexadecyloxy)-3-(octadecyloxy) propyl ester (Fig. 1).

The profile of the surface compounds extracted from the studied triticale showed a clear difference in content and composition. The most striking feature of the whole picture was represented by complete lack of ß-dicarbonyl compounds in waxy-less cultivar. Furthermore, such a total inhibition of the ß-diketones biosynthesis are passed on and maintained on leaves and also on the ears. When the individuals were studied some differences in content and presence of the surface chemicals were found as well (Table 2).

Results obtained in the performed experiments clearly suggest that epicuticular waxes play an important role in host plant selection by the grain aphid, especially during the first contact of winged migrant aphids with colonized plants. On the other hand, it is well documented that stimuli detected by herbivorous insects at the plant surface are important cues for host plant selection, particularly during the first contact with new plants (Stadler, 1986; Woodhead and Chapman, 1986). The surface waxes are responsible for the plant texture and color of the aerial plant organs that can appear "light green" (nonglaucosness), "dark green" "blue-gray" or even "white" according to the nature and content of the epicuticular lipid layer. Glaucosness, due to a superficial deposit of light - scattering crystallites of wax, is generally thought to be a result of the amount and chemical composition of the epicuticular waxes.

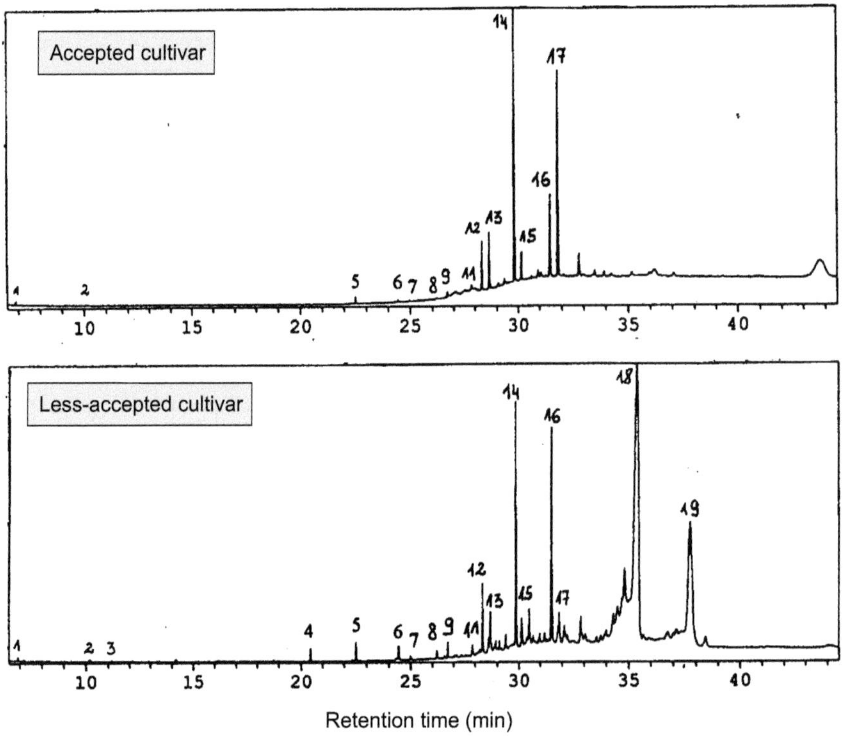

Fig. 1 Profiles of surface compounds extracted from the waxy and wax-less triticale cultivars. 1, propanoic acid; 2, nonanal; 3, octanoic acid; 4, tetradecanoic acid; 5, hexadecanoic acid; 6, octadecanoic acid; 7, 2-methyl eicosane; 8, eicosanoic acid; 9, tricosane; 10, pentacosane; 11, docosanoic acid; 12, hexacosane; 13, tetracosanoic acid; 14, octacosane; 15, hexacosanoic acid; 16, triacontane; 17, hentriacontane; 18, 14, 16-hentriacontanedione; 19, lauric acid 2-(hexadecyloxy)-3-(octadecyloxy) propyl ester

Thus, the surface compounds are related to the glaucousness or non-glaucousness and often play an important role as mediators of insect-plant interactions (Lowe *et al.*, 1985; Foster and Harri, 1992; Eigenbrode and Espelie, 1995; Eigenbrode, 1996; Storer *et al.*, 1996; Kanno and Harris, 1999). For the aphids, phytochemicals detected at the plant surface are especially important, since their initial plant selection clearly depends on the color and texture of plants (Prokopy and Owen, 1983; Niraz *et al.* 1985; Montlor, 1991).

The first step of the host-plant colonization by the aphids begins with secretion of their saliva on the plant surface, and is followed by minute punctures into the plant tissues. Thus, the presence of the 14,16-hentriacontanedione and lauric acid (2-(hexadecyloxy)-3-(octadecyloxy) propyl ester on the surface of the heavy waxy-covered cereal cultivars

Table 2 Content of the identified surface components of waxy and wax-less cereals (overview of the all studied cultivars)

Identified chemicals	*Cultivars*	
	Waxy	*Wax-less*
Propanoic acid	+	+
Nonanal	+	+
Octanoic acid	+	+
Tetradecanoic acid	+	+
Hexadecanoic acid	++	+
Octadecanoic acid	++	+
2-Methyl eicosane	+	+
Eicosanoic acid	+	+
Tricosane	++	+
Pentacosane	+	+
Docosanoic acid	+	+
Hexacosane	++	++
Tetracosanoic acid	+	++
Octacosane	++++	++++
Hexacosanoic acid	++	++
Triacontane	++++	+++
Hentriacontane	+	+++
14,16-Hentriacontanedione	+++++	-
Lauric acid 2-(hexadecyloxy)-3-(octadecyloxy) propyl ester	++++	-

absent; + - traces; ++ - low content; +++ - medium content; ++++ - high content; +++++ - very high content

might have a great impact on the selection of suitable plants by the grain aphid in respect of its feeding and reproduction.

Effect of Allelochemicals Occurring within Cereal Peripheral Tissues on the Grain Aphid Probing Behavior

Cereal allelochemicals present on the plant surface are also important in further stages of the host-plant selection such as: puncturing of the surface with the aphid probes, taste of the plant sap with the gustatory papillae and epipharyngeal organ and during further ingestion of the phloem sap.

Duration of peripheral tissues penetration (epidermis and mesophyll) of the studied cultivars was positively related to number of probes. Detailed analyses showed that the aphids spent much more time penetrating epidermis and mesophyll tissues of the less accepted waxy cultivars (Table 3). There was no significant difference in the duration of the first pathway event, but the total duration and number of the ABC patterns were significantly larger on both types of the studied cereals.

Table 3 Comparison of the grain aphid activity while penetration of the peripheral tissues (epidermis and mesophyll) with content of the studied cereal allelochemicals

Studied parameters	*Cultivars*	
	Less accepted	*Accepted*
Number of ABC patterns	13.70 a	9.70 ab
Total duration (min) of pattern ABC	181.90 a	110.40 b
Duration (min) of the first ABC pattern	7.50 a	8.70 a
Phenolics (mg/g fresh weight)	2.20 a	1.20 b
Hydroxamic acids (mg/g fresh weight)	0.24 a	0.13 b

Values in rows followed by different letters are significantly different at 0,05 (Duncan's test);
ABC patterns – peripheral tissues penetration (epidermis and mesophyll)

Peneration of peripheral tissues (epidermis and mesophyll) was mostly affected by allelochemicals located within numerous vacuoles. Aphids are pretty sensitive to such allelochemicals as phenolic compounds, hydroxamic acids and indole alkaloids. These secondary plant metabolites significantly modified the aphids feeding behavior during peripheral tissues penetration (Table 3). So far, the role many of them in the chemical interaction between plants and aphids is unknown, but some of them e.g. hydroxamic acid and phenolic derivatives are important cereal allelochemicals strongly influencing the grain aphid feeding behavior and physiology (Todd *et al.*, 1971; Dreyer and Jones, 1981; Leszczynski *et al.*, 1995; Leszczynski *et al.*, 1996). Detailed analyses of the EPG recordings showed that aphids performed on the less accepted cultivars tended to increase a number of the stylet contacts with peripheral cells, which was observed in the form of so called pd (potential drops) patterns. During such cell interruptions, the aphids have the possibility to taste plant sap and get a message of the host-plant value.

Effect of Cereal Nutrients and Allelochemicals on the Grain Aphid Growth and Development

There was no significant difference in pattern E1 (salivation into sieve elements) on both types of the studied cultivars. However, on accepted wax-less cultivars the grain aphid twice as long ingested the phloem sap from the sieve element and no difference in xylem phase between the studied triticale was observed (Table 4).

During the 8 h of the EPG recordings, the aphids fed on the accepted cultivars, were spending about 64% of that time in the phloem phase (salivation and ingestion of the phloem sap), instead the aphids fed on the less accepted triticale only 33% (Fig. 2).

Table 4 Duration (min) and number of the grain aphid, *S. avenae* activities during feeding on the studied triticale cultivars.

Aphid activity	*Cereal cultivars*	
	Less accepted	*Accepted*
Duration (min) of phloem phase (E1 + E2)	157.6 b	304.0 a
Number of E1 patterns	3.1 a	3.2 a
Total duration (min) of E1 pattern	8.0 b	16.0 a
Duration (min)of the first pattern E1	2.3 b	6.8 a
Number of E2 patterns	2.5 a	2.5 a
Total duration (min) of E2 pattern	149.6 b	288.0 a
Duration (min) of the first E2 pattern	52.6 b	146.9 a

Values in rows followed by different letters are significantly different at 0,05 level (Duncan's test);
E1 patterns - aphid salivation into sieve elements; E2 – phloem sap ingestion by the aphids

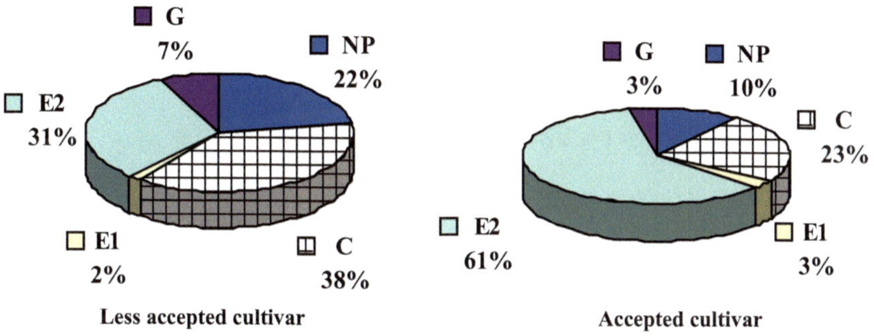

Fig. 2 Average percentage of the EPG patterns during 8 h recordings from apterous adult of the grain aphid fed on waxy and wax-less triticale cultivars. NP – Non-penetration; ABC(C) – penetration of the peripheral tissues (epidermis and mesophyll); E1 – aphid salivation into sieve elements; E2 – phloem sap ingestion by the aphids; G – xylem sap ingestion by the aphids.

Accepted cereals were characterized by a higher content of nutrients (amino acids) and phagostimuli (sucrose) in comparison into less-accepted cultivars (Table 5). As a result, the worst cultivars decreased the grain aphid feeding activity and in consequence prolonged its development time and reduced fecundity (Table 5). The obtained results suggest allelopathic influence of, the cereal allelochemicals towards the grain aphid include both resistance mechanisms: non-preference and antibiosis.

Non-preference mechanism is closely related to the preliminary stages of the host-plant selection by the grain aphid. Undoubtedly, during landing and testing of the texture of the plant surface the most important are allelochemicals occurring within the epicuticular wax layer.

Table 5 Antibiotic effect of less accepted cereals on the grain aphid biology at the phloem phase level

Studied parameters	*Cultivars*	
	Less accepted	*Accepted*
Prereproductive period (days)	11.20 a	10.30 b
Daily fecundity	2.20 b	3.10 a
Free amino acids (g/g freeze dry weight)	0.10 ab	0.12 a
Sucrose (g/g freeze dry weight.)	0.05 b	0.08 a
Phloem sap ingestion (%)	31.00 b	61.00 a

The second resistance mechanism - antibiosis is closely related to the level of nutrients and cereal allelochemicals located in phloem sap and within the peripheral tissues. The cultivars characterized by a high level of the cereal allelochemicals e.g. phenolic compounds, hydroxamic acids and/or indole alkaloids and lower concentration of nutrients and feeding stimulators seriously affect the grain aphid metabolism. Thus the chemical interaction between cereals and aphids determine their pest status and play a crucial role in the aphid control in the agrocenoses.

CONCLUSION

Cereal allelochemicals act at three levels towards the grain aphid during the chemical interactions going on between the host-plant and its herbivorous pest. The first level occurs at the plant surface where the wax layer and differences in its chemical composition is responsible for the host-plant selection by the grain aphid. The second level appears within peripheral tissues, when the aphid meets most of the cereal allelochemicals located within the vacuoles. When the aphids reach the vascular system, the third level occurs mostly connected with level of the nutrients and phagostimuli and traces concentration of the allelochemicals within the cereal phloem sap.

References

Bianchi, G., Figini, M.L. (1986) Epicuticular waxes of glaucous and nonglaucous durum lines. J. Agric. Food Chem. 34: 429-433.

Dabrowski, Z.T. (1988) Odpornosc roslin uprawnych na owady (Insecta). *In* Podstawy Odpornosci Roslin na Szkodniki, 2nd Edn. PWRiL Warszawa, pp. 30-63. (In Polish).

Dixon, A.F.G. (1987). Cereal aphids as an applied problem. Agric. Zool. Rev. 2: 1-57.

Dreyer, D.L., Jones, K.C. (1981) Feeding deterrency of flavonoids and related phenolics towards *Schizaphis graminum* and *Myzus persicae*, aphid feeding deterrents in wheat. Phytochemistry. 20: 2489-2493.

Eigenbrode, S.D. (1996) Plant surface waxes and insect behaviour. *In* Plant Cuticles. An Integrated Functional Approach. G. Kerstiens. (ed). Bios Scientific Publishers, Oxford. UK. pp. 201-221.

Eigenbrode, S.D., Espelie, K.E. (1995) Effect of plant epicuticular lipids on insect herbivores. Ann. Rev. Entomol. 40: 171-194.

Foster, S.P., Harris, M.O. (1992) Foliar chemicals of wheat and related grasses influencing oviposition by hessian fly, Mayetiola destructor (Say) (Diptera: Ccidomyiidae). J. Chem. Ecol. 18: 1965-1980.

Kanno, H., Harris, M. (1999) Plant surface substances as insect mediators. *In* the Proceddings of the 16th Annual Meeting of International Society of Chemical Ecology. p. 69.

Leszczyński, B. (1987) Mechanizmy odporności pszenicy ozimej na mszycę zbożową *Sitobion avenae F.* ze szczególnym uwzględnieniem roli związków fenolowych. Rozp. Nauk. Nr 21, Wyd. Nauk. WSRP Siedlce (in Polish).

Leszczynski, B., Wright, L.C., Bakowski, T. (1989) Effect of secondary plant substances on winter wheat resistance to grain aphid. Entomol. Exp. Appl. 52: 135-139.

Leszczyski, B., Dixon, A.F.G. (1990) Resistance of cereals to aphids: Interaction between hydroxamic acids and the aphid *Sitobion avenae* (Homoptera: Aphididae). Ann. Appl. Biol. 117: 21-30.

Leszczynski, B., Dixon, A.F.G., Bąkowski, T., Matok, H. (1995) Cereal allelochemicals in grain aphid control. Allelopathy J. 2: 31-36.

Leszczyński, B. (1996) Sacharoza. *In* Kurs praktyczny w zakresie chemicznych interakcji owady - rośliny na przykładzie mszyc (Aphidoidea). Wyd. Nauk WSRP Siedlce. pp. 194-196 (In Polish).

Leszczynski, B., Bakowski, T., Rozbicka, B., Matok, H., Urbanska, A., Dixon, A.F.G. (1996) Interaction between cereal phenolics and grain aphid (*Sitobion avenae* Fabr.). IOBC/ WPRS Bull. 19: 100-105.

Leszczynski, B., Jozwiak, B., Laskowska, I., Matok, H. (1997) Cereals naturally resistant to aphids - an approach to ecological agriculture. J. Appl. Genet. 38B: 179-183.

Leszczynski, B. (1999) Plant allelochemicals in aphids management. *In* Allelopathy, Vol . II, Basic and Applied Aspects, S.S. Narwal. (ed). Oxford & IBH Publishing Co. Pvt. Ltd. New Delhi, India. pp. 285-320.

Leszczynski, B., Jozwiak, B., Lukasik, I., Matok, H., Sempruch, C. (1999). Influence of nutrients and water content on host plant alternation of bird cherry-oat aphid *R. padi.* Aphids and Other Homopterous Aphids. 7: 223-229.

Lowe, H.J.B., Murphy, G.I.P., Parker, M.L. (1985) Non-glaucousness, a probable aphid-resistance character of wheat. Ann. Appl. Biol. 106: 555-560.

Lykouressis, D. (1984) A comparative study of different aphid population parameters in assessing resistance in cereals. Z ang. Ent. 97: 77-84.

Montlor, C.B. (1991) The influence of plant chemistry on aphid feeding behaviour. *In* Insect-Plant Interactions, Vol. III, E. Bernays (ed.) CRC Press, Boca Ranton. Fl, USA. pp. 125-173.

Niemeyer, H.M. (1990) Secondary plant chemicals in aphid-host interactions. *In* the Proceedings of Aphid-Plant Interactions: Populations to Molecules. August 12-17, Stillwater, Oklahoma, USA. pp. 101-111.

Niraz, S., Leszczynski, B., Ciepiela, A., Urbanska, A., Warchol, J. (1985) Biochemical aspects of winter wheat resistance to aphids. Insect Sci. Applic. 6: 253-257.

Niraz, S., Leszczynski, B., Urbanska, A., Matok, H., Ciepiela, A. (1996) Biochemical mechanisms of aphid resistance in cereals - 20 years of research. Plant Breed. Seed Sci. 40: 85-89.

Prokopy, R.J., Owens, E.D. (1983) Visual detection of plants by herbivorous insects. Ann. Rev. Ent. 28: 337-364.

Stadler, E. (1986) Oviposition and feeding stimuli in leaf surface waxes. *In* Insects and the Plant Surface. B.E. Juniper, T.R.E. Southwood (eds.). Edward Arnold, London. UK. pp. 105-121.

Storer, J.R., Powell, G., Hardie, J. (1996) Settling responses of aphids in air permeated with non-host plant volatiles. Entomol. Exp. Appl. 80: 76-78.

Tjallingii, W.F. (1988) Electrical recording of stylet penetration activities. *In* Aphids Their Biology, Natural Enemies and Control, Vol. B. A.K. Minks, P. Harrewijh (eds.) Elsevier, Amsterdam. The Netherlands. pp. 95-108.

Todd, G.W., Getahun, A., Cress, D.C. (1971) Resistance in barley to greenbug, *Schizaphis graminum*. I. Toxicity of phenolic and flavonoid compounds and related substances. Ann. Ent. Soc. Amer. 64: 718-722.

Tottman, D.R., Broad, H. (1987) The decimal code for the growth stages of cereals, with illustrations. Ann. Appl. Biol. 110: 441-454.

Woodhead, S., Chapman, R.F. (1986) Insect behaviour and the chemistry of plant surface waxes. *In:* Insects and the Plant Surface. B.E. Juniper, T.R.E. Southwood (eds.) Edward Arnold, London. UK. pp. 123-135.

Wratten, S.D., Lee, G., Stevens, D.J. (1979) Duration of cereal aphid populations and the effects on wheat yield and quality. *In the* Proceedings of British. Crop Protection Conference - Pest and Diseases. pp. 1-8.

24

Rat Sexual Behavior and Volatile Substance from Plants

Sadao Yamaoka[1], Teruyo Tomita[1] and Akikazu Hatanaka[2]

[1]*Department of Regulatory Physiology, Dokkyo University School of Medicine, Mibu, Tochigi 321-0293, Japan*

[2]*Department of Applied Biological Chemistry, Yamaguchi University, Yamaguchi 753-8511, Japan*

The sexual behavior of male and female rats (Sprague-Dawley) were observed 30 min after exposure of volatile odorant substances (α-pinene, green odors) from green plants. Male sexual behavior was divided in to mount, intromission and ejaculation. The mount or intromission latency and post-ejaculation interval was significantly shortened after odorant exposure compared with the non-exposed group. The lordosis quotient (LQ), the percent of lordosis reflex/ 20-30 mounts by sexually active male rats, was measured 30 min after odorant exposure for 2 μg EB (subliminal doses) pretreated OVX rats. These experiments were repeated three times for two groups, with changed odorant exposure and non-exposure at each experiment. The odorant exposure group showed significantly higher LQ at each experiment. From these results, it can be suggested that odorant exposure may activate the motivation of sexual behavior in rats of both sexes, and the phenomena may correlate with the olfactory-hypothalamic-limbic pathway.

Keywords: *α-pinene, green odor sexual behavior, intromission latency, post-ejaculation interval, lordosis*

INTRODUCTION

The sexual behavior of male and female rats is a series of complicated actions, which in males comes from ejaculation in the pursuit of and mounting estrous female rats, and in females is characterized by the lordosis reflex of the male mount. These rat sexual behaviors seem to be a series of neural reactions in the central nervous system (CNS) that the sexual excitement occurs by pheromones of each male and female, but in castrated animals, such sexual behavior does not appear. The non estrous female never accepts the male and continues to deny it. Therefore, the CNS must be sensitized by androgen in the male and estrogen in the female and pheromones may be involved in CNS sensitization of sex steroids. On the other hand, it is known that volatile substances from plants affect several animal behaviors, such as the effects of Actinidia on the madness behavior of cats, the effects of α-pinene on bark-deprivation behavior of the Japanese bear (Kamiyama, 1985). These phenomena, suggest that the very low concentration of plant-derived volatile substances of ppb has had a lot of influence on these animals. It is also suggested by this mechanism that these volatile substances are not absorbed from the nasal mucosa, but are transmitted through olfactory-central nervous system as odor. It is well known that animal sexual behavior correlate with the olfactory-central nervous system. However, there is little evidence to consider the relationship between plant derived-odor and animal sexual behavior. In this report, we studied the effects of α-pinene and green odors on the sexual behavior of both sexes of rats.

MATERIALS AND METHODS

Plants Derived Odorants and its Delivery System

The alpha-pinene [(±- α-pinene (Aldrich Chem. Co. Inc., US)] was purchased, and the green odorous substances (equal mixture of n-hexenol and n-hexenal,) were supplied from Soda Aromatic, Tokyo, Japan. Both odorous substances were diluted with triethyl citrate to 0.3% (w/w) (Fig. 1). The volatile odorants from both substances were introduced to animal cages through a Teflon tube by using odorants delivery system (Fig. 2).

Animals

The Sprague–Dawley (originally Charles River, international global standard, CD-IGS) 24 male rats weighing 450-550 g (sexually experienced) and 24 female rats weighing 200-300 g bred at the SPF system in Experimental Animal Center of Dokkyo University School of Medicine were used. Twelve males and 12 females were used for the experiment of α-pinene exposure, and the remaining rats (i.e., 12 male and 12 female) for the experiment of green mix odor exposure. The rats were housed under

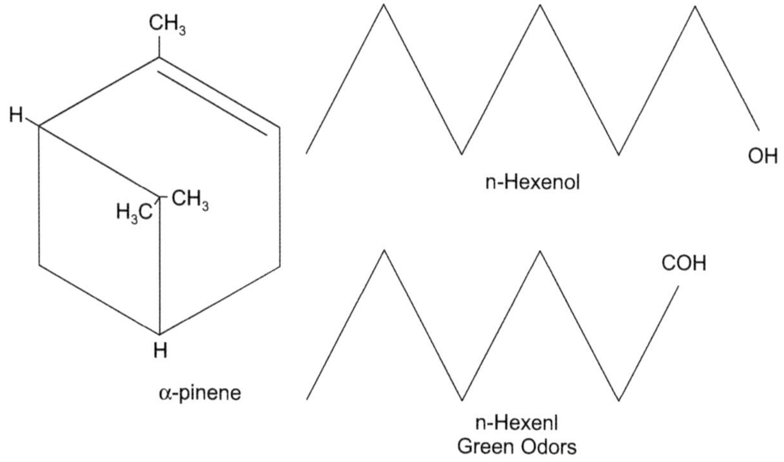

Fig. 1 Chemical structure of odors

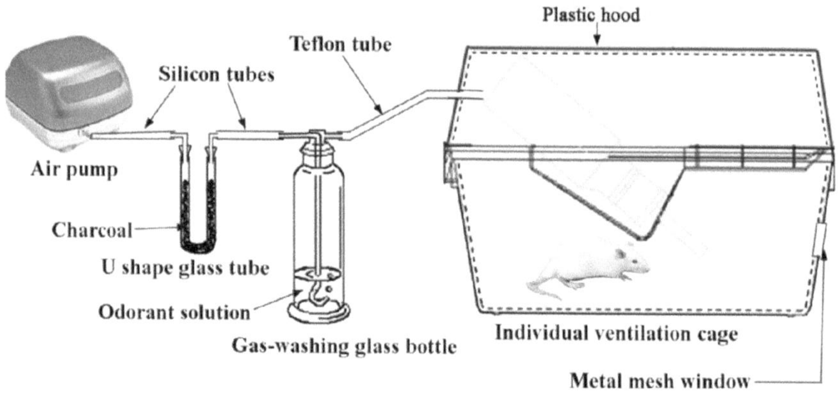

Fig. 2 The odor delivery system. Odorant solution (70 ml) (0.3% w/w in triethyl citrate) in a gas-washing bottle was bubbled by the charcoal filtered air and delivered into individual ventilation cage.

controlled temperature (24 ± 0.5°C), humidity (50 ± 5%) and illumination (light 0500-1900), and given food and water *ad libitum*.

For the male sexual behavior, male rats were moved 5 d before the experiment to the room of different lighting schedule (light 00:00-14:00). Six female partners of males were ovariectomized (OVX) at least one mon before, and were pretreated with 20 μg estradiol benzoate (EB) 3 d before and 2 mg progesterone 4 to 6 h before the experiments. These female partners showed maximum sexual activation during the experimental period. Twelve male rats were divided into two groups (A, B each in individual cages) at 9:00 am on the day of experiment. The observation of sexual behavior was conducted between 16:00 and 19:00 pm under dim red light. The difference of group A and B was taken as the existence of odor exposure. The second testing 1 wk later made the existence of odor exposure reverse (Fig. 3). Group A (group B in second testing period) was exposed to odor for 30 min before the experiment in a different room apart from females and group B using the odorant delivery system.

The mating tests were done during the dark phase of the cycle under dim red light. Rats were placed in acrylic cylindrical test arena ($\phi 40 \times 60$ cm), and after a 5-min adaptation period, each subject was presented with a receptive female partner. The number of mounts and intromissions as well as the mount, intromission, and ejaculation latencies, and the post-ejaculatory interval, were manually recorded and measured by a strip chart recorder (Fig. 3). The test ended in one of the following circumstances: 1) 15 min after the presentation of the female to the male if no intromission occurred, 2) 30 min after the first intromission if no ejaculation had occurred, or 3) after the first intromission following ejaculation.

For female sexual behavior, bilateral ovariectomy (ovx) was performed under light ether anesthesia. Three weeks after ovx, at 12:00 pm on the 3 d before the first testing day, all animals received subcutaneous (s.c.) injection of 2 μg EB, the subliminal dose which does not elicit female sexual behavior by only one injection. Twelve female rats were divided into two groups (A, B each in individual cage) at 9:00 am on the day of the first experiment. The difference of group A and B was taken as the existence of odor exposure. The second and third experiments on 3 d intervals, made the existence of odor exposure reverse every time (Fig. 4). The observation of sexual behavior was conducted between 14:00 and 17:00 pm under a dim red light. Odor exposure 30 min before the experiment was done in different rooms away from the non exposure group and males. Two sexually experienced stud male rats were used during the testing period, which consisted of 10-15 mounts. Sexual receptivity was quantified by lordosis quotient (LQ, lordosis responses/number of mounts $\times$ 100%). After the first and second experiments, all animals received s.c. injection of 2 μg EB. All experimental procedures followed "Guidelines for the Care and Use of Laboratory Animals, Dokkyo University School of Medicine" and "Guiding principles for the care and use of animals in the field of physiological sciences" of the Physiological Society of Japan.

Statistics

Statistical comparisons were performed using the two-tailed paired Student's *t* test when comparing two groups in all parameters of male

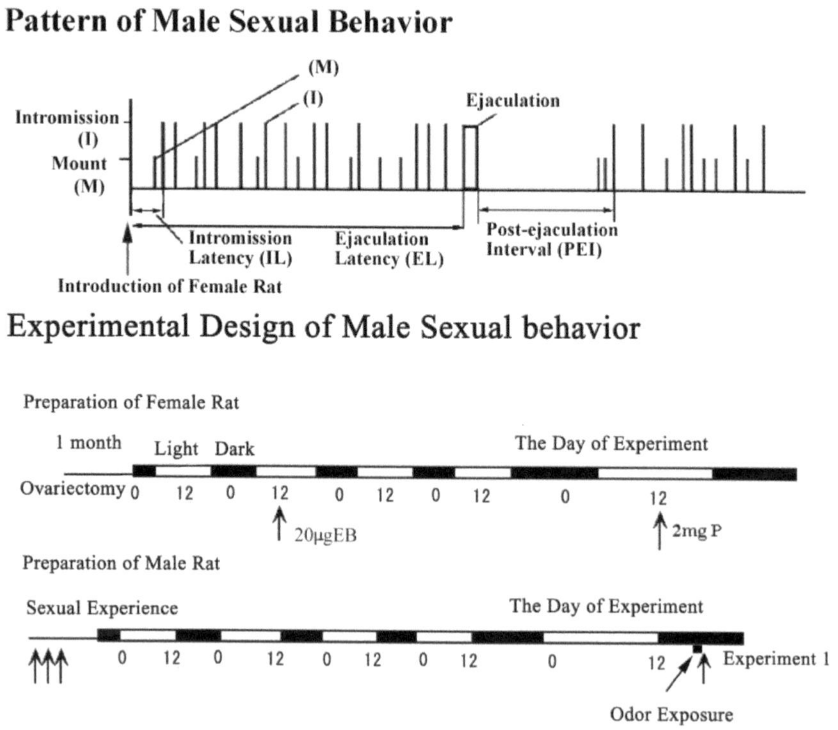

Fig. 3 The upper-part of the figure represents the illustrated male sexual pattern as recorded by the strip chart recorder. The lower-part of the figure represents the experimental design of male sexual behavior. The different lighting schedules were used for the male (light on at 00:00 and off at 14:00pm) and the female partner (light on at 05:00 and off at 19:00 pm). The female partner was ovariectomized (OVX) at least one mon before, and was pretreated with 20 μg estradiol benzoate (EB) at 12:00 pm 3 d before and 2 mg progesterone (P) 4 to 6 h before the experiments. The odor exposure for sexually experienced male rats was started 30 min before the experiment.

sexual behavior. Differences of the mean lordosis quotient were analyzed by one-way ANOVA with repeated measures and the Newman-Keuls multiple comparison test for the experiments repeated three times in each group and were compared with the two-tailed un-paired Student's *t* test in each experiment. The criterion for statistical significance was $P < 0.05$ for all tests.

RESULTS

Male Sexual Behavior

The male rats pursue the rear of sexually heated females, who are introduced into the testing arena marking the territory, and are shown

Experimental Schedule of Female Sexual Behavior

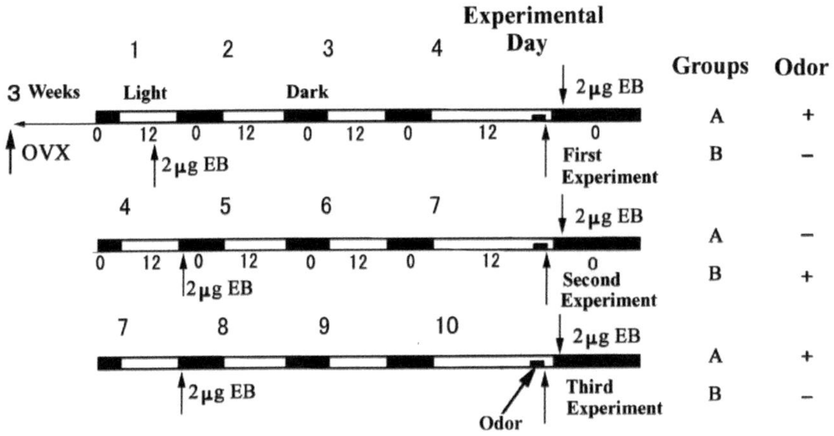

Fig. 4 Experimental schedule of female sexual behavior. All experimental female rats were ovariectomized over 3 wk before the experiment. Three days before the experiment, the subcutaneous injection of 2 μg EB was injected every other third day for all animals. The first EB injection was injected at 12:00 pm, and at the other three times, EB injections were injected after each experiment.

wiggling and hopping. They repeat this for several seconds, and then the male mounts the female. All males displayed a complete pattern of sexual behavior (mounting, intromission, and ejaculation) within 30 min after the female was introduced. Group A of the first testing period (A-1) and group B of the second testing period (B-2) did not show any significant difference in both the α-pinene and green mix exposure for the all parameters of male sexual behavior. Therefore, group A-1 and group B-2 were collectively made into the odor group, and B-1 and A-2 were made into the non-odor (control) group.

The mount frequency (MF) and the intromission frequency (IF) were 20-50 times and 20-40 times respectively in the a-pinene experiments, and 4-13 times and 7-22 times respectively in the green mix experiments. The intromission latencies (IL) of control and odor exposure were 85.1 ± 26.7 (MEAN ± S.E.) sec and 24.7 ± 7.8 sec respectively in the α-pinene experiments, and 31.4 ± 22.5 sec and 6.2 ± 1.2 sec respectively in the green mix experiment. The post ejaculation intervals (PEI) of control and odor exposure were 382 ± 127 sec and 290 ± 92 sec respectively in the α-pinene experiments and 404 ± 21.2 sec and 328 ± 22.8 sec respectively in the green mix experiment. The ejaculation latency (EL) of control and odor exposure were 846 ± 268 sec and 834 ± 264 sec respectively in α-pinene experiments and 516.3 ± 56.6 sec and 248.3 ± 52.3 sec respectively in the green mix experiment. The parameters except PEI were less and shorter in the green mix experiment than in the α-pinene experiment. The reason for this result may be that both experiments were done separately in different animal groups and the green mix experiment group was more active in copulatory behavior than the α-pinene experiment group. The mount frequency and intromission frequency in both the α-pinene and green mix experiment showed any significant difference between odor and non-odor groups. The intromission latency and post-ejaculation interval in both the α-pinene and green mix experiment were significantly shortened by odor exposure. The ejaculation latency was significantly shortened by only green odor exposure (Fig. 5).

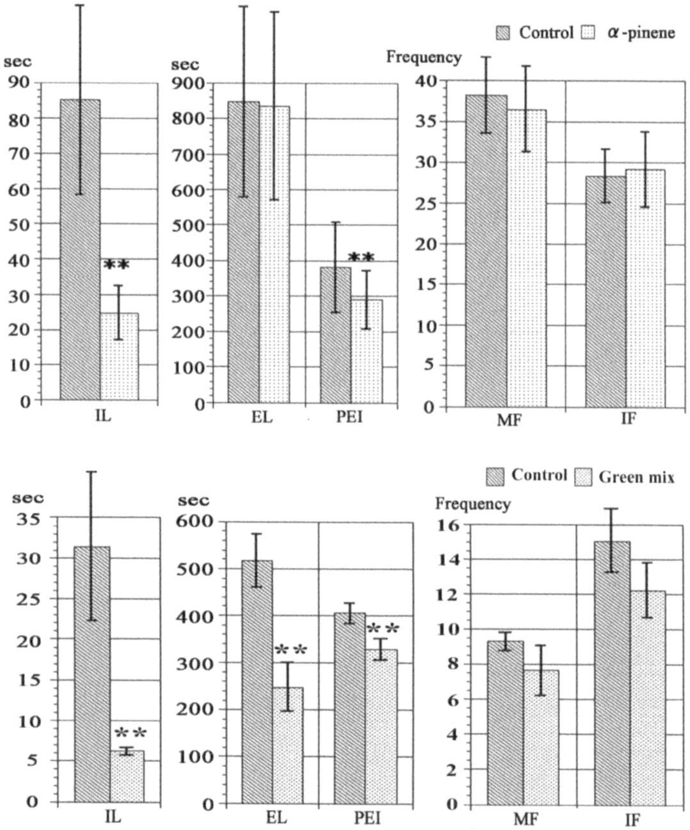

Fig. 5 Effects of odor exposure on male sexual behavior. Upper graphs represent α-pinene exposure and lower graphs represent green mix exposure. Values are mean ± SEM. **, $P < 0.001$ versus non-odor group (control) in each experiment. IL: intromission latency, EL: ejaculation latency, PEI: post-ejaculation interval, MF: mount frequency, IF: intromission frequency.

Female Sexual Behavior

Three days after the first s.c. injection of 2 μg EB alone (the first experiment), lordosis responses were 0% in non-odor groups of both the α-pinene and green mix experiment and in the green mix exposed group before the experiment, but the α-pinene exposed group showed the 35.2 ± 6.7% LQ. The second experiment showed significantly higher LQ in the α-pinene exposed group (84 ± 6.8%) than the non-odor group (30.5 ± 3.7%), but the green mix exposed group showed only 2.9 ± 1.7% LQ and did not show lordosis behavior in the non-odor group. The third experiment showed significantly higher LQ in the odor exposed groups (89.5 ± 5.4% and 88.14 ± 5.06% in the α-pinene and green mix exposed group respectively) than the non-odor groups (55 ± 5.6% and 45.4 ± 11.2%in in the α-pinene and green mix exposed group respectively) (Fig. 6). The difference of experiments repeated three times in each group was also analyzed by one-way ANOVA and the Newman-Keuls multiple comparison test and their results (the data were not shown) were mostly significant (P < 0.001) except first versus second test of B group in the green mix exposure experiment (both of them showed 0% LQ).

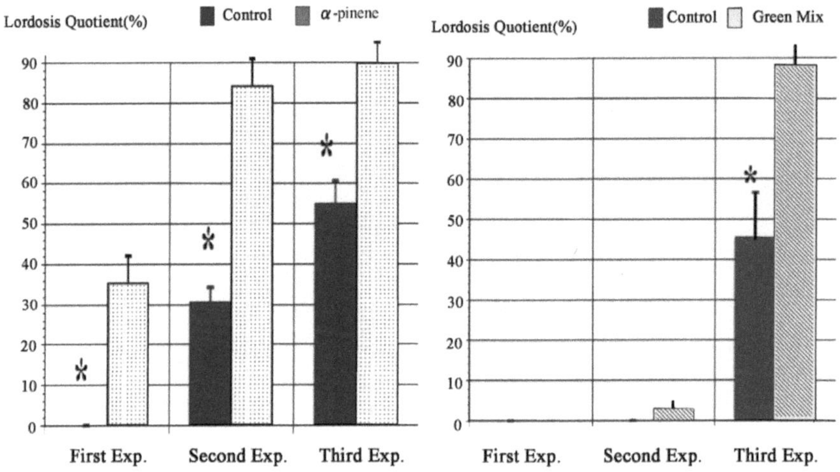

Fig. 6 Effect of odor exposure on lordosis quotient in subliminally sexual elicited female rat. Left graph represents the experiment of α-pinene exposure and right graph represents the experiment of green mix exposure. Values are mean ± SEM. *, P < 0.05 versus odor exposed group in each experiment.

CONCLUSION

The results reported here provide evidence that 30 min exposure of plant derived odors, α-pinene and green mix, represents the enhancement of sexual behavior in both the sexes of rats. Both the odors shortened the intromission latency; the period when the male interposes after being introduced to the female partner, and post-ejaculation latency, that is, the sexual refractory period after ejaculation. Both odors also enhanced the lordosis quotient under subliminal EB treatment in female rats.

The sexual behavior of the male rat is divided into four types of behavior, i.e., pursuit, mount, intromission, and ejaculation, which results in ejaculation after about 10 to 30 intromissions, and this behavior is again repeated after the sexual refractory periods for 5 to 8 min. This mechanism of male sex behavior in which various sensory information such as olfaction and visual sense and auditory sense and somatosensor is important, and there is the complicated central nervous process that a series of action is completed by reflecting these information for each central nervous area part place with the involvement of the androgen. It is also reported that the mount latency is extended, when olfaction is removed in the rat. However, olfaction and visual sense such tendency is dependent on the existence of the sexual experience of the rat, and there is some evidence that it is concerned in the female confirmation which came into heat without influencing in copulatory behavior itself directly. The most related central nervous system (CNS) in the male sexual behavior is the medial preoptic area of hypothalamus (MPOA). It is known that the lesion of MPOA suppress the male sexual behavior and the discharge rate of MPOA neuron increases during copulatory behavior (Sachs and Meisel, 1992).

There are numerous studies on the pheromone of rodents, such as the mouse on the relationship between the olfaction and sexual behavior (Johns *et al.*, 1978; Reynolds and Keverne, 1979; Lomas and Keverne, 1982). Several reports have postulated that the oral administration of plant extracts such as Mexican zoapatle (Carro-Juarez *et al.*, 2004), *Butea frondonsa* (Ramachandran *et al.*, 2004), *Tribulus terrestris* (Gauthaman *et al.*, 2003) enhance male sexual behavior similar to the aphrodisiac activity. Ginseng (one of herbals used for the treatment of sexual dysfunction in Asia) enhance libido and copulatory performance, penile erection by directly inducing the vasodilatation and relaxation of penile corpus cavernosum (Murphy and Lee, 2002).

However, we could not find any report concerning with the relationship between very weak smell of plant-derived odor and mammalian sexual behavior. From our present experiments, it is suggested that 30 min of odor exposure before sexual behavior may stimulate the CNS (including MPOA) related to male copulatory behavior through the olfactory system.

In the female rat, copulatory acts are limited to the display of lordosis. There are also numerous reports of female lordosis behavior and its central mechanisms. It is known that the female copulatory behavior is limited for a short time around the ovulatory period and is dependent mainly on estrogen conditions. The sexual behavior related to the nervous areas in females are also estrogen-dependent areas, such as MPOA, ventromedial hypothalamic nucleus, midbrain central gray (Meisel *et al.*, 1979; Pfaff and Sakuma, 1979; Sakuma and Pfaff, 1979; Yamanouchi *et al.*, 1990; Rajendren *et al.*, 1991; Calizo *et al.*, 2003).

Many reports have postulated that the olfactory stimulation affect the female reproduction in mammalians including humans (Keverne, 1978; Vandenbergh, 1983). There is some evidence of menstrual cycle-related changes in the emotional evaluation of odors (Hummel *et al.*, 1991; Graham *et al.*, 2000) and in olfactory sensitivity, with maximal sensitivity around ovulation (Doty *et al.*, 1981).

In this experiment, it was shown that 30 min of odor exposure (the same as the male experiment) before sexual behavior enhanced female receptivity with treatment of subliminal EB dosage. Especially in the experiment of α-pinene exposure, the first experiment has shown the enhancement of LQ after α-pinene exposure (group A) and the same group lowered LQ (no α-pinene exposure) in the second experiment. From this result, it may be suggested that α-pinene transiently enhanced the CNS effect of EB through olfactory-hypothalamic system. The green mix odor may have the same effects, but EB effect may be weaker than the α-pinene experiment.

Although this study was relatively small and exploratory, our findings suggest that plant-derived weak odors transiently enhanced the sexual behavior of rats. The precise neural mechanisms of these findings are problems for the future.

Reffrences

Calizo, L.H., Flanagan-Cato, L.M. (2003) Hormonal-Neural Integration in the Female Rat Ventromedial Hypothalamus: Triple Labeling for Estrogen Receptor-α, Retrograde Tract Tracing from the Periaqueductal Gray, and Mating-Induced Fos Expression. Endocrinology. 144: 5430-5440.

Carro-Juarez, M., Cervantes, E., Cervantes-Mendeza, M., Rodriguez-Manzob, G. (2004) Aphrodisiac properties of *Montanoa tomentosa* aqueous crude extract in male rats. Pharmacol. Biochem. Behavior. 78: 129-134.

Doty, R.L., Snyder, P.J., Huggins, G.R., Lowry, R.A. (1981) Endocrine, cardiovascular, and psychological correlates of olfactory sensitivity changes during the human menstrual cycle. J. Comp. Physiol. Psychol. 95: 45-60.

Gauthaman, K., Ganesan, A.P., Prasad, R.N. (2003) Sexual effects of puncturevine (*Tribulus terrestris*) extract (protodioscin): an evaluation using a rat model. J. Altern. Complement Med. 9: 257-65.

Graham, C.A., Janssen, E., Sanders, S.A. (2000) Effects of fragrance on female sexual arousal and mood across the menstrual cycle. Psychophysiology. 37: 76-84

Hummel, T., Gollisch, R., Wildt, G., Kobal, G. (1991) Changes in olfactory perception during the menstrual cycle. Experimentia. 47: 712-715.

Johns, M.A., Feder, H.H., Komisaruk, B.R., Mayer, A.D, (1978) Urine-induced reflex ovulation in anovulatory rats may be a vomero-nasal effect. Nature. 272: 446-448.

Kamiyama, K. (1985) The volatile substances in forest. The Amenity of forest. *In* The investigation report of the Forestry Agency" (In Japanese).

Keverne, E.B. (1978) Olfactory cues in mammalian sexual behaviour. *In* Biological Determinants of Sexual Behaviour. J.B. Hutchison (ed.). Wiley, UK.

Lomas, D.E., Keverne, E.B. (1982) Role of the vomeronasal organ and prolactin in the acceleration of puberty in female mice. J. Reprod. Fertil. 66: 101-107.

Mazaro, R., Di Stasi, L., De Grava Kempinas, K. (2002) Effects of the hydromethanolic extract of *Austroplenckia populnea* (Celastraceae) on reproductive parameters of male rats. Contraception. 66: 205-9.

Meisel, R.L., Dohanich, G.P., McEwen, B.S., Pfaff, D.W. (1987) Antagonism of sexual behavior in female rats by ventromedial hypothalamic implants of antiestrogen. Neuroendocrinology. 45: 201-207.

Murphy, L., Lee, T., Ginseng, J. (2002) Sex behavior, and nitric oxide. Ann NY Acad. Sci. 962: 372-377.

Pfaff, D.W., Sakuma, Y. (1979) Facilitation of the lordosis reflex of female rats from the ventromedial nucleus of the hypothalamus. J. Physiol. 288: 189-202.

Rajendren, G., Dudley, C.A., Moss, R.L. (1991) Role of the ventromedial nucleus of hypothalamus in the male-induced enhancement of lordosis in female rats. Physiol. Behavior. 50: 705-719.

Ramachandran, S., Sridhar, Y., Sam, S.K., Saravanan, M., Leonard, J.T., Anbalagan, N., Sridhar, S.K. (2004) Aphrodisiac activity of *Butea frondosa* Koen. ex Roxb. extract in male rats. Phytomedicine. 11:165-8.

Reynolds, J., Keverne, E.B. (1979) The accessory olfactory system and its role in the pheromonally mediated suppression of oestrus in grouped mice. J. Reprod. Fertil. 57: 31-35.

Runganga, A., Pitts, M., McMaster, J. (1992) The use of herbal and other agents to enhance sexual experience. Soc. Sci. Med. 35: 1037-42.

Sachs, B.D., Meisel, R.L. (1994) The physiology of male sexual behavior. *In* The Physiology of Reproduction, Vol. 2. J.D. Neill. and E. Knobil (eds.). Raven Press, NY, USA. pp. 3-106.

Sakuma, Y., Pfaff, D.W. (1979) Mesencephalic mechanisms for integration of female reproductive behavior in the rat. Am. J. Physiol. 237: R285-R290

Vandenbergh, J.G. (ed.). (1983) Pheromones and reproduction in mammals. Academic Press, NY, USA.

Yamanouchi, K., Nakano, Y., Arai, T. (1990) Roles of the pontine dorsomedial tegmentum and midbrain central gray in regulating female rat sexual behaviors: effects of p-chlorophenylalanine. Brain Res. Bull. 25: 381-5.

Color Plate Section

Chapter 9

Plate 1 White rocket (*Diplotaxis erucoides* L.)

Plate 2 White rocket (*Diplotaxis erucoides* L.) invading olive fields choking out other weed species

Plate 3 General view of white rocket (*Diplotaxis erucoides* L.) invasion of cultivated land by wheat and barley crops in Jordan

Plate 4 Allelopathic effect of soil incorporated dried residues (rates increased from left to right) of white rocket (*Diplotaxis erucoides* L.) on germination and growth of wheat grown under glasshouse conditions

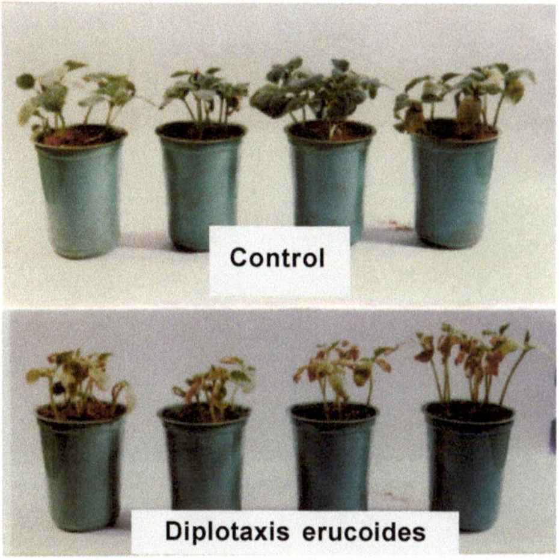

Plate 5 Allelopathic effect of decayed soil incorporated dried residues of white rocket (*Diplotaxis erucoides* L.) on germination and growth of cucumber grown under glasshouse conditions

Plate 6 Allelopathic effect of decayed soil incorporated dried residues of white rocket (*Diplotaxis erucoides* L.) on germination and growth of tomato grown under glasshouse conditions